BIOLOGICAL AND MEDICAL PHYSICS,
BIOMEDICAL ENGINEERING

BIOLOGICAL AND MEDICAL PHYSICS, BIOMEDICAL ENGINEERING

The fields of biological and medical physics and biomedical engineering are broad, multidisciplinary and dynamic. They lie at the crossroads of frontier research in physics, biology, chemistry, and medicine. The Biological and Medical Physics, Biomedical Engineering Series is intended to be comprehensive, covering a broad range of topics important to the study of the physical, chemical and biological sciences. Its goal is to provide scientists and engineers with textbooks, monographs, and reference works to address the growing need for information.

Books in the series emphasize established and emergent areas of science including molecular, membrane, and mathematical biophysics; photosynthetic energy harvesting and conversion; information processing; physical principles of genetics; sensory communications; automata networks, neural networks, and cellular automata. Equally important will be coverage of applied aspects of biological and medical physics and biomedical engineering such as molecular electronic components and devices, biosensors, medicine, imaging, physical principles of renewable energy production, advanced prostheses, and environmental control and engineering.

V.V. Tuchin   L.V. Wang   D.A. Zimnyakov

Optical Polarization in Biomedical Applications

With 102 Figures and 2 Tables

 Springer

Professor Dr.Sc. Ph.D. Valery V. Tuchin
Head of Optics and Biomedical Physics Chair
Department of Physics, Saratov State University
83, Astrakhanskya st., Saratov, 410012, Russia
E-mail: tuchin@sgu.ru

Professor Lihong V. Wang, Ph.D.
Royce E. Wisenbaker II Professor of Engineering
Department of Biomedical Engineering, Texas A&M University
3120 Tamu, College Station, TX 77843-3120, USA
E-mail: LWang@bme.tamu.edu

Professor Dr.Sc. Ph.D. Dmitry A. Zimnyakov
Optics and Biomedical Physics Chair
Department of Physics, Saratov State University
83, Astrakhanskya st., Saratov, 410012, Russia
E-mail: zimnykov@sgu.ru

Library of Congress Control Number: 2006925437

ISSN 1618-7210

ISBN-10 3-540-25876-0 Springer Berlin Heidelberg New York

ISBN-13 978-3-540-25876-6 Springer Berlin Heidelberg New York

Springer is a part of Springer Science+Business Media

springer.com

© Springer-Verlag Berlin Heidelberg 2006

The use of general descriptive names, registered names, trademarks, etc. in this publication does not imply, even in the absence of a specific statement, that such names are exempt from the relevant protective laws and regulations and therefore free for general use.

Cover concept by eStudio Calamar Steinen

Typesetting: Camera-ready by SPI Publisher Services, Pondicherry
Final layout: SPI India
Cover production: *design & production* GmbH, Heidelberg

Printed on acid-free paper   SPIN 10884375   57/3100/SPI - 5 4 3 2 1 0

Preface

This book introduces some of the recent key developments in polarization optical methods that have made possible the quantitative studies of tissues and biological cells essential to biomedical diagnostics. It also presents a number of advanced novel polarimetric technologies that have future potential for laboratory and clinical medical diagnosis.

In the book, the theory of polarization transfer in a random medium is used as the basis for a quantitative description of the interaction of polarized light with tissues and fluids. This theory employs the modified transfer equation for Stokes parameters to predict the polarization structure of single and multiple scattered optical fields. With our ultimate goal being the design of new noninvasive medical diagnostic methods, we introduce the backscattering polarization matrices (Jones matrix and Mueller matrix) that describe strongly scattering tissues. Experimental 2-D polarization patterns and Monte Carlo simulations of matrix elements are also presented. Optical coherence polarization tomography is described as a new tool for the measurement of polarization in tissues. Jones vector measurements, imaging by optical coherence tomography, and the problem of conversion from a Jones matrix to a Mueller matrix are discussed. A number of diagnostic techniques based on polarized light detection are presented, including CW polarization imaging, multiwavelength polarization imaging, polarization correlometry of tissues with expressed birefringence, partially coherent polarization-sensitive speckle-spectroscopy, polarization spectroscopy, microscopy and cytometry. Examples of biomedical applications of these techniques for cataract and glaucoma early diagnostics, glucose sensing in the human body, hematological and skin disease prediction, and bacteria detection are presented.

The audience for which this book is written consists of researchers, postgraduate students, biomedical engineers, and medical doctors who are interested in the design and application of optical and laser methods and instruments for medical science. Investigators who are deeply involved in the field will find up-to-date results on the topics discussed. Physicians and biomedical engineers may be interested in clinical applications of the presented techniques and in the instrumentation described. Laser and fiber optics engineers may also be interested in the book because acquaintance with the extensive potential for laser and fiber optics medical applications may stimulate new ideas for laser and fiber optics design. The large number of fundamental concepts and basic research on light-tissue interactions presented in the book should make it useful for a broad audience that includes students and physicians.

The authors are grateful to Dr. Habil Claus E. Ascheron for his valuable suggestions and help on preparation of this book. It should be mentioned that some of the original materials included in the book were arrived at while working on the following grants: 25.2003.2 the Russian Federation President's grant "Supporting of Scientific Schools" of the Russian Ministry for Industry, Science and Technologies, 2.11.03 "Leading Research-Educational Teams" of the Russian Ministry of Education, 04-02-16533 of the Russian Foundation for Basic Research, and REC-006 "Nonlinear Dynamics and Biophysics" of CRDF and the Russian Ministry of Education, as well as EB00319, CA71980 and CA092415 of the U.S. National Institutes of Health, and BES-9734491 of the U.S. National Science Foundation.

The authors are thankful to their colleagues, especially Prof. Irina L. Maksimova, Prof. Yurii P. Sinichkin, and Dr. Georgy V. Simonenko, for their cooperation in this endeavor.

They are also thankful to Prof. P. Gupta, Prof. K. Meek, and Dr. A. Kishen for their comments on sections of this book. Finally, the authors are grateful to Mary Ann Dickson for grammatically editing the entire manuscript.

Valery V. Tuchin   March 2006
Lihong V. Wang
Dmitry A. Zimnyakov

Contents

1

Introduction

1.1 Light Interaction with Tissues

Utilizing the interaction of light with biological tissues and fluids for practical purposes depends on understanding the properties of two large classes of biological media. One of them comprises weakly scattering (transparent) tissues and fluids like cornea, crystalline lens, vitreous humor, and aqueous humor of the front chamber of eye. The other class includes strongly scattering (opaque or turbid) tissues and fluids like skin, brain, vessel wall, eye sclera, blood, and lymph [1–7]. The interaction of light with biological media of the first class can be described by a model of single (or low-step) scattering in an ordered medium with closely packed scatterers that have a complex refractive index. Light propagation in tissues of the second class can be described by a model of multiple scattering of scalar or vector waves in a random or ordered low-absorbing medium.

The optical transparency of tissues is maximal in the near infrared (NIR) region, which is due to the absence, in this spectral range, of strong intrinsic chromophores that would absorb radiation in living tissues [1–7]. However, these tissues are characterized by rather strong scattering of NIR radiation, which prevents the attainment of clear images of localized inhomogeneities arising due to various pathologies, e.g., tumor formation or the growth of microvessels. Thus, this volume devotes special attention in the sections on tissue optical tomography and spectroscopy to the development of methods for the selection of image-carrying photons and the detection of photons providing spectroscopic information. Often the vector nature of light transport in scattering media, such as tissues, is ignored because of its rapid depolarization during propagation in a randomly inhomogeneous medium. However, in certain tissues (transparent eye tissues, cellular monolayers, mucous membrane, superficial skin layers, etc.), the degree of polarization of the transmitted or reflected light is measurable even when the tissue has considerable thickness.

Many tissues – such as eye cornea, sclera, tendon, cartilage, which are classified as fibrous tissues, and other structured tissues such as retina, tooth enamel and dentin – show a wide variety of polarization properties: linear birefringence, optical activity, and diattenuation. These properties are primarily defined by the tissue structure – anisotropy of form – or by the intrinsic anisotropic character of the tissue components or metabolic molecules – anisotropy of material. Collagen, muscle fibers, keratin, and glucose belong to this latter group.

The propagation of polarized light in a birefringent turbid medium is complicated because both the birefringent and the scattering effects can change the polarization state of light. Information about the structure of a tissue and the birefringence of its components can be extracted from the registered depolarization degree of initially polarized light, the polarization state transformation, or the appearance of a polarized component in the scattered light.

Since incident polarized light is rapidly depolarized in turbid tissues by light scattering, polarization-sensitive detection of reflected or transmitted light selects only the early escaping photons and rejects the multiply scattered light [1–7]. Thus, for a light beam reflected from a tissue, the polarization properties of light can be employed as a selector of the photons coming from different depths in the tissue. For transmitted light, they can act as a selector of ballistic or quasi-ballistic photons. Such polarization gating can, therefore, provide novel contrast mechanisms for tissue imaging and spectroscopy. As for practical implications, polarization techniques are expected to lead to simpler schemes of optical medical tomography than those used in existing diagnostic methods and also to provide additional information about the structure of tissues.

Since a variety of optical medical techniques employ lasers, light coherence is very important for the analysis of light interaction with tissues and cells. The problem can be viewed in terms of a loss of coherence due to the scattering of light in a random medium with multiple scattering and/or a change in the statistics and polarization states of speckles in a scattered field. Similarly, the coherence of light is of fundamental importance for the selection of photons that have experienced no, or a small number, of scattering events, as well as for the generation of speckle-modulated fields from scattering phase objects with single and multiple scattering. Such approaches are important for coherence tomography, diffractometry, holography, photon-correlation spectroscopy, and the speckle-interferometry of tissues and biological flows. The use of optical sources with a short coherence length opens up new opportunities in coherent interferometry and the tomography of tissues and blood flows.

To understand the general formalism for the scattering and absorption of light by arbitrarily shaped and arbitrarily oriented particles in tissue components, to learn exact and approximate theoretical methods and computer codes for calculating the scattering and polarization properties of these arbitrary shaped small particles, the following literature is recommended [8–20].

1.2 Definitions of Polarized Light

Definitions of polarized light, its properties, as well as production and detection techniques are described in a voluminous literature on this topic [21–28].

Polarization refers to the pattern described by the electric field vector as a function of time at a fixed point in space. When the electrical field vector oscillates in a single, fixed plane all along the beam, the light is said to be linearly polarized. A linearly polarized wave can be resolved into two mutually orthogonal components. If the plane of the electrical field rotates, the light is said to be elliptically polarized, because the electrical field vector traces out an ellipse at a fixed point in space as a function of time. If the ellipse happens to be a circle, the light is said to be circularly polarized. The connection between phase and polarization can be understood as follows: circularly polarized light consists of equal quantities of linear mutually orthogonal polarized components that oscillate exactly 90° out of phase. In general, light of arbitrary elliptical polarization consists of unequal amplitudes of linearly polarized components where the electrical fields of the two polarizations oscillate at the same frequency but have some constant phase difference. Light of arbitrary polarization can be represented by four numbers known as Stokes parameters [21–28].

In polarimetry, the Stokes vector $\mathbf{S}$ of a light beam is constructed based on six flux measurements obtained with different polarization analyzers in front of the detector

$$\mathbf{S} = \begin{pmatrix} I \\ Q \\ U \\ V \end{pmatrix} = \begin{pmatrix} I_\mathrm{H} + I_\mathrm{V} \\ I_\mathrm{H} - I_\mathrm{V} \\ I_{+45°} - I_{-45°} \\ I_\mathrm{R} - I_\mathrm{L} \end{pmatrix}, \tag{1.1}$$

where I_H, I_V, $I_{+45°}$, $I_{-45°}$, I_R, and I_L are the light intensities measured with a horizontal linear polarizer, a vertical linear polarizer, a +45° linear polarizer, a 45° linear polarizer, a right circular analyzer, and a left circular analyzer in front of the detector, respectively. Because of the relationship $I_\mathrm{H} + I_\mathrm{V} = I_{+45°} + I_{-45°} = I_\mathrm{R} + I_\mathrm{L} = I$, where I is the intensity of the light beam measured without any analyzer in front of the detector, a Stokes vector can be determined by four independent measurements, for example, $I_\mathrm{H}, I_\mathrm{V}, I_{+45°}$, and I_R,

$$\mathbf{S} = \begin{pmatrix} I_\mathrm{H} + I_\mathrm{V} \\ I_\mathrm{H} - I_\mathrm{V} \\ 2I_{+45°} - (I_\mathrm{H} + I_\mathrm{V}) \\ 2I_\mathrm{R} - (I_\mathrm{H} + I_\mathrm{V}) \end{pmatrix}. \tag{1.2}$$

From the Stokes vector, the degree of polarization (DOP), the degree of linear polarization (DOLP), and the degree of circular polarization (DOCP) are derived as

$$\mathrm{DOP} = \sqrt{Q^2 + U^2 + V^2} \Big/ I, \tag{1.3}$$

$$\mathrm{DOLP} = \sqrt{Q^2 + U^2} \Big/ I,$$

$$\mathrm{DOCP} = \sqrt{V^2} \Big/ I.$$

If the DOP of a light field remains unity after transformation by an optical system, this system is nondepolarizing; otherwise, the system is depolarizing.

The Mueller matrix ($\mathbf{M}$) of a sample transforms an incident Stokes vector $\mathbf{S}_{\mathrm{in}}$ into the corresponding output Stokes vector $\mathbf{S}_{\mathrm{out}}$:

$$\mathbf{S}_{\mathrm{out}} = \mathbf{M}\mathbf{S}_{\mathrm{in}}. \tag{1.4}$$

Obviously, the output Stokes vector varies with the state of the incident beam, but the Mueller matrix is determined only by the sample and the optical path. Conversely, the Mueller matrix can fully characterize the optical polarization properties of the sample. The Mueller matrix can be experimentally obtained from measurements with different combinations of source polarizers and detection analyzers. In most general cases, a 4×4 Mueller matrix has 16 independent elements; therefore, at least 16 independent measurements must be acquired to determine a full Mueller matrix.

The normalized Stokes vectors for the four incident polarization states, H, V, $+45°$, and R, are, respectively,

$$\mathbf{S}_{\mathrm{Hi}} = \begin{pmatrix} 1 \\ 1 \\ 0 \\ 0 \end{pmatrix}, \ \mathbf{S}_{\mathrm{Vi}} = \begin{pmatrix} 1 \\ -1 \\ 0 \\ 0 \end{pmatrix}, \ \mathbf{S}_{+45°\mathrm{i}} = \begin{pmatrix} 1 \\ 0 \\ 1 \\ 0 \end{pmatrix}, \ \mathbf{S}_{\mathrm{Ri}} = \begin{pmatrix} 1 \\ 0 \\ 0 \\ 1 \end{pmatrix}, \tag{1.5}$$

where H, V, $+45°$, and R, represent horizontal linear polarization, vertical linear polarization, $+45°$ linear polarization, and right circular polarization, respectively. We may express the 4×4 Mueller matrix as $\mathbf{M} = [\mathbf{M}_1 \ \mathbf{M}_2 \ \mathbf{M}_3 \ \mathbf{M}_4]$, where $\mathbf{M}_1$, $\mathbf{M}_2$, $\mathbf{M}_3$, and $\mathbf{M}_4$ are four column vectors of four elements each. The four output Stokes vectors corresponding to the four incident polarization states, H, V, $+45°$ and R, are denoted, respectively, by $\mathbf{S}_{\mathrm{Ho}}$, $\mathbf{S}_{\mathrm{Vo}}$, $\mathbf{S}_{+45°\mathrm{o}}$, and $\mathbf{S}_{\mathrm{Ro}}$. These four output Stokes vectors are experimentally measured based on (1.2) and can be expressed as

$$\begin{cases} \mathbf{S}_{\mathrm{Ho}} = \mathbf{M}\mathbf{S}_{\mathrm{Hi}} = \mathbf{M}_1 + \mathbf{M}_2 \\ \mathbf{S}_{\mathrm{Vo}} = \mathbf{M}\mathbf{S}_{\mathrm{Vi}} = \mathbf{M}_1 - \mathbf{M}_2 \\ \mathbf{S}_{+45°\mathrm{o}} = \mathbf{M}\mathbf{S}_{+45°\mathrm{i}} = \mathbf{M}_1 + \mathbf{M}_3 \\ \mathbf{S}_{\mathrm{Ro}} = \mathbf{M}\mathbf{S}_{\mathrm{Ri}} = \mathbf{M}_1 + \mathbf{M}_4. \end{cases} \tag{1.6}$$

The Mueller matrix can be calculated from the four output Stokes vectors [29]:

$$\mathbf{M} = \frac{1}{2} \times$$
$$\left[S_{\text{Ho}} + S_{\text{Vo}}, \, S_{\text{Ho}} - S_{\text{Vo}}, \, 2S_{+45°\text{o}} - (S_{\text{Ho}} + S_{\text{Vo}}), \, 2S_{\text{Ro}} - (S_{\text{Ho}} + S_{\text{Vo}}) \right].$$

(1.7)

In other words, at least four independent Stokes vectors must be measured to determine a full Mueller matrix, and each Stokes vector requires four independent intensity measurements with different analyzers.

For an overview of light polarization properties, its fundamentals and applications, the reader is referred to [21–28].

2

Tissue Structure and Optical Models

2.1 Introduction

As biological tissue is an optically inhomogeneous and absorbing medium, light propagation within a tissue depends on the scattering and absorption properties of its components – the cells, cell and fiber structures, and cell organelles [1–7]. In particular, the parameters such as particle size, shape, and density, the properties of the ground substance around the scattering particles, and the polarization states of the incident light play important roles in the propagation of light in turbid media [1–20].

The vector nature of propagating light waves [21–28] is especially important in transparent tissues, such as anterior eye tissues [2, 4, 5], but much attention has also been focused recently on investigation of the polarization properties of light propagation in strongly scattering media [4–7]. In scattering media, the vector nature of light waves is manifested as the polarization of an initially unpolarized light or as the depolarization (generally, a change in the character of the polarization) of an initially polarized beam that has been propagated in a medium.

Tissue polarization anisotropy exhibits primarily linear birefringence caused by fibrous structures – common constituents of many connective tissues. The refractive index of the medium is higher along the length of the fibers than across the width. This is called anisotropy of form. Due to their chirality, some molecules of tissue structures, for example keratin, or metabolic molecules like glucose, are responsible for what is called the material anisotropy of tissues.

Propagation of polarized light in a birefringent turbid medium is rather complicated, because both the birefringence and the scattering effect the polarization states of the light. The scattered radiation contains information about the sizes and shapes of the tissue structural elements, their orientation, optical constants, and other parameters. To extract this information and to interpret experimental results on light scattering, the researcher must develop an appropriate optical model for the particular tissue being considered and describe the light propagation in the medium based on this model.

In view of the great diversity and structural complexity of tissues [1–7], the development of adequate optical models accounting for the scattering and absorption of light is often the most complex step of a study. Therefore, in this chapter we will consider (1) tissue structure; (2) the optical properties of tissue components; (3) continuous and particle tissue models, including eye tissues; (4) the origin of tissue anisotropy; and (5) the fractal properties of tissues and cell aggregates.

2.2 Continuous and Discrete Tissue Models

Two approaches are currently used for tissue modeling. They are (1) tissue modeled as a medium with a continuous random spatial distribution of optical parameters [2, 4, 5, 29, 30] and (2) tissue modeled as a discrete ensemble of scatterers [1–7, 31]. The choice of approach is dictated by both the structural specificity of the tissue under study and the kind of light scattering characteristics that are to be obtained.

Most tissues are composed of structures with a wide range of sizes, and most can be described as a random continuum of the inhomogeneities of the refractive index with a varying spatial scale [29, 30]. Phase contrast microscopy has been used, in particular, to show that the structure of the refraction index inhomogeneities in mammalian tissues is similar to the structure of frozen turbulence in a number of cases [29]. This fact is of fundamental importance for understanding the peculiarities of radiation transfer in tissue, and it may be a key to the solution of the inverse problem of tissue structure reconstruction. This approach is applicable for tissues with no pronounced boundaries between elements that feature significant heterogeneity. The process of scattering in these structures may be described under certain conditions using the model of a phase screen [30, 32, 33].

The second approach to tissue modeling is its representation as a system of discrete scattering particles. This model has been advantageously used to describe the angular dependence of the polarization characteristics of scattered radiation [16, 19, 20, 34]. Blood is the most important biological example of a disperse system that entirely corresponds to the model of discrete particles.

Biological media are often modeled as ensembles of homogeneous spherical particles, since many cells and microorganisms, particularly blood cells, are close in shape to spheres or ellipsoids. A system of noninteracting spherical particles is the simplest tissue model. Mie theory rigorously describes the diffraction of light in a spherical particle [16, 24]. The development of this model involves taking into account the structures of the spherical particles, namely, the multilayered spheres and the spheres with radial nonhomogeneity, anisotropy, and optical activity [19, 20].

As connective tissue consists of fiber structures, a system of long cylinders is the most appropriate model for it. Muscular tissue, skin dermis, dura mater, eye cornea, and sclera belong to this type of tissue formed essentially by collagen fibrils. The solution of the problem of light diffraction in a single homogeneous or multilayered cylinder is also well understood [16].

2.3 Scatterer Size Range and Distribution

The sizes of cells and tissue structure elements vary in size from a few dozen nanometers to hundreds of micrometers [4, 5–7, 35–48]. The size of bacteria is usually a few micrometers. Blood cells(erythrocytes, leukocytes, and platelets) exhibit the following parameters. A normal erythrocyte in plasma has the shape of a concave–concave disc with a diameter varying from 7.1 to 9.2 μm, a thickness of 0.9–1.2 μm in the center and 1.7–2.4 μm on the periphery, and a volume of 90 μm³. Leukocytes are formed like spheres with a diameter of 8–22 μm. Platelets in the blood stream are biconvex disk-like particles with diameters ranging from 2 to 4 μm. Normally, blood has about ten times as many erythrocytes as platelets and about 30 times as many platelets as leukocytes.

Most other mammalian cells have diameters in the range of 5–75 μm. In the epidermal layer, the cells are large (with an average cross-sectional area of about 80 μm²) and quite uniform in size. Fat cells, each containing a single lipid droplet that nearly fills the entire cell and therefore results in eccentric placement of the cytoplasm and nucleus, have a wide range of diameters from a few microns to 50–75 μm. Fat cells may reach a diameter of 100–200 μm in pathological cases.

Additionally, there are a wide variety of structures within cells that determine tissue light scattering. Cell nuclei are on the order of 5–10 μm in diameter, mitochondria, lysosomes and peroxisoms have dimensions of 1–2 μm, ribosomes are on the order of 20 nm in diameter, and structures within various organelles can have dimensions up to a few hundred nanometers. In reality, the scatterers in cells are not spherical. The models of prolate ellipsoids with a ratio of the ellipsoid axes between 2 and 10 are more typical.

In fibrous tissues or tissues containing fiber layers (cornea, sclera, dura mater, muscle, myocardium, tendon, cartilage, vessel wall, retinal nerve fiber layer, etc.) and composed mostly of microfibrils and/or microtubules, typical diameters of the cylindrical structural elements are 10–400 nm. Their length is in a range from 10–25 μm to a few millimeters.

The dominant scatterers in an artery may be the fibers, the cells, or the subcellular organelles. Muscular arteries have three main layers. The inner intimal layer consists of endothelial cells with a mean diameter of less than 10 μm. The medial layer consists mostly of closely packed smooth muscle cells with a mean diameter of 15–20 μm; small amounts of connective tissue, including elastic, collagenous, and reticular fibers as well as a few fibroblasts, are also located in the media. The outer adventitial layer consists of dense fibrous connective tissue that is largely made up of 1–$12\,\mu$m in diameter collagen fibers and thinner, 2–$3\,\mu$m in diameter, elastin fibers.

Another two examples of complex scattering structures are myocardium and the retinal nerve fiber layer. The myocardium consists mostly of cardiac muscle which comprises myofibrils (about $1\,\mu$m in diameter) that in turn consist of cylindrical myofilaments (6–$15\,$nm in diameter) and aspherical mitochondria (1–$2\,\mu$m in diameter). The retinal nerve fiber layer comprises bundles of unmyelinated axons that run across the surface of the retina. The cylindrical organelles of the retinal nerve fiber layer are axonal membranes, microtubules, neurofilaments, and mitochondria. Axonal membranes, like all cell membranes, are thin (6–$10\,$nm) phospholipid bilayers that form cylindrical shells enclosing the axonal cytoplasm. Axonal microtubules are long tubular polymers of the protein tubulin with an outer diameter of $\approx 25\,$nm, an inner diameter of $\approx 15\,$nm, and a length of 10–$25\,\mu$m. Neurofilaments are stable protein polymers with a diameter of $\approx 10\,$nm. Mitochondria are ellipsoidal organelles that contain densely involved membranes of lipid and protein. They are 0.1–$0.2\,\mu$m thick and 1–$2\,\mu$m long.

Gaussian, gamma, or power size distributions are typical in optics of dispersed systems [49]. For some tissues, the size distribution of the scattering particles may be essentially monodispersive and for others it may be quite broad. Two opposite examples are transparent eye cornea stroma which has a sharply monodispersive distribution and turbid eye sclera which has a rather broad distribution of collagen fiber diameters [4,5]. There is no universal distribution size function that would describe all tissues with equal adequacy. Polydispersion for randomly distributed scatterers can be accounted for by using the gamma distribution or the skewed logarithmic distribution of scatterers' diameters, cross sections, or volumes [4,5,29,31,44,50,51]. For turbid tissues such as eye sclera, the gamma radii distribution function is applicable [50,51]:

$$\eta(a) = a^{\mu}\exp(-\mu\beta), \tag{2.1}$$

where $\sigma/a_{\mathrm{m}} = 2.35\,\mu^{-0.5}$, $\beta = a/a_{\mathrm{m}}$, σ is the half-width of the distribution, and a_{m} is the more probable scatterer radius.

A two-phase system made up of an ensemble of equally sized small particles and a minor fraction of larger ones provides a good model of pathological tissue, e.g., a cataractous lens [4].

For epithelial cells and their nuclei scattering structures, log-normal size distributions of spherical or slightly prolated ellipsoidal particles are characteristic [45]:

$$\eta(a) = (1/a\sigma\sqrt{2\pi})\exp\{-[(\ln(a) - \ln(a_{\mathrm{m}})]^2/2\sigma^2\}. \tag{2.2}$$

In particular, for epithelial cells and their nucleus components, two log-normal size distributions for small and big spherical scatterers with the following

parameters were found in a certain line of rat prostate carcinoma cells [45]:
$a_{m1} = 0.012\,\mu m$, $\sigma_1 = 1.15\,\mu m$ and $a_{m2} = 0.59\,\mu m$, $\sigma_2 = 0.43\,\mu m$.

When a description of scattering by particles of complex shape is required, different procedures, e.g., the method of T-matrices, can be applied [19, 20, 45]. Complexly shaped scatterers, like cells themselves, may be modeled as aggregates of spherical particles [46].

2.4 Refractive-Index Variations and Absorption

The tissue components that contribute most to the local refractive-index variations are the connective tissue fibers (bundles of elastin and collagen), cytoplasmic organelles (mitochondria, lysosoms, and peroxisomes), cell nuclei, and melanin granules [29, 31, 40–45, 47, 48, 50–57]. Figure 2.1 shows a hypothetical index profile formed by measuring the refractive index along a line in an arbitrary direction through a tissue. The widths of the peaks in the actual index profile are proportional to the diameters of the elements, and their heights depend on the refractive index of each element relative to that of its surroundings. This is the origin of the tissue discrete particle model. In accordance with this model, the index variations may be represented by a statistically equivalent volume of discrete particles having the same index but different sizes.

In inhomogeneous materials, such as tissues, the refractive indices of the fibrils, the interstitial medium, and the tissue itself can be derived using the law of Gladstone and Dale, which states that the resulting value represents an average of the refractive indices of the components related to their volume fractions [4]:

$$\bar{n} = \sum_{i=1}^{N} n_i f_i, \quad \sum_i f_i = 1, \tag{2.3}$$

where n_i and f_i are the refractive index and volume fraction of the individual components, respectively, and N is the number of components.

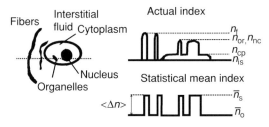

Fig. 2.1. A hypothetical index profile formed by measuring the refractive index along a line in an arbitrary direction through a volume of soft tissue [31]

The statistical mean index profile in Fig. 2.1 illustrates the nature of the approximation implied by this model. According to (2.3), the average background index is defined as the weighted average of the refractive indices of the cytoplasm and the interstitial fluid, n_{cp} and n_{is}, as

$$\bar{n}_0 = f_{cp} n_{cp} + (1 - f_{cp}) n_{is}, \tag{2.4}$$

where f_{cp} is the volume fraction of the fluid in the tissue contained inside the cells. Since approximately 60% of the total fluid in soft tissue is contained in the intracellular compartment, in accordance with [40–42, 52] $n_{cp} = 1.37$ and $n_{is} = 1.35$, it follows that $\bar{n}_0 = 1.36$.

The refractive index of a particle can be defined as the sum of the background index and the mean index variation:

$$\bar{n}_s = \bar{n}_0 + \langle \Delta n \rangle, \tag{2.5}$$

which can be approximated by another volume–weight average,

$$\langle \Delta n \rangle = f_f(n_f - n_{is}) + f_{nc}(n_{nc} - n_{cp}) + f_{or}(n_{or} - n_{cp}). \tag{2.6}$$

Here subscripts f, is, nc, cp, and or refer to fibers, interstitial fluid, nuclei, cytoplasm, and organelles, respectively, which are identified as the major contributors to the index variations. The terms in parentheses in this expression are the differences between the refractive indices of the three types of tissue components and their respective backgrounds. The multiplying factors are the volume fractions of the elements in the solid portion of the tissue. The refractive index of the connective-tissue fibers is about 1.47, which corresponds to approximately 55% hydration of collagen, its main component [4]. The nucleus and the cytoplasmic organelles in mammalian cells that contain similar concentrations of proteins and nucleic acids, such as the mitochondria and the ribosomes, have refractive indices that fall within a relative narrow range (1.38–1.41) [40, 41]. The measured index for the nuclei is $n_{nc} = 1.39$ [42, 52]. Accounting for this and supposing that $n_{or} = n_{nc} = 1.39$, the mean index variation can be expressed in terms of the fibrous-tissue fraction f_f only:

$$\langle \Delta n \rangle = f_f(n_f - n_{is}) + (1 - f_f)(n_{nc} - n_{cp}). \tag{2.7}$$

Collagen and elastin fibers comprise approximately 70% of the fat-free dry weight of the dermis, 45% of the heart, and 2–3% of the nonmuscular internal organs [31]. Therefore, depending on the tissue type, f_f may be as small as 0.02 or as large as 0.7. For $n_f - n_{is} = 1.47 - 1.35 = 0.12$ and $n_{nc} - n_{cp} = n_{or} - n_{cp} = 1.39 - 1.36 = 0.03$, the mean index variations that correspond to these two extremes are $\langle \Delta n \rangle = 0.03$–0.09.

For example, the nucleus and cell membrane of fibroblasts have an index of refraction of 1.48, the cytoplasm has an index of 1.38, and the averaged index of a cell is 1.42 [53]. The collagenous fibrils of cornea and sclera have an index of refraction of 1.47, and the refraction index of the ground matter is 1.35 [55]. The relative index of human lymphocytes in respect to plasma varies from $1.01 < m < 1.08$ [56]. Additional information on the refractive indices of biological cells and tissues may be found in [4, 57, 58].

The matter surrounding the scatterers (intercellular liquid and cytoplasm), the so-called ground substance, is composed mainly of water with salts and organic components dissolved in it. The ground matter index is usually taken as $n_0 = 1.35$–1.37. The scattering particles themselves (organelles, protein fibrils, membranes, protein globules) exhibit a higher density of proteins and lipids in comparison with the ground substance and thus a greater index of refraction $n_1 = 1.39$–1.47. This implies that the simplest way to model tissue is to consider the binary fluctuations in the index of refraction of the various tissue structures.

Absorption for most tissues in the visible region is insignificant except for the absorption bands of blood hemoglobin and some other chromophores (see Fig 2.2) [1–7]. The absorption bands of protein molecules are mainly in the near UV region. Absorption in the IR region is essentially defined by water contained in tissues. For example, the index for *Bacillus subtilis* spores has

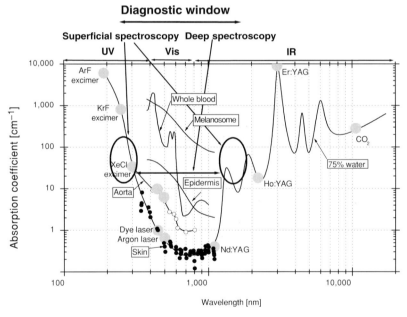

Fig. 2.2. Absorption spectra of skin and aorta; spectra of tissue components – water (75%), epidermis, melanosome, and whole blood – are also presented; diagnostic lasers and their wavelengths as well as diagnostic window and wavelength ranges suitable for superficial and deep spectroscopy are shown (Adapted from S. Jacques, "Strengths and weaknesses of various optical imaging techniques," Saratov Fall Meeting'01, Internet Plenary Lecture, Saratov, Russia, 2001, http://optics.sgu.ru/SFM)

a value of about 1.5, and its imaginary part is smaller than 0.01 in a wide spectral range [54]. The real part of the erythrocyte index with respect to plasma is $m = 1.041$–1.067 ($\lambda = 600\,nm$), and its imaginary part is varied within 10^{-2}–10^{-5} in the wavelength range $\lambda = 350$–$1,000\,nm$.

The above examples provide evidence that tissue inhomogeneities have sizes comparable to, or smaller than, visible or NIR wavelengths and small relative indices of refraction. Hence, they must be considered as optically soft. For most tissues the size parameter of the particles $x = 2\pi r/\lambda$ (where r is the particle radius, λ is the light wavelength in a tissue) varies in a wide range $0.1 < x < 100$ for the visible/IR region. The absorption of particles and ground medium is rather small in this wavelength range. This enables different approximation methods, described in [1–7, 12–14, 16, 19, 20], to be used for calculations.

2.5 Tissue Anisotropy

Many biological tissues are optically anisotropic [59–96] Tissue birefringence results primarily from the linear anisotropy of fibrous structures, which forms the extracellular media. The refractive index of a medium is higher along the length of a fiber than along the cross section. A specific tissue structure is a system composed of parallel cylinders that create an uniaxial birefringent medium with the optic axis parallel to the cylinder axes. This is called birefringence of form. A large variety of tissues, such as eye cornea, tendon, cartilage, eye sclera, dura mater, testis, muscle, nerve, retina, bone, teeth, myelin, etc., exhibit form birefringence. All of these tissues contain uniaxial and/or biaxial birefringent structures. For instance, in bone and teeth, these are mineralized structures originating from hydroxyapatite crystals. In particular, enamel prisms, fairly well-oriented hexagonal crystals of hydroxyapatite of 15–20 nm in diameter and up to 160 nm in length, packed into an organic matrix with an overall cross section of 4–6 µm, and dentin tubules, shelled organic cylinders with a highly mineralized shell with diameters of 1–5 µm, play an important role in tooth birefringence.

Tendon consists mostly of parallel, densely packed collagen fibers. Interspersed between the parallel bundles of collagen fibers are long, elliptical fibroblasts. In general, tendon fibers are cylindrical in shape with diameters ranging from 20 to 400 nm [59, 60]. The ordered structure of collagen fibers running parallel to a single axis makes tendon a highly birefringent tissue.

Arteries have a more complex structure than tendons. The medial layer consists mostly of closely packed smooth muscle cells with a mean diameter of 15–20 µm. Small amounts of connective tissue, including elastic, collagenous, and reticular fibers, as well as a few fibroblasts, are also located in the media. The outer adventitial layer consists of dense fibrous connective tissue. The adventitia is largely made up of collagen fibers, 1–12 µm in diameter, and thinner elastin fibers, 2–3 µm in diameter. As with tendon, the cylindrical collagen and elastin fibers are ordered mainly along one axis, thus, causing the tissue to be birefringent.

Myocardium, on the other hand, contains fibers oriented along two different axes. Myocardium consists mostly of cardiac muscle fibers arranged in sheets that wind around the ventricles and atria. In pigs, the myocardium cardiac muscle is comprised of myofibrils (about $1\,\mu$m in diameter) that in turn consist of cylindrical myofilaments (6–15 nm in diameter) and aspherical mitochondria (1–2 μm in diameter). Myocardium is typically birefringent since the refractive index along the axis of the muscle fiber is different from that in the transverse direction [59,60].

Form birefringence arises when the relative optical phase between the orthogonal polarization components is nonzero for forwardly scattered light. After multiple forward scattering events, a relative phase difference accumulates and a delay (δ) similar to that observed in birefringent crystalline materials is introduced between orthogonal polarization components. For organized linear structures, an increase in phase delay may be characterized by a difference (Δn) in the effective refractive index for light polarized along, and perpendicular to, the long axis of the linear structures. The effect of tissue birefringence on the propagation of linearly polarized light is dependent on the angle between the incident polarization orientation and the tissue axis. Phase retardation, δ, between orthogonal polarization components, is proportional to the distance (d) traveled through the birefringent medium [82]

$$\delta = \frac{2\pi\Delta nd}{\lambda}. \qquad (2.8)$$

A medium of parallel cylinders is a positive uniaxial birefringent medium $[\Delta n = (n_{\mathrm{e}} - n_{\mathrm{o}}) > 0]$ with its optic axis parallel to the cylinder axes (see Fig. 2.3a). Therefore, a case defined by an incident electrical field directed parallel to the cylinder axes will be called "extraordinary," and a case with the incident electrical field perpendicular to the cylinder axes will be called "ordinary." The difference $(n_{\mathrm{e}} - n_{\mathrm{o}})$ between the extraordinary index and the

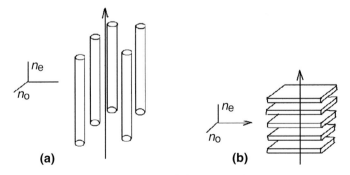

Fig. 2.3. Models of tissue birefringence: (**a**) system of long dielectric cylinders, (**b**) system of thin dielectric plates

ordinary index is a measure of the birefringence of a medium comprised of cylinders. For the Rayleigh limit ($\lambda \gg$ cylinder diameter), the form birefringence becomes [63, 66]

$$\Delta n = (n_e - n_o) = \frac{f_1 f_2 (n_1 - n_2)^2}{f_1 n_1 + f_2 n_2}, \tag{2.9}$$

where f_1 is the volume fraction of the cylinders, f_2 is the volume fraction of the ground substance, and n_1, n_2 are the corresponding indices. For a given index difference, maximal birefringence is expected for approximately equal volume fractions of thin cylinders and ground material. For systems with large diameter cylinders ($\lambda \ll$ cylinder diameter), the birefringence goes to zero [66].

For a system of thin plates (see Fig. 2.3b), the following equation is obtained [24]

$$n_e^2 - n_o^2 = -\frac{f_1 f_2 (n_1 - n_2)}{f_1 n_1^2 + f_2 n_2^2}, \tag{2.10}$$

where f_1 is the volume fraction occupied by the plates, f_2 is the volume fraction of the ground substance, and n_1, n_2 are the corresponding indices. This implies that the system behaves like a negative uniaxial crystal with its optical axis aligned normally with the plate surface.

Form birefringence is used in biological microscopy as an instrument for studying cell structure. The sign of the observed refractive index difference points to the particle shape closest to that of the rod or the plate, and if n_1 and n_2 are known, one can then assess the volume fraction occupied by the particles. To separate the birefringence of the form and the particle materials, the refractive indices of the particles and the ground substance should be matched, because form birefringence vanishes with $n_1 = n_2$.

Linear dichroism (diattenuation), i.e., different wave attenuation for two orthogonal polarizations, in systems formed by long cylinders or plates is defined by the difference between the imaginary parts of the effective indices of refraction. Depending on the relationship between the sizes and the optical constants of the cylinders or plates, this difference can take both positive and negative values [24].

Reported birefringence values for tendon, muscle, coronary artery, myocardium, sclera, cartilage, skin are on the order of 10^{-3} (see, for instance, [64, 67, 68, 75, 76, 78–82]). The measured refractive index variations for the fast and slow axes of rabbit cornea show that its birefringence varies within the range of 0 at the apex, or top of the cornea, to 5.5×10^{-4} at the base of the cornea where it attaches to the sclera [62, 70]. The predominant orientation of collagenous fibers in different regions of the cornea results in birefringence and dichroism [69]. Based on experimental results, it has been assumed that the birefringent portions of the corneal surface all have a relatively universal fast axis located approximately 160° from the vertical axis, defined as a line that runs from the apex of the cornea through the pupil [70].

A new technique – polarization-sensitive optical coherence tomography (PS OCT) – allows for the measurement of linear birefringence in turbid tissue with high precision [78–82, 84]. The following data have been reported using this technique: for rodent muscle, 1.4×10^{-3} [81,82]; for normal porcine tendon, $(4.2 \pm 0.3) \times 10^{-3}$ and for thermally treated ($90°$C, 20 s), $(2.24 \pm 0.07) \times 10^{-3}$; for porcine skin, 1.5×10^{-3}–3.5×10^{-3}; for bovine cartilage 3.0×10^{-3} [84]; and for bovine tendon, $(3.7 \pm 0.4) \times 10^{-3}$ [79]. Such birefringence provides 90% phase retardation at a depth on the order of several hundred micrometers.

The magnitude of birefringence and diattenuation are related to the density and other properties of the collagen fibers, whereas the orientation of the fast axis indicates the orientation of the collagen fibers. The amplitude and orientation of birefringence of the skin and cartilage are not as uniformly distributed as in tendon. In other words, the densities of collagen fibers in skin and cartilage are not as uniform as in tendon, and the orientation of the collagen fibers are not distributed in as orderly a fashion [84].

Because many components in biological tissues contain intrinsic, and/or form, birefringence, polarization-sensitive optical technologies are of great interest for application in ophthalmology [4–7, 65, 67, 69–72, 85, 86], dermatology [81, 87, 88], and dentistry [89, 90]. Functional information in some biological systems is associated with transient changes in birefringence. For instance, photothermal injury during laser surgery is associated with birefringence changes in subsurface tissue components. Therefore, changes in birefringence may indicate changes in the functionality, structure, or viability of tissues.

In addition to linear birefringence and dichroism (diattenuation), many tissue components show optical activity. In polarized light research, the molecule's chirality that stems from its asymmetric molecular structure, results in a number of characteristic effects generically called optical activity [15,93]. A well-known manifestation of optical activity is the ability to rotate the plane of linearly polarized light about the axis of propagation. The amount of rotation depends on the chiral molecular concentration, the pathlength through the medium, and the light wavelength. For instance, chiral asymmetrically encoded in the polarization properties of light transmitted through a transparent media enables very sensitive and accurate determination of glucose concentration. Tissues containing chiral components display optical activity [70,91,92]. Interest in chiral turbid media is driven by the attractive possibility of noninvasive in situ optical monitoring of the glucose in diabetic patients. Within turbid tissues, however, where the scattering effects dominate, the loss of polarization information is significant and the chiral effects due to the small amount of dissolved glucose are difficult to detect.

In complex tissue structures, chiral aggregates of particles, in particular spherical particles, may be responsible for tissue optical activity (see Fig. 2.4). More sophisticated anisotropic tissue models can also be constructed. For example, eye cornea can be represented as a system of plane anisotropic layers (plates, i.e., lamellas), each of which is composed of densely packed long cylinders (fibrils) (see Fig. 2.3) with their optical axes oriented along a spiral (see Fig. 2.5). This fibrilar—lamellar structure of the cornea is responsible for the

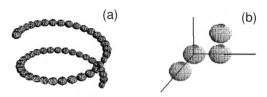

Fig. 2.4. Examples of chiral aggregates of spherical particles

Fig. 2.5. A schematic illustration of the lamellar organization of the cornea stroma [96]

linear and circular dichroism and its dependence on the angle between the lamellas [69].

2.6 Volume Fraction and Spatial Ordering of Particles

A discrete particle ensemble is characterized by the packing density or, in other words, by the volume fraction occupied by particles. Evidently, in addition to particle size, the volume fraction of particles defines the optical properties of an ensemble by changing the refractive index (see (2.3–2.7)), optical anisotropy (see (2.9, 2.10)), and other characteristics. The volume fraction of particles for a certain tissue may be experimentally found using electron micrographs of tissue slices. This is a straightforward approach based on the measurement of the area occupied by an element of a particular size for a certain slice. Unfortunately, systematic errors, caused by cross sectioning of 3D particles within the examined slice, may occur. Such errors lead to distortion of the volume fractions of the different tissue elements [94]. Estimations of a volume fraction occupied by scattering particles may also be accomplished by the weighting of a native tissue and dry rest.

The volume fraction occupied by the scattering particles in tissues, such as muscle, cornea, sclera and eye lens, covers from 20 to 40%. Conventionally, whole blood contains $(4–5) \times 10^6$ erythrocytes, $(4–9) \times 10^3$ leukocytes, and $(2-3) \times 10^5$ platelets in $1\,\mathrm{mm^3}$. Cells make up 35–45% of the blood volume. The volume fraction f of erythrocytes in the blood is called the hematocrit H. For normal blood, $H = 0.4$. The remaining 60% of the blood volume is mostly the plasma – an essentially transparent water solution of salts.

Most tissues are comprised of cellular and subcellular structures located in close proximity to each other. In general, densely packed structures are likely to exhibit correlation scattering, an effect that has been observed, for instance, in cornea stroma [4, 5, 95–99]. Cornea is comprised of individual collagen fibrils that are closely packed parallel to one another in a lamella. If each fibril in the lamella scattered light independently, then the scattering cross section of the lamella should be the product of the cross section of a single fibril and the number of fibrils in the lamella. If all of the cornea fibers scattered light independently, the cornea would scatter 90% of the light incident on it, and we would see essentially nothing. However, the fibrils do not scatter independently and the coherent scattering (interference) effects cannot be neglected. Accordingly, correlated polarization effects can be observed [98–101]. For example, in spherical particle suspensions, as the particle concentration increases beyond a concentration at which independent scattering can be assumed, the degree of polarization increases (rather than decreases) as the scatterer concentration increases [100, 101].

Thus, the spatial organization of the particles forming a tissue plays a substantial role in the propagation of polarized light. As mentioned above, with very small packing densities, incoherent scattering by independent particles occurs. If the volume fraction occupied by the particles is equal to, or more than, 0.01–0.1, coherent concentration effects appear. The concentration of scattering particles is adequate in most tissues to allow spaces between individual scatterers that are comparable to their sizes. If, however, the particle-size distribution is rather narrow, then dense packing entails a certain degree of order in the arrangement of the particles.

Spatial ordering is of utmost importance in optical eye tissue [4, 5, 69, 94–99, 102–104]. In a large variety of other tissues, spatial ordering is also more-or-less inherent, particularly in tendon, cartilage, dura mater, skin or muscle. The high degree of order in densely packed scatterers ensures high transmission in the cornea and eye lens. Tissue structures with statistically ordered periodic variations in the index at characteristic scales of light wavelength, like photon crystals [104], exhibit high transmission spectral regions and bands for which the propagation of electromagnetic waves is forbidden. The position and depth of these bands depends on the size, refractive index and spatial arrangement of the scattering particles.

To account for the interparticle correlation effects which are important for systems with volume fractions of scatterers higher than 1–10% (dependent on particle size), the following expression is valid for the packing factor ω_p of a medium filled with a volume fraction f_s of scatterers with different shapes [16]:

$$\omega_{\mathrm{p}} = \frac{(1 - f_{\mathrm{s}})^{p+1}}{[1 + f_{\mathrm{s}}(p - 1)]^{p-1}}, \tag{2.11}$$

where p is a packing dimension that describes the rate at which the empty space between scatterers diminishes as the total density increases. The packing of spherical particles is described well by packing dimension $p = 3$. The packing of sheet-like and rod-shaped particles is characterized by a p that approaches 1 and 2, respectively. Since the elements of tissue have all of these different shapes and may exhibit cylindrical and spherical symmetry simultaneously, the packing dimension may lie anywhere between 1 and 5. When one calculates optical coefficients at high concentrations of particles, the size distribution $\eta(2a)$ [(2.1) and (2.2)] should be replaced by the correlation-corrected distribution [31]

$$\eta'(2a) = \frac{[1 - \eta(2a)]^{p+1}}{[1 + \eta(2a)(p - 1)]^{p-1}} \eta(2a). \tag{2.12}$$

Most of the observed scattering properties of soft tissue that are explained in the model treat tissue as a collection of scattering particles, whose volume fractions are distributed according to a skewed log-normal distribution modified by a packing factor, to account for correlated scattering among densely packed particles [31].

2.7 Eye Tissue Optical Models

Healthy tissues of the anterior human eye chamber, e.g., the cornea and lens, are highly transparent to visible light due to their ordered structure and the absence of strongly absorbing chromophors. Scattering is an important feature of light propagation in eye tissues. The size of the scatterers and the distances between them are smaller than, or comparable with, the wavelength of visible light. The relative refractive index of the scattering matter is equally small (soft particles). Typical eye tissue models are long, round dielectric cylinders (corneal and scleral collagen fibers) or spherical particles (lens protein structures) that have a refractive index n_{s} and are distributed in the isotropic ground matter with a refractive index $n_0 \leq n_{\mathrm{s}}$ in an orderly (transparent cornea and lens) or quasiorderly (sclera, opaque lens) manner [2,4,5,50,51,61,63,66,69,95–99,102–127]. Light scattering analysis in eye tissue often is possible using a single scattering model owing to the small scattering cross-section (soft particles).

Let's consider the structure of the cornea and the sclera in more detail to demonstrate tissues with different size distributions and spatial ordering of scatterers [2,4,5,95,105–114]. The cornea is the frontal section of the eye's fibrous capsule; its diameter is about $\approx 10\,\mathrm{mm}$. The sclera is a turbid opaque tissue that covers nearly 80% of the eye and serves as a protective membrane to provide, along with the cornea, for counteraction against internal and external forces and to retain eye shape. Both tissues are composed of collagen fibrils immersed in a ground substance [95, 105–114]. The fibrils have a shape similar to that of a cylinder. They are packed in bundles like lamellae. Within each lamella, all of the fibers are nearly in parallel with each other and with the lamella plane. Fibrils and lamellar bundles are immersed within an amorphous ground (interstitial) substance containing water, glycosaminoglycans, proteins, proteoglycans, and various salts. The glycosaminoglycans play a key role in regulating the assembly of the collagen fibrils as well as in tissue permeability to water and other molecules [110]. The indices of refraction for the fibers and the ground substance differ markedly.

The structural elements that give the cornea the strength to preserve its proper curvature while withstanding intraocular pressure (14–18 mm Hg) are located within its stromal layer, which constitutes 0.9 of the cornea's thickness [95–97, 105]. The stroma is composed of several hundred successively stacked lamellae, each about 2 μm in thickness (three sequential lamellae are shown in Fig. 2.6a). Human corneal thickness averages 0.52 mm. A few flat cells (keratocytes) are dispersed between the lamellae, and these occupy only 0.03–0.05 of the stromal volume. Each lamella is composed of a parallel array of collagen fibrils.

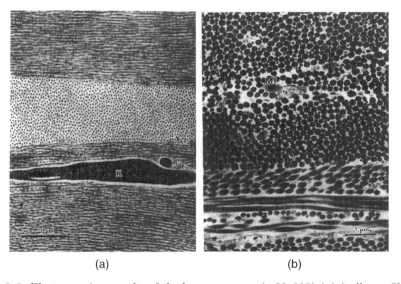

(a)   (b)

Fig. 2.6. Electron micrographs of the human cornea ($\times 32,000$) (**a**) (collagen fibrils have a uniform diameter and are arranged in the same direction within the lamellae) and sclera ($\times 18,000$) (**b**) (collagen fibrils display various diameters; they are much larger than those in the cornea) [105]. K is the keratocyte; Mf is the microfibril.

Although the cornea fibril diameters vary from 25 to 39 nm in different mammals, the fibrils are quite uniform in diameter within each species [105, 110, 114]. Spacing between fibril centers is equal to 45–65 nm; intermolecular spacing within fibrils is in the range of 1.56–1.63 nm [114]. The fibrils in the human cornea have a uniform diameter of about 30.8 ± 0.8 nm with a periodicity close to two diameters, 55.3 ± 4.0 nm, and rather high regularity in the organization of fibril axes about one another (see Fig. 2.6a), The intermolecular spacing is 1.63 ± 0.10 nm [114]. Thus, the stroma has at least three levels of structural organization: the lamellae that lie parallel to the cornea's surface; the fibrillar structure within each lamella that consists of small, parallel collagen fibrils with uniform diameters that have some degree of order in their spatial positions; and the collagen molecular ultrastructure.

The sclera contains three layers: the episclera, the stroma and the lamina fusca [108]. The stroma is the thickest layer of the sclera. The thickness of the sclera and the arrangement of the collagen fibers show regional (limbal, equatorial, and posterior pole region) and aging differences. In the stroma, the collagen fibrils exhibit a wide range of diameters from 25 to 230 nm (see Fig. 2.6b) [105]. The average diameter of the collagen fibrils increases gradually from about 65 nm in the innermost part to about 125 nm in the outermost part of the sclera [109]; the mean distance between fibril centers is about 285 nm [112]. Collagen intermolecular spacing is similar to that in the cornea; in bovine sclera, particularly, it is equal (1.61 ± 0.02 nm) [110].

The fibrils are also arranged in individual bundles in a parallel fashion but more randomly than those in the cornea. Moreover, within each bundle, the groups of fibers are separated from each other by large empty lacunae randomly distributed in space [105]. Collagen bundles show a wide range of widths (1–50 μm) and thicknesses (0.5–6 μm) and tend to be wider and thicker toward the inner layers. These ribbon-like structures are multiply cross-linked; their length can be a few millimeters [108]. They cross each other in all directions but remain parallel to the scleral surface. The episclera has a similar structure with more randomly distributed, and less compact, bundles than the stroma. The lamina fusca contains a larger quantity of pigments, mainly melanin, which is generally located between the bundles. The sclera itself does not contain blood vessels but has a number of channels that allow arteries, veins, and nerves to enter into or leave the eye [108]. The thickness of the sclera is variable. It is thicker at the posterior pole (0.9–1.8 mm); it is thinnest at the equator (0.3–0.9 mm); and at the limbus, it is in the range of 0.5–0.8 mm [108]. Hydration of the human sclera can be estimated at 68%. About 75% of its dry weight is due to collagen; 10% is due to other proteins; and 1% to mucopolysaccharides [108].

In designing an optical model of a tissue, in addition to the form, size, and density of the scatterers as well as tissue thickness, it is important to have information on the refractive indices of the tissue components. Following

[51, 108, 113, 114], we can estimate the refractive index of the corneal and scleral fibrils (hydrated collagen), n_c, using (2.4) which was written for the average refractive index of the tissue, $\bar{n}_t$:

$$n_c = \frac{\bar{n}_t - (1 - f_c)n_{is}}{f_c}, \qquad (2.13)$$

where f_c is the volume fraction of the hydrated collagen, and n_{is} is the refractive index of the interstitial fluid.

The refractive indices measured for the dry corneal collagen and for the interstitial fluid are: $n_c^{dry} = 1.547$ and $n_{is} = 1.345$–1.357 [95–97, 108, 113, 114]. The refractive index of the corneal stroma measured for many species is $\bar{n}_t = 1.375 \pm 0.005$ [114]. Therefore, for $n_{is} = 1.356$ and $f_c = 0.32$, corresponding to a tissue hydration of 76.2% and a collagen content of 61.3% of the dry weight [114], on the basis of (2.13), it is easy to obtain the refractive index of the hydrated fibrils as $n_c = 1.415$.

The direct measurement of the average refractive index of sclera using an Abbe refractometer gives $\bar{n}_t = 1.385 \pm 0.005$ for $\lambda = 589$ nm. Because of the similarly fibrous nature of the cornea and the sclera, it is expected that at equal hydration the refractive indices of scleral collagen and its interstitial fluid should be equal to these indices in the cornea. For $\bar{n}_t = 1.385, n_{is} = 1.345$, and $f_c = 0.31$, corresponding to a tissue hydration of 68% and a collagen content of 75% of the dry weight, it follows from (2.13) that for the refractive index of the scleral fibrils, $n_c = 1.474$. Changes of n_c and f_c with hydration can be evaluated from measurements of the refractive index and the thickness of the collagen films [128].

While both tissues are composed of similar molecular components, they have different microstructures and thus very different physiological functions. The cornea is transparent, allowing for more than 90% of the incident light to be transmitted. The collagen fibrils in the cornea have a much more uniform size and spacing than those in the sclera, resulting in a greater degree of spatial order in the organization of the fibrils in the cornea compared with the sclera. The sclera of the eye is opaque to light; it scatters almost all wavelengths of visible light and thus appears white.

Light propagation in a densely packed disperse system can be analyzed using the radial distribution function $g(r)$, which statistically describes the spatial arrangement of particles in the system. The function $g(r)$ is the ratio of the local number density of the fibril centers at a distance r from a reference fibril at $r = 0$ to the bulk number density of fibril centers [96]. It expresses the relative probability of finding two fibril centers separated by a distance r; thus $g(r)$ must vanish for values of $r \leq 2a$ (a is the radius of a fibril; fibrils cannot approach each other closer than touching). The radial distribution function of scattering centers $g(r)$ for a certain tissue may be calculated on the basis of tissue electron micrographs (see Fig 2.6).

The technique for the experimental determination of $g(r)$ involves counting the number of particles, placed at a specified spacing from an arbitrarily chosen initial particle, followed by statistical averaging over the whole ensemble. In a two dimensional case, the particle number ΔN at the spacing from r to $r + \Delta r$ is related to function $g(r)$ by the following equation:

$$\Delta N = 2\pi\rho g(r)r\Delta r, \tag{2.14}$$

where ρ is the mean number of particles for a unit area.

The radial distribution function $g(r)$ was first found for the rabbit cornea by Farrell *et al.* [96]. Figure 2.7a depicts a typical result for one of the cornea regions, which was obtained by determining the ratio of the local mean density of the centers as a function of radii taken from 700 fibril centers. The function $g(r) = 0$ for $r \leq 25$ nm, which is consistent with a fibril radius of 14 ± 2 nm, can be calculated from the electron micrograph [96]. The first peak in the distribution gives the most probable separation distance, which is approximately 50 nm. The value of $g(r)$ is essentially unity for $r \geq 170$ nm, indicating that the fibril positions are correlated over no more than a few of their nearest neighbors. Therefore, a short-range order exists in the system.

Similar calculations for several regions of the human eye sclera [103] are illustrated in Fig. 2.7b. Electron micrographs from [105], averaged for 100 fibril centers, were processed (see Fig 2.6b). Function $g(r)$ for the sclera was obtained on the basis of the spatial distribution of the fibril centers, neglecting discrepancy in their diameters. Some noise is due to the small volume of statistical averaging. The obtained results present evidence of the presence of a short-range order in the sclera, although the degree of order is less pronounced than in the cornea. The function $g(r) = 0$ for $r \leq 100$ nm, which is consistent with the mean fibril diameter of ≈ 100 nm derived from the electron

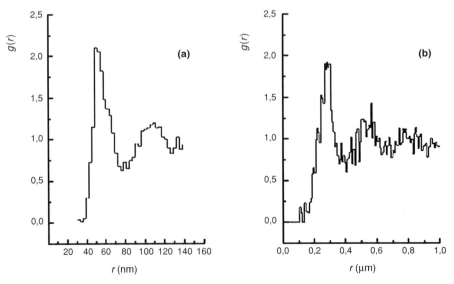

Fig. 2.7. Histograms of radial distribution functions $g(r)$ obtained from electron micrographs of the rabbit cornea [96] (**a**) and the human sclera [103] (**b**).

micrograph (see Fig 2.6b) [105]. The first peak in the distribution gives the most probable separation distance, which is approximately 285 nm. The value of $g(r)$ is essentially unity for $r \geq 750$ nm, indicating a short-range order in the system. The short-range order, being characterized by a ratio of this specific distance (decay of spatial correlation) to the most probable particle separation distance, $(750/285) \approx 2.7$, is smaller than the similar ratio for the cornea, $(170/50) = 3.4$.

The spatial density (refractive index) fluctuations of a tissue can also be analyzed by resolving 2D-profiles of refractive index variations into Fourier components, which provide a basis for a detailed and quantitative description of the microstructure [111, 112]. These Fourier components represent the predominant spatial density fluctuations and the structural ordering. A comparable study of the human cornea and sclera has shown that the cornea reveals much less collagen fibril spacing and greater spatial order than the sclera [111, 112].

The eye lens is also an example of a tissue in which the short-range spatial order is of crucial importance. Because of its high index of refraction and transparency, a lens focuses light to form an image at the retina. The eye lens material exhibits a certain viscosity that is capable of altering its radius of curvature and thus its focal length through the action of accommodating muscles. The healthy human lens is a coherent structure containing about 60% water and 38% protein [119–127]. The lens consists of many lens fiber cells. The predominant dry components of a mammalian lens are three kinds of structural proteins named α -, β -, and γ - crystallins, and their combined weight accounts for about 33% of the total weight of the lens [129]. The crystalline lens grows throughout life and in addition undergoes a variety of biochemical changes as a person ages. The potential changes include age-related cataract formation, which can lead to greatly increased light scattering and coloration and eventually to lens opacity. The light scattering is caused by random fluctuations in the refractive index. These fluctuations can be density or optical anisotropy fluctuations [61,65,66,75,102,119,121,125]. Fluctuations in the refractive index due to density may arise because of (1) an aggregation of lens proteins, (2) a microphase separation (cold-induced cataract), or (3) syneresis (water releases from the bound state in the hydration layers of lens proteins and becomes bulk water which increases the refractive index difference between the lens proteins and the surrounding fluid). Analyses of polarized light scattering by human cataracts have shown that 15–30% of the turbidity results from optical anisotropy fluctuations.

Eye lens transparency can be explained by a short-range ordering in the packing structure of the lens proteins. This idea was first suggested by Benedek [130]. The primary role among the ocular lens proteins is played by the water-soluble α-crystallin which has a shape that is close to spherical with a diameter of about 17 nm. Studies of lens transparency, birefringence and optical activity are of importance to the facilitation of early diagnosis of cataracts [119–127, 129–133].

The types of fiber cell disruption due to cataract formation include intracellular globules, clusters of globules, vacuoles with the contents wholly or partially removed, clusters of highly curved cell membranes, and odd-shaped domains of high or low density [127]. These spherical objects are variable in size (often in the range 100–250 nm) and occur in clusters that create potential scattering centers.

2.8 Fractal Properties of Tissues and Cell Aggregates

Since a biological tissue is generally composed of complex structures with dimensions ranging from several dozens of nanometers to several millimeters, a fractal-like structure may be employed to investigate the relationship between optical (polarization) properties and structural features [4, 5, 29–31, 44, 77, 134–152]. Particle aggregation is assumed to be a model of physiological and rheological dense media, for example, red blood cell aggregation, blood coagulation, and gelation and aggregation of gastric mucin, etc. [145, 146].

Fractals or fractal objects are either self-similar structures or scale invariant ones [134–138]. Fractals that are found in nature are called random fractals, and their structure shows self-similarity only in a statistical sense. Random fractals are better described by the term "scale invariant" rather than self-similar. The fractal concept enables one to describe such random systems as polymers, colloidal aggregates and tissues [29–31, 44, 77, 134–152]. The fractal dimension is a measure of how the fractal object fills up space. There is some correspondence between the observed complexity or roughness of a random object and its fractal dimensions. The fractal properties of random systems strongly affect their light scattering capability [29–31, 44, 77, 136–151]. The same mass of particles may induce small scattering in a dense cluster and significantly greater scattering in a fractal one. The peculiarities that occur in the multiple scattering of fractals are caused by a slowly falling correlation of particle density [137, 145]. Fractal effects at multiple scattering are observed even for fractal clusters whose sizes are smaller than the wavelength; they are sensitive to light polarization.

Since the spatial distributions of the constituents of many types of tissues appear to satisfy the conditions of statistical self-similarity [140, 143], fractal analysis may potentially provide a much simpler basis for the analysis of tissue. Statistical self-similarity implies that an object is composed of building blocks with inherent statistical regularities that can be described by a power law. The correlations of a variety of tissues in the refractive indices exhibit characteristics of a random fractal with a Hurst coefficient between 0.3 and 0.5 [44].

The tissue structure can be represented as a multifractal composed of various fractal formation types [77, 140]. For bone tissue, the main fractal elements are trabeculae (formations with flatly lying mineralized fibers), and osteons (a region with a spiral-like orientation of fibers raised at an angle of 30–60°) [77]. The above fractal types form an architectonic multifractal network. The geometric dimensions of biofractals may be rather large (100–1000 μm). In many cases, fractal geometry provides a key to understanding the scattering peculiarities of these objects.

The structure of various biological aggregates may be described in terms of statistical (irregular) fractal clusters [152], i.e., statistically self-similar objects with the fractal dimension $D_f < 3$ defined by power relations

$$G(r) \sim (r/R)^{D_f - 3}, \quad N \sim (R/2a)^{D_f}, \tag{2.15}$$

where $G(r)$ is the binary density–density correlation function, N is the number of particles in the aggregate, R is the average size of the aggregates (r.m.s. radius R, gyration radius R_g, etc.), and $2a$ is the size of the monomers. From (2.15), one can see the essential property of fractal aggregates – low average density and large density fluctuations within short-range distances. A direct consequence of this property of the binary density correlations of monomers inside a cluster is the power law for the angular dependence of the static structure factor (normalized intensity) of light, X-ray, or neutron scattering $S(q) \sim (qR)^{-D_f}$, where $q = 2k/\sin(\theta/2), k = 2\pi n/\lambda_0$, and θ is the scattering angle [147].

The above-presented power laws for $G(r)$ and $S(q)$ are observed in the asymptotic sense only, when the value of the scattering vector of the probing irradiation q satisfies the strong inequality $2a \ll q^{-1} \ll R$ [148]. For real objects within the visible and IR range, the condition $qR \gg 1$ is usually not fulfilled rigorously, since the average size of the aggregates does not, as a rule, exceed 1 μm. In these cases, the character of the decrease in the density correlation when approaching the cluster boundary becomes important. This decrease is described by using the so-called cutoff function $h(r/R)$, which is included in the complete correlation function $r^{D_f - 3}h(r/R)$ [149, 150]. Several forms of $h(x = r/R)$, including a single exponential model, $h(x) \sim \exp(-bx)$, have been proposed in the literature (see discussion and relevant citations in [150]). Based on experimental data [136, 150, 151] and computer simulations [149], the following approximation seems to be the most appropriate for fractal aggregates

$$h(x) \approx \exp(-bx^\nu), \quad \nu \approx D_f \approx 2, \tag{2.16}$$

where $b \sim 1$ for reaction limited aggregates, and $b \sim 1/2$ for diffusion limited aggregates [147, 152].

Thus, the scattering centers in tissue have a wide range of dimensions and tend to aggregate into complex forms suggestive of fractal objects. The skewed logarithmic distribution function, which is the most plausible on physical grounds, is used extensively in particle-size analysis. The skewed logarithmic distribution function for the volume fraction of particles of diameter $2a$ has the view [31]:

$$\eta(2a) = \frac{F_v}{C_m} (2a)^{3-D_f} \exp\left[-\frac{\{\ln(2a) - \ln(2a_m)\}^2}{2\sigma_2}\right], \tag{2.17}$$

where

$$C_m = \sigma\sqrt{2\pi}(2a_m)^{4-D_f} \exp[(4 - D_f)^2\sigma^2/2]$$

is the normalizing factor;

$$F_v = \int_0^\infty \eta(2a)\mathrm{d}(2a)$$

is the total volume fraction of the particles; and the quantities $2a_m$ and σ set the center and width of the distribution, respectively; D_f is the (volumetric) fractal dimension.

At the limit of an infinitely broad distribution of particle sizes,

$$\eta(2a) \approx (2a)^{3-D_f}. \tag{2.18}$$

For $3 < D_f < 4$, this power-law relationship describes the dependence of the volume fractions of the subunits of an ideal mass fractal on their diameter $2a$. These size distributions expand the size distributions, described by (2.1) and (2.2), to account for the fractal properties of tissues.

For calculations of the optical coefficients at a high concentration of particles, the size distribution $\eta(2a)$ (2.17) and (2.18) should be replaced by the correlation-corrected distribution, described by (2.12) [31].

Scatterers in the epidermal layer of the skin also exhibit a log-normal size distribution, whereas the spatial fluctuations in the index of refraction of dense fibrous tissues, such as the dermis, follow a power law [44].

2.9 Summary

Biological tissues and cells are optically inhomogeneous and slightly absorbing media in the visible and the NIR ranges. Light propagation and interactions with tissue depend on the optical and structural properties of cells, fibers, and other structural elements making up the tissue. The size range, typical shapes, values of refractive indices, densities, and arrangement of tissue components that are overviewed in this chapter are important for the development and application of adequate theories or approximations for describing polarized light interactions with particular types of tissue. Theoretical approaches that are valid for both weakly and strongly scattering media and corresponding experimental protocols will be considered in Chaps. 3–5.

Tissue models that will be studied further include virtually all areas of dispersion-media optics including (1) simple single scattering approximation; (2) incoherent multiple scattering, described by the radiation transfer equation; and (3) multiple-wave scattering in condensed systems of electrodynamically interacting scatterers.

3

Polarized Light Interactions with Weakly Scattering Media

3.1 Introduction

As demonstrated in Chap. 2, the majority of tissues and cells can be represented as particle systems composed of optically soft scattering particles (with a low degree of refractive index mismatch between the scatterers and ground medium). That allows one to restrict the description of light propagation to a single scattering approximation in a number of cases [1–7]. Such media are considered to be weakly scattering ones. Polarized light interactions with a scattering medium are displayed as a transformation of the polarization state (linear, circular, or elliptical) when the light beam propagates within the medium. To correctly exploit a single scattering approximation, the optical thickness of the object under study must be quite small. In strongly scattering structures, like turbid tissues or blood, this means it is necessary to restrict the technique to thin histological tissue sections and blood monolayers.

The majority of mammalian tissues are structured as densely packed particle systems with characteristic dimensions on the order of the wavelength. Therefore, a certain correlation, which can be accounted for theoretically, should exist between waves scattered by adjacent particles. To arrive at this correlation, it is necessary to sum the amplitudes of the scattered waves with regard to their phase relations. Such interference interaction is expected to result in an essential alteration of the scattered intensity and the polarization characteristics of the scattered light when compared to similar quantities for a system of noninteracting particles.

Models described in this chapter include only certain elements of the theoretical apparatus used in dispersion-media optics, including simple single scattering approximation and scattering in condensed systems of electrodynamically interacting scatterers. More complete theoretical descriptions may be found in [8–20].

Definitions of polarized light, its properties, as well as production and detection techniques are described in a voluminous literature on this topic

[21–28]. Some important definitions, such as Stokes vector and Mueller matrix formalism for polarized light characterization, are presented in Chap. 1.

In this chapter we will consider (1) the fundamentals of polarized light propagation in scattering media and (2) transformation light polarization in scattering anisotropic media using Mueller matrix formalism and the derivation of the Stokes parameters of scattered light.

3.2 Noninteracting Particles

Let us consider the transformation of polarization (linear, circular, or elliptical) in a scattering medium composed of noninteracting particles. The optical softness of the tissue scattering particles makes it possible to utilize a single scattering approximation in a number of cases. To correctly exploit this approximation, the optical thickness of the object under study must be small, $\tau < 0.1$ [14]. In strongly scattering structures, this means a restriction to thin histological sections.

The geometry needed to describe the scattering of light by a particle is shown in Fig. 3.1 [16]. The incident monochromatic plane wave comes from below and travels along the positive z-axis. Some of the light is scattered by the particle along the direction indicated by the vector, $\vec{S}_1$, toward a detector located at a distance r from the particle. The scattering direction is defined by the scattering angle, θ, and azimuthal angle, φ. The scattering plane is originated by the vector $\vec{S}_1$ and the z-axis. The electrical field of the incident

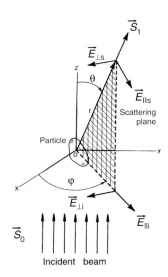

Fig. 3.1. Geometry of the scattering of light by a particle located at the origin [16]. The incident light beam is parallel to the z-axis. A detector is located at a distance r from the origin along the vector $\vec{S}_1$

light is in the $x - y$ plane and can be resolved into components parallel, $\vec{E}_{\|\mathrm{i}}$, and perpendicular, $\vec{E}_{\perp\mathrm{i}}$, to the scattering plane. The electrical field vector and the intensity of the incident light beam are given by

$$\vec{E}_{\mathrm{i}} = \vec{E}_{\|\mathrm{i}} + \vec{E}_{\perp\mathrm{i}}, \tag{3.1}$$

$$I_{\mathrm{i}} = \left\langle E_{\|\mathrm{i}} E_{\|\mathrm{i}}^* + E_{\perp\mathrm{i}} E_{\perp\mathrm{i}}^* \right\rangle, \tag{3.2}$$

where the asterisk denotes complex conjugation and the brackets denote a time average.

The electrical field of the scattered light wave is perpendicular to $\vec{S}_1$ and can be resolved into components $E_{\|\mathrm{s}}$ and $E_{\perp\mathrm{s}}$, which are parallel and perpendicular, respectively, to the scattering plane. The scattered electrical field vector is given by

$$\vec{E}_{\mathrm{s}} = \vec{E}_{\|\mathrm{s}} + \vec{E}_{\perp\mathrm{s}}. \tag{3.3}$$

There is a linear relationship between the incident and scattered field components, defined by (3.1) and (3.3) [16, 30, 31]:

$$\begin{bmatrix} E_{\|\mathrm{s}} \\ E_{\perp\mathrm{s}} \end{bmatrix} = \frac{e^{ik(r-z)}}{-ikr} \begin{bmatrix} S_2 & S_3 \\ S_4 & S_1 \end{bmatrix} \begin{bmatrix} E_{\|\mathrm{i}} \\ E_{\perp\mathrm{i}} \end{bmatrix}, \tag{3.4}$$

where $k = 2\pi/\lambda$, $\lambda = \lambda_0/\bar{n}$ is the wavelength in the scattering medium, $\bar{n}$ is the mean refractive index of the scattering medium, λ_0 is the wavelength of the light in the vacuum, $\mathrm{i} = \sqrt{-1}$, r is the distance from the scatterer to the detector, and z is the position coordinate of the scatterer. The complex numbers S_{1-4} are the elements of the amplitude scattering matrix (S-matrix) or the Jones matrix (J) [16, 21–31]. They each depend on scattering and azimuthal angles θ and φ, and contain information about the scatterer. Both amplitude and phase must be measured to quantify the amplitude scattering matrix. The measurements of the matrix elements can be done using a two-frequency Zeeman laser, which produces two laser lines with a small frequency separation (about 250 kHz) and orthogonal linear polarizations [31], or by the coherence optical tomography (OCT) technique [32].

In terms of the electrical field components the Stokes parameters from (1.1) and (1.2) are given by

$$I = \left\langle E_{\|} E_{\|}^* + E_{\perp} E_{\perp}^* \right\rangle,$$

$$Q = \left\langle E_{\|} E_{\|}^* - E_{\perp} E_{\perp}^* \right\rangle, \tag{3.5}$$

$$U = \left\langle E_{\|} E_{\perp}^* + E_{\perp} E_{\|}^* \right\rangle,$$

$$V = \left\langle \mathrm{i}(E_{\|} E_{\perp}^* - E_{\perp} E_{\|}^*) \right\rangle.$$

All Stokes parameters have the same dimension – energy per unit area per unit time per unit wavelength. For an elementary monochromatic plane or spherical electromagnetic wave [19],

$$I^2 \equiv Q^2 + U^2 + V^2. \tag{3.6}$$

For an arbitrary light beam, as in the case of a partially polarized quasimono-chromatic light that is due to the fundamental property of additivity, the Stokes parameters for the mixture of the elementary waves are sums of the respective Stokes parameters of these waves. Equation (3.6) is replaced by the inequality [16, 19]:

$$I^2 \geq Q^2 + U^2 + V^2. \tag{3.7}$$

The degree of polarization (DOP), the degree of linear polarization (DOLP), and the degree of circular polarization (DOCP) for the incident and scattered light are defined by (1.3). In particular, for the DOLP (P_{L}) and the DOCP (P_{C}) of the scattered light, we have:

$$P_{\mathrm{L}} = (I_\| - I_\perp)/(I_\| + I_\perp) = \sqrt{Q_{\mathrm{s}}^2 + U_{\mathrm{s}}^2}/I_{\mathrm{s}}, \tag{3.8}$$

$$P_{\mathrm{C}} = \sqrt{V_{\mathrm{s}}^2}/I_{\mathrm{s}}. \tag{3.9}$$

The values of the normalized Stokes parameters, which correspond to a certain polarization, are described by (1.5).

In the far field, the polarization of the scattered light is described by the Stokes vector $\mathbf{S}_{\mathrm{s}}$ connected with the Stokes vector of the incident light $\mathbf{S}_{\mathrm{i}}$ (see (1.4) [16]

$$\mathbf{S}_{\mathrm{s}} = \mathbf{M} \cdot \mathbf{S}_{\mathrm{i}}, \tag{3.10}$$

where $\mathbf{M}$ is the normalized 4×4 scattering matrix (intensity or Mueller's matrix):

$$\mathbf{M} = \begin{bmatrix} M_{11} & M_{12} & M_{13} & M_{14} \\ M_{21} & M_{22} & M_{23} & M_{24} \\ M_{31} & M_{32} & M_{33} & M_{34} \\ M_{41} & M_{42} & M_{43} & M_{44} \end{bmatrix}. \tag{3.11}$$

Elements of the light-scattering matrix (LSM) depend on the scattering angle θ, the wavelength, and also the geometrical and optical parameters of the scatterers.

Element M_{11} is what is measured when the incident light is unpolarized, the scattering angle dependence of which is the phase function of the scattered light. It provides only a fraction of the information theoretically available from scattering experiments. M_{11} is much less sensitive to chirality and long-range structure than some of the other matrix elements [16, 31]. M_{12} refers to the degree of linear polarization of the scattered light, M_{22} displays the ratio of depolarized light to the total scattered light (a good measure of the scatterers' nonsphericity), M_{34} displays the transformation of 45° obliquely polarized incident light into circularly polarized scattered light (uniquely characteristic for different biological systems); the difference between M_{33} and M_{44} is a good measure of the scatterers' nonsphericity.

In general, all 16 elements of the LSM are nonzero. However, during scattering by a single particle with a fixed orientation, only 7 elements (out of 16) of the LSM are independent resulting from the 4 complex elements of the amplitude matrix minus an irrelevant phase.

The following nine relations connect the LSM elements of a single particle, or nondepolarizing system of particles, whose Mueller matrix is obtained by coherent summation of the matrices of the individual particles of the system [19, 33, 34]:

$$(M_{11} + M_{22})^2 - (M_{12} + M_{21})^2 = (M_{33} + M_{44})^2 + (M_{34} - M_{43})^2 , \quad (3.12)$$

$$(M_{11} - M_{22})^2 - (M_{12} - M_{21})^2 = (M_{33} - M_{44})^2 + (M_{34} + M_{43})^2 , \quad (3.13)$$

$$(M_{11} + M_{21})^2 - (M_{12} + M_{22})^2 = (M_{13} + M_{23})^2 + (M_{14} + M_{24})^2 , \quad (3.14)$$

$$(M_{11} - M_{21})^2 - (M_{12} - M_{22})^2 = (M_{13} - M_{23})^2 + (M_{14} - M_{24})^2 , \quad (3.15)$$

$$(M_{11} + M_{12})^2 - (M_{21} + M_{22})^2 = (M_{31} + M_{32})^2 + (M_{41} + M_{42})^2 , \quad (3.16)$$

$$(M_{11} - M_{12})^2 - (M_{21} - M_{22})^2 = (M_{31} - M_{32})^2 + (M_{41} - M_{42})^2 , \quad (3.17)$$

$$(M_{13}M_{14} - M_{23}M_{24}) \left(M_{33}^2 - M_{34}^2 + M_{43}^2 - M_{44}^2\right) \quad (3.18)$$
$$= (M_{33}M_{34} + M_{43}M_{44}) \left(M_{13}^2 - M_{14}^2 - M_{23}^2 + M_{24}^2\right) ,$$

$$(M_{31}M_{41} - M_{32}M_{42}) \left(M_{33}^2 - M_{43}^2 + M_{34}^2 - M_{44}^2\right) \quad (3.19)$$
$$= (M_{33}M_{43} + M_{34}M_{44}) \left(M_{31}^2 - M_{41}^2 - M_{32}^2 + M_{42}^2\right) ,$$

$$(M_{14}M_{23} - M_{32}M_{42}) \left(M_{33}^2 - M_{43}^2 + M_{34}^2 - M_{44}^2\right) \quad (3.20)$$
$$= (M_{42}M_{31} + M_{41}M_{32}) \left(M_{14}^2 - M_{24}^2 - M_{13}^2 + M_{23}^2\right) .$$

For particle systems with depolarization, the first six equalities rearrange to inequalities, i.e., in (3.12)–(3.17) we have to change "=" to "≥." During scattering by a collection of randomly oriented scatterers, there are ten independent parameters.

Another important characteristic of LSM is the $\|M\|^2$ quantity [34]:

$$\|M\|^2 = \sum_{i,j}^{4} \overline{M}_{ij}^2, \quad (3.21)$$

where $\overline{M}_{ij}$ is the LSM element normalized to the first one, i.e., M_{11}. The equality $\|M\|^2 = 4$ is the necessary and sufficient condition for a given matrix $\mathbf{M}$ to describe a nondepolarizing biological object. For depolarizing objects, $\|M\|^2$ takes a value from 1 to 4. This quantity serves as a test of the consistency of the experimental data from a light scattering experiment and is especially important in tissue measurements. For example, the square of the norm of the experimental matrices of the human normal and cataract eye lenses is minimal for scattering angles close to 90°, and, correspondingly, equal to 3.3 and 2.5 [34]. This result agrees well with the theoretical estimate [35].

In addition to the degree of light polarization, defined by (1.3), (3.8) and (3.9), diattenuation (linear dichroism) is introduced as

$$D = (P_1^2 - P_2^2)/(P_1^2 + P_2^2) = \sqrt{M_{12}^2 + M_{13}^2 + M_{14}^2}/M_{11}, \qquad (3.22)$$

where P_1 and P_2 are the principal coefficients of the amplitude transmission for the two orthogonal polarization eigenstates.

The LSM for macroscopically isotropic and symmetric media has the well-known block-diagonal structure [14]

$$\mathbf{M}(\theta) = \begin{bmatrix} M_{11}(\theta) & M_{12}(\theta) & 0 & 0 \\ M_{12}(\theta) & M_{22}(\theta) & 0 & 0 \\ 0 & 0 & M_{33}(\theta) & M_{34}(\theta) \\ 0 & 0 & -M_{34}(\theta) & M_{44}(\theta) \end{bmatrix}. \qquad (3.23)$$

In general, only eight LSM elements are nonzero and only six of these are independent. Moreover, there are special relationships for two specific scattering angles 0 and π [19]:

$$M_{22}(0) = M_{33}(0), \; M_{22}(\pi) = -M_{33}(\pi),$$

$$M_{12}(0) = M_{34}(0) = M_{12}(\pi) = M_{34}(\pi) = 0,$$

$$M_{44}(\pi) = M_{11}(\pi) - 2M_{22}(\pi). \qquad (3.24)$$

Rotationally symmetric particles have an additional property [19]:

$$M_{44}(0) = 2M_{22}(0) - M_{11}(0). \qquad (3.25)$$

The structure of the LSM further simplifies for spherically symmetric particles, which are homogeneous or radially inhomogeneous (composed of isotropic materials with a refractive index depending only on the distance from the particle center), because in this case [19]

$$M_{11}(\theta) \equiv M_{22}(\theta), \quad M_{33}(\theta) \equiv M_{44}(\theta). \qquad (3.26)$$

The phase function, i.e., the M_{11} element, satisfies the normalization condition [16, 19]:

$$2\pi \int_0^\pi M_{11}(\theta) \sin\theta d\theta = 1. \qquad (3.27)$$

The quantity

$$g \equiv \langle \cos\theta \rangle = 2\pi \int_0^\pi M_{11}(\theta) \cos\theta \sin\theta d\theta \qquad (3.28)$$

is called the scattering anisotropy parameter (mean cosine of the scattering angle θ) or the asymmetry parameter of the phase function. The g-factor varies from -1 to $+1$. It is positive for particles that scatter predominantly in the forward direction, negative for backscattering particles, and zero for symmetric phase functions with $M_{11}(\pi - \theta) = M_{11}(\theta)$.

The average scattering cross section per particle is given by [16]

$$\sigma_{\text{sca}} = (\lambda^2/2\pi)(1/I_0) \int_0^\pi I(\theta) \sin \theta \mathrm{d}\theta, \qquad (3.29)$$

where I_0 is the intensity of the incident light and $I(\theta)$ is the angle distribution of the scattered light. For macroscopically isotropic and symmetric media, the average scattering cross section is independent of the direction and polarization of the incident light. The average extinction, σ_{ext}, and absorption, σ_{abs}, coefficients are also independent of the direction and polarization state of the incident light:

$$\sigma_{\text{ext}} = \sigma_{\text{sca}} + \sigma_{\text{abs}}. \qquad (3.30)$$

The probability that a photon incident on a small volume element will survive is equal to the ratio of the scattering and extinction cross sections and is called the albedo for single scattering, Λ:

$$\Lambda = \frac{\sigma_{\text{sca}}}{\sigma_{\text{ext}}}. \qquad (3.31)$$

If a particle is small with respect to the wavelength of the incident light, its scattering can be described as the re-emission of a single dipole. This Rayleigh theory is applicable under the condition that $m(2\pi a/\lambda) \ll 1$, where m is the relative refractive index of the scatterers, $(2\pi a/\lambda)$ is the size parameter, a is the radius of the particle, and λ is the wavelength of the incident light in a medium [16]. For NIR light and scatterers with a typical (for biological tissue) refractive index relative to the ground m=1.05–1.11, the maximum particle radius must be about 12–14 nm for Rayleigh theory to remain valid. With this theory, the scattered intensity is inversely proportional to λ^4 and increases as a^6; the angular distribution of the scattered light is isotropic.

The Rayleigh–Gans or Rayleigh–Debye theory addresses the problem of calculating scattering by a special class of arbitrary shaped particles. It requires $|m - 1| \ll 1$ and $(2\pi a'/\lambda)|m - 1| \ll 1$, where a' is the largest dimension of the particle [5, 19, 20, 31]. These conditions mean that the electrical field inside the particle must be close to that of the incident field and that the particle can be viewed as a collection of independent dipoles that are all exposed to the same incident field. A biological cell might be modeled as a sphere of cytoplasm with a higher refractive index ($n_{\text{cp}} = 1.37$) relative to that of the surrounding water medium ($n_{\text{is}} = 1.35$); then $m = 1.015$ for the NIR light. This theory is valid for particle dimensions up to $a' = 0.8$–$1.0\,\mu$m. This approximation has been applied extensively to calculations of light scattering from suspensions of bacteria [31]. It is also applicable for describing light scattering from cell components (mitochondria, lysosomes, peroxisomes, etc.) in tissues with small dimensions and refraction [4, 36–38].

The Fraunhofer diffraction approximation is useful [31] for describing forward direction scattering caused by large particles (on the order of $10\,\mu m$). According to this theory, scattered light has the same polarization as incident light and the scatterer pattern is independent of the refractive index of the object. For small scattering angles, the Fraunhofer diffraction approximation accurately represents a change in intensity as a function of particle size. That is why this approach is applicable in laser flow cytometry.

Mie or Lorenz-Mie scattering theory is an exact solution of Maxwell's electromagnetic field equations for homogeneous spheres [16,24]. In the general case, light scattered by a particle becomes elliptically polarized. The single-scattering Jones and Mueller matrices for a spherical particle of an optically inactive material are presented in Appendix.

3.3 Densely Packed Correlated Particles

A certain correlation exists between waves scattered by adjacent particles in a densely packed medium that has characteristic dimensions on the order of a wavelength. Therefore, it is necessary to sum the amplitudes of scattered waves with regard to their phase relations. The interference interaction may result in an essential alteration of the total scattered intensity, of its angular dependence, or of the polarization characteristics of the scattered light as compared with similar quantities for a system of noninteracting particles.

To illustrate light scattering in a correlated disperse system, we will use a radial distribution function $g(r)$, which is a statistical characteristic of the spatial arrangement of the scatterers [5, 39] (see Fig. 2.6). Let us consider N spherical particles in a finite volume. The pair distribution function $g_{ij}(r)$ is proportional to the conditional probability of finding a particle of type j at distance r from the origin given that there is a particle of type i at the origin (Fig. 3.2). In a model of mutually impenetrable (hard) spheres, the interparticle forces are zero, except for the fact that two neighbor particles cannot interpenetrate each other.

The arrangement of particles in a densely packed system is not entirely random. A short-range order can be observed which is more ordered when the density of the scattering centers is greater and their size distribution is narrower. Near the origin of the coordinates, in the region within the effective particle diameter, the function $g(r) = 0$, which points to the impenetrability of a particle. Function $g(r)$ has a few maxima whose positions correspond to distances from the chosen particle to its first, second, etc. neighbors. Nonzero values of minima are indicative of a particle distribution between various coordination spheres. It is obvious that the correlation between the pairs of particles should be degraded with r; hence, $\lim_{r \to \infty} g(r) = 1$. Function $g(r)$ is the ratio of the local number density of the scattering centers at a distance r from an arbitrary center to the bulk number density.

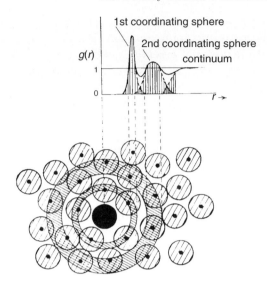

Fig. 3.2. Diagram of radial distribution function $g(r)$ that is proportional to the probability of particle displacement at a certain distance r from an arbitrarily fixed particle [40]

Analogous problems are characteristic for statistical mechanics considering the dynamics and positions of particles with regard to interparticle interactions [40]. The pair distribution function was obtained in the framework of various approximate theories for one, two, or L species in a mixture [41–43]. In a model of hard spheres distributed in three-dimensional space, $g(r)$ exists as the analytical solution of the Percus–Yevick (PY) integral equation. The Monte Carlo statistical simulation is also sometimes used.

For monodispersing systems of spherical particles with a diameter of $2a$, $g(r)$ is represented by an approximation of the hard spheres as follows [44]:

$$g(r) = 1 + \frac{1}{4\pi f} \int_0^\infty \frac{H_3^2(z)}{1 - H_3(z)} \frac{\sin zx}{zx} z^2 dz, \text{ for } x > 1, \qquad (3.32)$$

where $x = r/2a$,

$$H_3(z) = 24f \int_0^1 c_3(x) \frac{\sin zx}{zx} x^2 dx, \quad c_3(x) = -\alpha - \beta x - \delta x^3 \qquad (3.33)$$

$$\alpha = \frac{(1 + 2f)^2}{(1 - f)^4}, \quad \beta = -6f \frac{(1 + 0.5f)^2}{(1 - f)^4}, \quad \delta = \frac{1}{2} f \frac{(1 + 2f)^2}{(1 - f)^4}, \qquad (3.34)$$

f is the volume fraction of particles.

Let us consider light scattering by a system of N spherical particles [45]. In general, the field affecting a particle differs from the field of the incident wave E_i since the latter also contains the total field of adjacent scatterers. Within the single scattering approximation (Born's approximation), the field affecting the particle does not essentially differ from that of initial wave. In cases where double scattering of the field affects the particle, one needs to take the sum of the initial field plus the single-scattered field, and so on [46]. For transparent tissues composed of optically soft quasiregularly packed particles, the use of the single-scattering approximation yields quite satisfactory results [4, 5, 39, 47–62].

A field scattered by a particle with the center defined by radius-vector $\vec{r}_j$ differs from one scattered by a particle placed at the origin of the coordinates by a phase multiplier characterizing the phase shift of the waves. The phase difference is equal to $(2\pi/\lambda)(\vec{S}_0 - \vec{S}_1)\vec{r}_j$, where $\vec{S}_0$ and $\vec{S}_1$ are the unit vectors of the directions of the incident and the scattered waves (see Fig. 3.1). The difference between these vectors is called the scattering vector $\vec{q}$:

$$\vec{q} = \frac{2\pi}{\lambda}(\vec{S}_0 - \vec{S}_1). \tag{3.35}$$

Taking into account that the wavevector module is invariable with elastic scattering, the value of the scattering vector is found as follows:

$$|\vec{q}| \equiv q = \frac{4\pi}{\lambda}\sin(\theta/2), \tag{3.36}$$

where θ is the angle between directions $\vec{S}_0$ and $\vec{S}_1$, i.e., it is the scattering angle. The amplitude of a wave scattered by a system of N particles will be

$$E_{\mathrm{s}} = \sum_{j=1}^{N} E_{\mathrm{s}j} = \sum_{j=1}^{N} E_{0j} e^{\mathrm{i}\vec{q}\vec{r}_j}, \tag{3.37}$$

where E_{0j} is the scattering amplitude of an isolated particle. The single scattering intensity for the given spatial realization of the N particle arrangement is

$$I = |E_{\mathrm{s}}|^2 = \sum_{j=1}^{N} E_{0j} \sum_{i=1}^{N} E_{0i}^* e^{\mathrm{i}\vec{q}(\vec{r}_j - \vec{r}_i)}. \tag{3.38}$$

For real systems, the mean scattering intensity of an ensemble of particles can be only detected, because of thermal particle motion, finite measuring time, and a finite area of a photodetector, such as

$$\langle I \rangle = \left\langle \sum_{j=1}^{N} \sum_{i=1}^{N} E_{0j} E_{0i}^* e^{\mathrm{i}\vec{q}(\vec{r}_j - \vec{r}_i)} \right\rangle. \tag{3.39}$$

The brackets show the averaging over all possible configurations of the particle arrangement in the system. This equation represents the sum of the two contributions to noncoherent scattered intensity. One defines the light distribution on the assumption that there is no interference of light scattered by various particles. The other term regards the interference affect on the light field structure and depends on the degree of order in the particle arrangement that is characterized by the radial distribution function $g(r)$. For an isotropic system of identical spherical particles, we may write [52]

$$\langle I \rangle = |E_0|^2 \, N S_3(\theta), \tag{3.40}$$

$$S_3(\theta) = \left\{ 1 + 4\pi\rho \int_0^R r^2 \, [g(r) - 1] \, \frac{\sin qr}{qr} dr \right\}, \tag{3.41}$$

where q is defined by (3.36), ρ is the mean density of particles, and R is the distance for that $g(r) \rightarrow 1$. Quantity $S_3(\theta)$ is the so-called structure factor. This factor describes the alteration of the angle dependence of the scattered intensity which appears with a higher particle concentration. To approximate the hard spheres used for the derivation of (3.41), the structure factor is equal to

$$S_3(\theta) = 1/[1 - H_3(q)], \tag{3.42}$$

where $H_3(q)$ is defined by (3.33).

For small particle concentrations, the approximation of excluded volume is applicable: $g(r) = 0$ for r that are shorter than the particle diameter and have unity over long distances. In this approximation, the structure factor for a system of spherical particles takes the form:

$$S_3(\theta) = 1 - f\Phi(qa), \tag{3.43}$$

where a is the particle radius and $\Phi(qa)$ is the function defined by the following equation:

$$\Phi(qa) = \frac{3 \, (\sin qa - qa \cos qa)}{(qa)^3}. \tag{3.44}$$

Function $\Phi(qa)$ modulates the angular dependence of the scattering intensity by diminishing its value at small angles and generating a diffusion ring at ten-degree angles for particle dimensions comparable with the wavelength.

For a very small concentration of particles, the structure factor is nearly a unit, and the intensity of scattering by a disperse system is essentially a sum of the contributions of the independent scatterers.

For systems of small soft particles, the structure factor only changes slightly as a function of the scattering angle. Therefore, the particle interaction reveals itself mainly by a uniform decrease in scattering intensity in all directions for linearly polarized and unpolarized incident light (see Fig. 3.3 (calculated by I. L. Maksimova) and Fig. 3.4). For systems of large particles, the structure factor is noticeably less than a unit only in the region of small scattering angles (see Figs. 3.3 and 3.5). In general, particle interaction makes

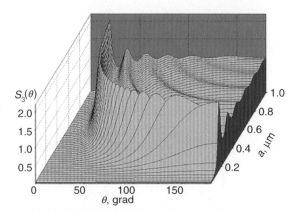

Fig. 3.3. The structure factor $S_3(\theta)$ (3.41) as a function of the scattering angle θ and particle radius a, the wavelength 633 nm, the volume fraction $f = 0.4$, the relative refraction index $m = 1.105$ (calculated by I. L. Maksimova)

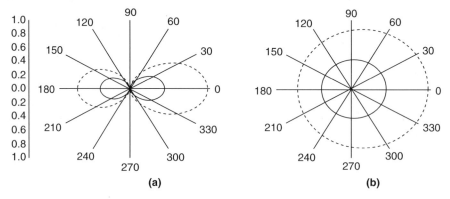

Fig. 3.4. The calculated angular dependencies of the scattered intensity for a system of small spherical particles, 0.02 µm radius, the incident wave is linearly polarized in parallel (**a**) or perpendicular (**b**) to the scattering plane; *dotted line* – independent particles, the wavelength 633 nm, the volume fraction $f = 0.1$, the relative refraction index $m = 1.105$ [45]

the angular dependence of the scattering intensity more symmetric with less overall scattered intensity, and, therefore, allows much more collimated transmittance for both small and large soft particles.

For the case of infinitely long identically aligned cylinders with a radius a and a light that is incident normally to their axes, the structure factor is defined within the approximation of a single scattering, as follows:

$$S_2(\theta) = \left\{ 1 + 8\pi a^2 \rho \int_0^R [g(r) - 1] J_0 \left(\frac{2\pi a}{\lambda} r \sin \frac{\theta}{2} \right) dr \right\}, \qquad (3.45)$$

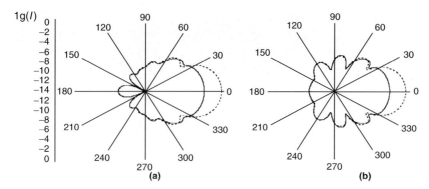

Fig. 3.5. The calculated angular dependences of the scattered intensity for a system of large spherical particles, $0.5\,\mu$m radius, the incident wave is linearly polarized in parallel (**a**) or perpendicular (**b**) to the scattering plane; *dotted line* – independent particles, the wavelength 633 nm, the volume fraction $f = 0.4$, the relative refraction index $m = 1.105$ [45]

where R is the distance for that $g(r) \to 1$. As the light is incident perpendicularly to the cylinder axis, the scattered light propagates only in the direction perpendicular to the axis.

The light scattering intensity angular dependences for systems of spherical and cylindrical particles in the single scattering approximation are described by (3.40), (3.41), and (3.45). The structure factor, which transforms these dependences, is defined by the spatial particle arrangement, and it is independent of the state of light polarization. Therefore, for systems of identical particles, when the single scattering approximation is valid, the angular dependences of all of the elements of LSM are multiplied by the same quantity, accounting for interference interaction (see (3.40)):

$$M_{ij}(\theta) = M_{ij}^0(\theta)NS_3(\theta), \qquad (3.46)$$

where $M_{ij}^0(\theta)$ is the LSM elements for an isolated particle. Consequently, the LSM for the system of monodisperse interacting particles coincides with that of the isolated particle (see (A.10)) if normalization to the magnitude of its first element M_{11} is used.

Unlike in monodispersing systems, in differently sized densely packed particle systems, the normalization of the matrix elements to M_{11} does not eliminate the influence of the structure factor on the angular dependences of the matrix elements. In the simplest case of a bimodal system of scatterers, an expression analogous to (3.41) and (3.45) can be found using four structural functions $g_{11}(r), g_{22}(r), g_{12}(r)$, and $g_{21}(r)$, which characterize the interaction between particles of similar and different sizes [52]. A bimodal system formed by a great number of equally sized small particles, and a minor fraction of coarse ones, provides a good model of pathological tissue, e.g., a cataract eye lens.

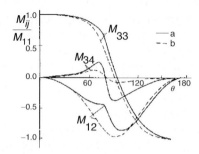

Fig. 3.6. The LSM angular dependences of binary mixture of spherical particles [52]. (a) Calculated by taking into account the interparticle interaction, (b) with independent scatterers. Particle diameters: $2a_1 = 0.06\,\mu$m and $2a_2 = 0.5\,\mu$m, volume fractions: $f_1 = 0.3$ and $f_2 = 0.02$, relative index of refraction $m = 1.07$, the wavelength $0.63\,\mu$m. Calculations done for the random arrangement of scattering particles (neglecting cooperative effects) are shown by a *dashed line*

Fig. 3.6 depicts the calculated results for the LSM of a binary mixture of spherical particles with different diameters and volume fractions. For comparison, the LSM angular dependences of the same binary mixture, neglecting cooperative effects, have also been calculated. It can be seen that the normalized LSM of a dense binary mixture is substantially altered due to the interference interaction. The high concentration of small particles is responsible for the order in their arrangement followed by a lower intensity of scattering in all directions and the higher intensity of straightforwardly propagating light (ballistic component). As a consequence, for the dense mixtures under study, the results of the solution of the inverse problem that were obtained for the experimental dense mixture LSM, neglecting the cooperative effects, should yield an overestimated value of the relative fraction of large particles. The LSM variations due to cooperative effects are of a more complicated nature for a binary system whose two components are sized on the order of the wavelength of incident light and they cannot be interpreted so uniquely as those in the preceding case. Numerical estimates for binary systems of different compositions show [52] the considered effects to be of the most crucial importance for the LSM in the visible region for the mixtures of particles with $2a_1 < 0.2\,\mu$m and $2a_2 > 0.25\,\mu$m.

3.4 Summary

In this chapter, Stokes vector and Mueller matrix formalism for polarized light characterization is described. Such formalism is general and, therefore, applicable for the analysis of polarized light propagation in complex optically inhomogeneous media. As a first step in describing polarized light propagation in tissues, a single scattering approximation that is valid for weakly scattering media is discussed.

Based on Mie theory, a single scattering Mueller matrix for a spherical particle of an optically inactive material is derived. More complex tissue structures, containing particles with a specific shape and composition, can be modeled using the Mie theory concept that the electromagnetic fields of the incident, internal, and scattered waves are each expanded in a series. In particular, Mie theory has been extended to arbitrary coated spheres and to arbitrary cylinders [19, 20, 31]. A linear transformation can be made between the fields in each of the regions. This approach can also be used for nonspherical objects such as spheroids [19, 20]. The linear transformation is called the transition matrix (T-matrix). The T-matrix for spherical particles is diagonal.

Previously, Stokes vectors were defined for the case of a monochromatic plane wave, and the Mueller matrix for single scattering. These concepts are generalized in this chapter to more complicated situations. Stokes vectors were defined for a quasimonochromatic wave [14] and the Mueller matrix for an ensemble of interacting particles [4, 5]. In systems of small soft particles, interaction reveals itself mainly in a uniform decrease of scattering intensity in all directions for linearly polarized and unpolarized incident light. For systems with large particle interactions, it is important only in the region of small scattering angles. In general, particle interaction makes angular dependence of scattering intensity more symmetric with less overall scattered intensity, which, therefore, provides much more collimated transmittance for both small and large soft particles.

For systems with identical particles where a single scattering approximation is valid, the angular dependences of all of the elements of LSM are multiplied by the same quantity to account for interference interaction. Unlike monodisperse systems, the normalization of matrix elements to M_{11} does not eliminate the influence of the structure factor (interaction) on the angular dependences of the matrix elements for different size densely packed particle systems.

4

Polarized Light Interactions with Strongly Scattering Media

4.1 Introduction

The majority of biological tissues are turbid media that display strong scattering characteristics and low absorption rates (up to two orders less absorption coefficient than scattering coefficient for visible and NIR wavelengths). Moreover, in their natural state, (non-sliced) tissues are rather thick. Therefore, multiple scattering is a specific feature of a wide class of tissues [1–7].

The polarization effects of light propagation through various multiply-scattering media, including biological tissues, are fully described by the vector radiative transfer equation [8–10]. Radiative transfer theory (RTT) originated as a phenomenological approach based on the consideration of the transport of energy through a medium filled with a large number of particles which ensures energy conservation [11–14]. This medium, composed of discrete, sparsely, and randomly distributed particles is treated as continuous and locally homogeneous. As discussed in Chap. 3, the concept of single scattering and absorption by an individual particle can be replaced by the concept of single scattering and absorption by a small homogeneous volume element. It was shown that under certain simplifying assumptions, RTT follows logically from the electromagnetic theory of multiple wave scattering in discrete random media (see [9]).

In the framework of RTT, the scattering and absorption of small volume elements follows from the Maxwell equations which are given by the incoherent sums of the respective characteristics of the constituent particles; the result of scattering is not the transformation of a plane incident wave into a spherical scattered wave but, rather, the transformation of the specific intensity vector (Stokes) of the incident light into the specific intensity vector of the scattered light [9].

In this chapter, we will consider the vector radiative transfer equation (VRTE) and its scalar approximation, as well as principles and results of the Monte Carlo modeling of light scattering matrix (LSM) elements for polarized light propagation within multiple scattering media. We will also analyze the specificity of densely packed particle systems.

4.2 Multiple Scattering and Radiative Transfer Theory

4.2.1 Vector Radiative Transfer Equation

For macroscopically isotropic and symmetric plane-parallel scattering media, the VRTE can be substantially simplified as follows [9]:

$$\frac{d\mathbf{S}(\bar{r}, \vartheta, \varphi)}{d\tau(\bar{r})} = -\mathbf{S}(\bar{r}, \vartheta, \varphi)$$

$$+\frac{\Lambda(\bar{r})}{4\pi} \int_{-1}^{+1} d(\cos \vartheta') \int_{0}^{2\pi} d\varphi' \, \bar{\mathbf{Z}}(\bar{r}, \vartheta, \vartheta', \varphi - \varphi')\mathbf{S}(\bar{r}, \vartheta', \varphi'), \qquad (4.1)$$

where $\mathbf{S}$ is the Stokes vector defined by (3.1); $\bar{r}$ is the position vector; ϑ and φ are the angles characterizing the incident direction, the polar (zenith) and the azimuth angles;

$$d\tau(\bar{r}) = \rho(\bar{r}) \langle \sigma_{\text{ext}}(\bar{r}) \rangle \, ds \qquad (4.2)$$

is the optical path-length element; ρ is the local particle number density; $\langle \sigma_{\text{ext}} \rangle$ is the local ensemble-averaged extinction coefficient;, ds is the path-length element measured along the unit vector of the direction of the light propagation; Λ is the single scattering albedo; ϑ' and φ' are the angles that characterize the scattering direction, the polar (zenith) and the azimuth angles, respectively; $\bar{\mathbf{Z}}$ is the normalized phase matrix

$$\bar{\mathbf{Z}}(\bar{r}, \vartheta, \vartheta', \varphi - \varphi') = \mathbf{R}(\Phi)\mathbf{M}(\theta)\mathbf{R}(\Psi); \qquad (4.3)$$

$\mathbf{M}(\theta)$ is the single scattering Mueller matrix, defined by (3.18); θ is the scattering angle, and $\mathbf{R}(\phi)$ is the Stokes rotation matrix for angle ϕ (see Appendix, (A.11)–(A.15), $\phi \equiv \varphi$),

$$\mathbf{R}(\phi) = \begin{bmatrix} 1 & 0 & 0 & 0 \\ 0 & \cos 2\phi & -\sin 2\phi & 0 \\ 0 & \sin 2\phi & \cos 2\phi & 0 \\ 0 & 0 & 0 & 1 \end{bmatrix}. \qquad (4.4)$$

The phase matrix, (4.3), links the Stokes vectors of the incident and scattered beams, specified relative to their respective meridional planes. To compute the Stokes vector of a scattered beam with respect to its meridional plane, one must calculate the Stokes vector of the incident beam with respect to the scattering plane, multiply it by the scattering matrix (to obtain the Stokes vector of the scattered beam with respect to the scattering plane), and then compute the Stokes vector of the scattered beam with respect to its meridional plane. This procedure involves two rotations of the reference plane, shown in Fig. 4.1: $\Phi = -\phi$; $\Psi = \pi - \phi$ and $\Phi = \pi + \phi$; $\Psi = \phi$. The scattering angle θ and the angles Φ and Ψ are expressed via the polar and the

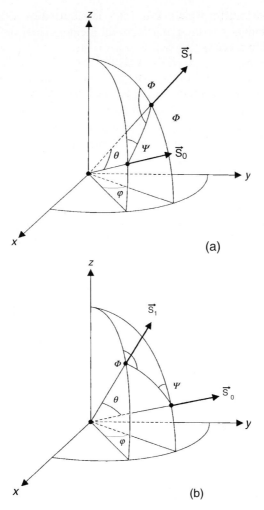

Fig. 4.1. Geometry for rotations of the Stokes vector during a scattering event, adapted from [9]

azimuth incident and scattering angles:

$$\cos\theta = \cos\vartheta'\cos\vartheta + \sin\vartheta'\sin\vartheta\cos(\varphi' - \varphi),$$

$$\cos\Phi = \frac{\cos\vartheta - \cos\vartheta'\cos\theta}{\sin\vartheta'\sin\theta},$$

$$\cos\Psi = \frac{\cos\vartheta' - \cos\vartheta\cos\theta}{\sin\vartheta\sin\theta}. \tag{4.5}$$

The first term on the right-hand side of (4.1) describes the change in the specific intensity vector over the distance ds that is caused by extinction and dichroism; the second term describes the contribution of light illuminating a small volume element centered at $\bar{r}$ from all incident directions and scattered into the chosen direction.

For real systems, the form of the VRTE, (4.1), tends to be rather complex and often intractable. Therefore, a wide range of analytical and numerical techniques have been developed to solve the VRTE. Because of an important property of the normalized phase matrix, (4.3), being dependent on the difference of the azimuthal angles of the scattering and incident directions rather than on their specific values [9], an efficient analytical treatment of the azimuthal dependence of the multiply scattered light, using a Fourier decomposition of the VRTE, is possible. The following techniques and their combinations can be used to solve VRTE: the transfer matrix method, the singular eigenfunction method, the perturbation method, the small-angle approximation, the adding-doubling method, the matrix operator method, the invariant embedding method, and the Monte Carlo (MC) method [5, 6, 8–10, 15, 16].

4.2.2 Scalar Radiative Transfer Equation

When the medium is illuminated by unpolarized light and/or only the intensity of multiply-scattered light needs to be computed, the VRTE can be replaced by its approximate scalar counterpart. In that case, in (4.1) the Stokes vector is replaced by its first element (i.e., radiance) (see (3.1)) and the normalized phase matrix by its (1,1) element (i.e., the phase function) (see (3.34)). The scalar approximation provides poor accuracy when the size of the scattering particles is much smaller than the wavelength but acceptable results for particles comparable to and larger than the wavelength [9, 11]. There is an ample literature on analytical and numerical solutions of the scalar radiative transfer equation [1–5, 7, 12, 14, 16–18].

The description of a continuous wave (CW) of unpolarized light propagation in a scattering medium is possible in the framework of stationary RTT. This theory has been successfully used to work out some practical aspects of tissue optics [1–5, 7]. Usually, in tissue optics, the main stationary equation of RTT for monochromatic light is used in the form that follows from (4.1):

$$\frac{\partial I(\bar{r}, \bar{s})}{\partial s} = -\mu_{e} I(\bar{r}, \bar{s}) + \frac{\mu_{s}}{4\pi} \int_{4\pi} I(\bar{r}, \bar{s}') p(\bar{s}, \bar{s}') d\Omega', \qquad (4.6)$$

where $I(\bar{r}, \bar{s})$ is the radiance (or specific intensity) – average power flux density at a point $\bar{r}$ in the given direction $\bar{s}$ ($\mathrm{W\,cm^{-2}\,sr^{-1}}$);

$$\mu_{e} = \mu_{s} + \mu_{a} \qquad (4.7)$$

is the extinction (interaction or total attenuation) coefficient, $\mathrm{cm^{-1}}$; μ_{s} is the scattering coefficient, $\mathrm{cm^{-1}}$; μ_{a} is the absorption coefficient, $\mathrm{cm^{-1}}$; $p(\bar{s}, \bar{s}')$ is the scattering phase function, $\mathrm{sr^{-1}}$; $d\Omega'$ is the unit solid angle about the direction $\bar{s}'$, sr. It is assumed that there are no radiation sources inside the medium. The extinction coefficient is connected with the extinction cross-section σ_{ext}, defined by (3.37) for a system of scattering and absorption particles as $\mu_e = \rho\sigma_{ext}$; ρ is the particle density.

To characterize the relationship between the scattering and absorption properties of a tissue, a parameter such as the single scattering albedo is usually introduced $\Lambda = \mu_s/\mu_e$ (see (3.38)). The albedo ranges from zero, for a completely absorbing medium, to unity for a completely scattering medium.

If radiative transport is examined in a domain $G \subset R^3$, and ∂G is the domain boundary surface, then the boundary conditions for ∂G can be written in the following general form:

$$I(\bar{r}, \bar{s})\Big|_{(\bar{s}\bar{N})<0} = I_0(\bar{r}, \bar{s}) + \tilde{R}I(\bar{r}, \bar{s})\Big|_{(\bar{s}\bar{N})>0}, \qquad (4.8)$$

where $\bar{r} \in \partial G$, $\bar{N}$, is the outside normal vector to ∂G, $I_0(\bar{r}, \bar{s})$ is the incident light distribution at ∂G, and $\tilde{R}$ is the reflection operator. When both absorption and reflection surfaces are present in domain G, conditions analogous to (4.8) must be given at each surface.

For practical purposes, the integral of the function $I(\bar{r}, \bar{s})$ over certain phase space regions $(\bar{r}, \bar{s})$ is of greater value than the function itself. Specifically, optical probing of tissues frequently measures the outgoing light distribution function at the medium surface, which is characterized by the radiant flux density or irradiance (W cm^{-2}):

$$F(\bar{r}) = \int_{(\bar{s}\bar{N})>0} I(\bar{r}, \bar{s})(\bar{s}\bar{N})d\Omega, \qquad (4.9)$$

where $\bar{r} \in \partial G$.

Often in tissue optics the measured quantity is actually the total radiant energy fluence rate $U(\bar{r})$. It is the sum of the radiance over all angles at a point $\bar{r}$ and is measured by W cm^{-2}:

$$U(\bar{r}) = \int_{4\pi} I(\bar{r}, \bar{s})d\Omega. \qquad (4.10)$$

The phase function $p(\bar{s}, \bar{s}')$ describes the scattering properties of the medium and is, in fact, the probability density function for scattering in the direction $\bar{s}'$ of a photon traveling in the direction $\bar{s}$. In other words, it characterizes an elementary scattering act. If scattering is symmetric relative to the direction of the incident wave, then the phase function depends only on the scattering angle θ (angle between directions $\bar{s}$ and $\bar{s}'$) (see (3.34)), i.e.,

$$p(\bar{s}, \bar{s}') = p(\theta) \equiv M_{11}(\theta). \qquad (4.11)$$

The assumption of a random distribution of scatterers in a medium (i.e., the absence of spatial correlation in the tissue structure) leads to normalization as described by (3.34).

In practice, the phase function is usually well approximated with the aid of the postulated Henyey–Greenstein function [1–5, 7]:

$$p(\theta) = \frac{1}{4\pi} \frac{1 - g^2}{(1 + g^2 - 2g \cos \theta)^{3/2}}, \tag{4.12}$$

where g is the scattering anisotropy parameter (mean cosine of the scattering angle θ), described by (3.35). In application to biological tissues and liquids, the value of g varies in a range from 0 to 1 : $g = 0$ corresponds to isotropic (Rayleigh) scattering and $g = 1$ to total forward scattering (Mie scattering of large particles).

In general, the integro-differential equation (4.6) is too complicated to be employed for the analysis of light propagation in scattering media. Therefore, it is frequently simplified by representing the solution in the form of spherical harmonics. Such simplification leads to a system of $(N + 1)^2$ connected differential partial derivative equations known as the P_N approximation. This system is reducible to a single differential equation of the order $(N + 1)$. For example, four connected differential equations reducible to a single diffusion-type equation are necessary for $(N = 1)$ [1–5]. It has the following form for an isotropic medium:

$$(\nabla^2 - \mu_d^2) U(\bar{r}) = -Q(\bar{r}), \tag{4.13}$$

where

$$\mu_d = [3\mu_a(\mu_s' + \mu_a)]^{1/2} \tag{4.14}$$

is the inverse diffusion length, cm^{-1};

$$Q(\bar{r}) = D^{-1}q(\bar{r}), \tag{4.15}$$

$q(\bar{r})$ is the source function (i.e., the number of photons injected into the unit volume),

$$D = \frac{c}{3(\mu_s' + \mu_a)} \tag{4.16}$$

is the photon diffusion coefficient, cm^2 c^{-1};

$$\mu_s' = (1 - g)\mu_s \tag{4.17}$$

is the reduced (transport) scattering coefficient, cm^{-1}; and c is the velocity of light in the medium. The transport mean free path (MFP) of a photon (cm) is defined as

$$l_t = \frac{1}{\mu_s' + \mu_a}. \tag{4.18}$$

It is worthwhile to note that the transport MFP l_t is the distance over which the photon loses its initial direction and, in a medium with anisotropic single scattering, significantly exceeds the MFP in a medium with isotropic single scattering, l_{ph}:

$$l_t >> l_{ph} \equiv \frac{1}{\mu_e}. \tag{4.19}$$

Diffusion theory provides a good approximation in the case of a small scattering anisotropy factor $g \leq 0.1$ and large albedo $\Lambda \rightarrow 1$. For many tissues, $g \approx 0.6$–0.9 and can be as large as 0.990–0.999, in blood for example [1–5]. This significantly restricts the applicability of the diffusion approximation. It is argued that this approximation can be used at $g < 0.9$ when the optical thickness τ of an object is of the order 10–20:

$$\tau = \int\limits_0^s \mu_e ds. \tag{4.20}$$

The so-called first order solution is realized for optically thin and weakly scattering media ($\tau < 1, \Lambda < 0.5$) when the intensity of a transmitted (coherent) wave is described by expression [19]:

$$I(s) = (1 - R_F)I_0 \exp(-\tau), \tag{4.21}$$

where the incident intensity $I_0(\mathrm{W\,cm^{-2}})$ is defined by the incident radiant flux density or irradiance (see (4.9)) F_0 and a solid angle delta function points in the direction $\overline{\Omega}_0 : I_0 = F_0 \delta(\overline{\Omega} - \overline{\Omega}_0)$; R_F denotes the Fresnel reflection on the tissue boundary. Given a narrow beam (e.g., a laser), this approximation may be applied to denser tissues ($\tau > 1, \Lambda < 0.9$). However, certain tissues have $\Lambda \approx 1$ in the diagnostic and therapeutic wavelength window range (see Fig. 2.2) which makes the first order approximation inapplicable even at $\tau \ll 1$.

A more strict solution of the transport equation is possible using the discrete ordinates method (multiflux theory) in which (4.6) is converted into a matrix differential equation for illumination along many discrete directions (angles) [13]. The solution approximates an exact one as the number of angles increases. It has been shown above that the fluence rate can expand in powers of the spherical harmonics, separating the transport equation into the components for spherical harmonics. This approach also leads to an exact solution provided the number of spherical harmonics is sufficiently large. For example, in a study of tissues that made use of up to 150 spherical harmonics [20], the resulting equations were solved by the finite-difference method [21]. However, this approach requires tiresome calculations if a sufficiently exact solution is to be obtained. Moreover, it is not very suitable for δ-shaped phase scattering functions [19].

Tissue optics extensively employs simpler methods for the solution of transport equations, e.g., the two-flux Kubelka–Munk theory or the three, four, and seven-flux models. Such representations are natural and very fruitful for laser tissue probing. Specifically, the four-flux model [13,22] is actually two diffuse fluxes traveling to meet each other (Kubelka–Munk model) and two collimated laser beams, the incident one and the one, and the one reflected from the rear boundary of the sample. The seven-flux model is the simplest three-dimensional representation of scattered radiation and an incident laser beam in a semi-infinite medium [23]. Of course, the simplicity and the possibility of expeditious calculation of the radiation or the rapid determination of tissue optical parameters (solution of the inverse scattering problem) are achieved at the expense of accuracy.

4.3 Monte Carlo Simulation Technique

The MC method that is widely used for the numerical solution of the RTT equation [10, 24, 25] in different fields (astrophysics, atmosphere and ocean optics, etc.) appears to be especially promising, in particular for the purposes of medical optical tomography and spectroscopy [1–5, 22, 26–36], for the solution of direct and inverse radiation transfer problems for media with arbitrary configurations and boundary conditions. The method is based on the numerical simulation of photon transport in scattering media. Random migrations of photons inside a sample can be traced from their input until absorption or output.

A straightforward simulation using the MC method has the following advantages:

– One can employ any scattering matrix; there are no obstacles to the use of strongly forward-directed phase functions or experimental single scattering matrices;
– The polarization calculation requires a computation time that is only twice that needed for the evaluation of the intensity;
– Any reasonable number of detectors can be accounted for without a noticeable increase in the computation time; there are no difficulties in determining the radiation parameters inside the medium;
– It is possible to model media with complex geometries where radiance depends not only on the optical depth, but also on the transverse coordinates.

The liability of the obtained results to statistical variations on the order of a few percentages with acceptable computation times is the main disadvantage of the MC technique. For a twofold increase in accuracy, one needs a fourfold increase in computation time. The MC method is also impractical for great optical depths ($\tau > 100$).

A few MC codes for the modeling of polarized light propagation through a scattering layer are available in the literature (see, for example, [10, 24, 28, 32–36]). To illustrate the MC simulation technique applied to modeling the angular dependencies of the scattering matrix elements in this section, the algorithm described in [35] is discussed. Another example demonstrating the application of a time-resolved MC technique to tissue imaging is discussed in Chap. 9.

Let a flux of photons within an infinitely narrow beam be incident exactly upon the center of the spherical volume filled up by the scattering particles [35]. The path of a single photon migration in the medium is accounted for in a process of computer simulation. The photons are considered in this case to be ballistic particles. Different evens possible in the course of the photon migration are estimated by the appropriate probability distributions. In the model under study, the photons would either be elastically scattered or absorbed during their collisions with the medium particles. A certain outcome of every event is found by a set of uniformly distributed random numbers. The probability of scattering in a given direction is determined in accordance with the scattering at a single particle. One is able to specify the cross-section of the scattering and the values of the scattering matrix elements for every photon interaction with a scatterer.

When an incident photon enters a scattering layer, it is allowed to travel a free pathlength, l. The l value depends on the particle concentration, ρ, and extinction cross-section, σ_{ext}. The free pathlength l is a random quantity that takes any positive value with the probability density, $p(l)$ [25]:

$$p(l) = \rho \sigma_{\text{ext}} e^{-\rho \sigma_{\text{ext}} l}. \tag{4.22}$$

The particular realization of the free pathlength l is dictated by the value of a random number γ that is uniformly distributed over the interval [0], [1]:

[0,1]

$$\int_0^l p(l)\, \mathrm{d}l = \gamma. \tag{4.23}$$

Substituting (4.22) into (4.23) yields the value l of the certain realization in the form

$$l = \frac{1}{\rho \sigma_{\text{ext}}} \ln \gamma. \tag{4.24}$$

If the distance l is larger than the thickness of the scattering system, then this photon is detected as transmitted without any scattering. If, having passed the distance l, the photon remains within the scattering volume, then the possible events of the photon–particle interaction (scattering or absorption) are randomly selected.

Within the spherical system of coordinates, the probability density of photon scattering along the direction specified by the angle of scattering θ between the directions of the incident and scattered photons and by the angle ϕ between the previous and new scattering planes, $p(\theta, \phi)$:

$$p(\theta, \phi) = \frac{I_s(\theta, \phi) \sin \theta}{\int_0^{2\pi} \int_0^{\pi} I_s(\theta, \phi) \sin \theta\, \mathrm{d}\theta\, \mathrm{d}\phi}, \tag{4.25}$$

where $I_s(\theta, \phi)$ is the intensity of the light scattered in the direction (θ, ϕ) in respect to the previous direction of the photon, defined by angles ϑ and φ (see (4.1) and (4.5)). For spherical particles, this intensity is given by Mie formulas with allowances for the state of polarization of each photon. An integral $I_s(\theta, \phi)$ over all scattering directions, similar to (3.36), determines the scattering cross-section

$$\sigma_{sca} = \int\limits_0^{2\pi} \int\limits_0^{\pi} I_s(\theta, \phi) \sin\theta \; d\theta \; d\phi. \qquad (4.26)$$

The probability density of the photon scattering along the specified direction, $p(\theta, \phi)$, depends on the Mueller matrix of the scattering particle $\mathbf{M}(\theta, \phi)$ (a single scattering matrix) and the Stokes vector $\mathbf{S}$ associated with the photon ((3.4) and (3.17)). The single scattering Mueller matrix $\mathbf{M}(\theta, \phi)$ links the Stokes vectors of the incident $[\mathbf{S}_i\,(0,0)]$ and scattered $[\mathbf{S}_s(\theta, \phi)]$ light (see Appendix, (A.1)–(A.10)). For spherical scatterers, the elements of this matrix may be factorized:

$$\mathbf{M}(\theta, \phi) = \mathbf{M}(\theta)\mathbf{R}(\phi). \qquad (4.27)$$

The single scattering matrix $\mathbf{M}(\theta)$ of spherical particles has the form, described by (3.30) and (3.33). The elements of this matrix are given by Mie formulas (A.2)–(A.10), which are functions of the scattering angle θ and the diffraction parameter $x = 2\pi a/\lambda$, where a is the radius of the spherical particle, and λ is the wavelength in the medium.

The matrix $\mathbf{R}(\phi)$ describes the transformation of the Stokes vector under rotation of the plane of scattering through the angle ϕ, which is defined by (4.4). Thus, the intensity of the light scattered by spherical particles is determined by the expression

$$I_s(\theta, \phi) = [M_{11}(\theta)I_i + (Q_i \cos 2\phi + U_i \sin 2\phi)M_{12}(\theta)], \qquad (4.28)$$

where Q_i and U_i are components of the Stokes vector of the incident light (see (3.4) and (3.17)). As follows from this equation, the probability $p(\theta, \phi)$ (4.25), unlike the scattering matrix (4.27), cannot be factorized. It appears to be parametrized by the Stokes vector associated with the scattered photon. In this case, one should use a rejection method to evaluate $p(\theta, \phi)$.

The following method of generating pairs of random numbers with the probability density $p(\theta, \phi)$ may be used [35,36]. In a three-dimensional space, the function $p(\theta, \phi)$ specifies some surface. The values (θ, ϕ) corresponding to the distribution $p(\theta, \phi)$ are chosen using the following steps:

(1) A random direction $(\theta_\gamma, \phi_\gamma)$ with a uniform spatial distribution is selected; the values of the random quantities θ_γ, and ϕ_γ distributed over the intervals $(0, \pi)$ and $(0, 2\pi)$, respectively, are found from the equations

$$\cos\theta_\gamma = 2\gamma - 1, \quad \phi_\gamma = 2\pi\gamma, \qquad (4.29)$$

where γ is a random number uniformly distributed over the interval (0,1).

(2) The surface specified by the function $p(\theta, \phi)$ is surrounded by a sphere of radius $\widehat{R}$, equal to the maximum value of the function $p(\theta, \phi)$; a random quantity $r_\gamma = \gamma \widehat{R}$ is generated.

(3) The direction $(\theta_\gamma, \phi_\gamma)$ is accepted as the random direction of the photon scattering at this stage, provided the condition $r_\gamma \leq p(\theta_\gamma, \phi_\gamma)$ is satisfied. In the opposite case, steps 1 and 2 are repeated.

The migration of the photon in the scattering medium can be described by a sequence of transformations for the related coordinate system. Each scattering event is accompanied by a variation of the Stokes vector associated with the photon. The new Stokes vector $\mathbf{S}_{n+1}$ is a product of the preceding Stokes vector, transformed to the new scattering plane, and the Mueller matrix (or LSM) $\mathbf{M}_k(\theta)$ of the scattering particle:

$$\mathbf{S}_{n+1} = \mathbf{M}_k(\theta)\mathbf{R}_n(\phi)\mathbf{S}_n, \qquad (4.30)$$

where the matrix $\mathbf{R}_n(\phi)$ (see (4.4)) describes the rotation of the Stokes vector around the axis specifying the direction of propagation of the photon before the interaction.

For the chosen scattering direction, the Stokes vector is recalculated using (3.17), (3.30), (3.33), (A.2)–(A.10), and (4.29). The value thus obtained is renormalized so that the intensity remains equal to unity. Thus, the Stokes vector associated with the photon contains information only about the variation of the state of polarization of the scattered photon. Real intensity is determined by measuring the number of detected photons in the chosen direction within the detector aperture.

The above procedure is repeated as long as the photon appears to be outside the scattering volume. In this case, if the photon propagation direction intersects the surface of the detector, the photon is detected. Upon detection, the Stokes vector is rotated from the current plane of the last scattering to the scattering plane of the laboratory coordinate system. The values obtained are accumulated in the appropriate cells of the detector whose number is defined by the photon migration direction. Furthermore, with registering, the photon is classified in accordance with the scattering multiplicity and the length of the total path. For every non-absorbed photon, the direction and the coordinates of the point at which it escapes the scattering volume, as well as the number of scattering acts it has experienced, are also recorded. The spatial distribution of the radiation scattered by the scattering volume can be obtained with regard to polarization by analyzing the above data for a sufficiently large number of photons.

To find the full LSM of an object, we detect the light scattering for four linearly independent states of polarization of incident light, $\mathbf{S}_{1i}$, $\mathbf{S}_{2i}$, $\mathbf{S}_{3i}$, and $\mathbf{S}_{4i}$. This allows us to construct the following system of linear equations

$$\mathbf{CM}' = \mathbf{S}', \qquad (4.31)$$

where $\mathbf{M}'$ is the column matrix composed of the found matrix elements of the LSM of the object and $\mathbf{S}'$ is the 16-element vector containing the Stokes vector elements recorded upon light scattering for the four independent states of incident light polarization. The transformation matrix $\mathbf{C}$ is determined by the choice of the initial set of the Stokes vectors of incident light. Having solved this system of equations for the following set of Stokes vectors $\mathbf{S}_{1i} = (1, 1, 0, 0)$, $\mathbf{S}_{2i} = (1, -1, 0, 0)$, $\mathbf{S}_{3i} = (1, 0, 1, 0)$, and $\mathbf{S}_{4i} = (1, 0, 0, 1)$, one can find the desired LSM of the object, $\mathbf{M}' = \mathbf{M}$:

$$\mathbf{M} = \frac{1}{2} \begin{bmatrix} I_1 + I_2 & I_1 - I_2 & 2I_3 - (I_1 + I_2) & 2I_4 - (I_1 + I_2) \\ Q_1 + Q_2 & Q_1 - Q_2 & 2Q_3 - (Q_1 + Q_2) & 2Q_4 - (Q_1 + Q_2) \\ U_1 + U_2 & U_1 - U_2 & 2U_3 - (U_1 + U_2) & 2U_4 - (U_1 + U_2) \\ V_1 + V_2 & V_1 - V_2 & 2V_3 - (V_1 + V_2) & 2V_4 - (V_1 + V_2) \end{bmatrix},$$

$$(4.32)$$

where the elements of the Stokes vectors of the scattered light obtained in each of the four cases are denoted as $\mathbf{S}_n = (I_n, Q_n, U_n, V_n), n = 1, 2, 3, 4$. As a result, we can calculate the angular dependencies for all of the elements of LSM with allowances for the contributions of multiple scattering.

A simulation was performed for the systems of spherical particles with a relative index of refraction of $m = 1.2$; these are uniformly distributed within a spherical volume at volume fraction $f = 0.01$ [35]. In the calculations, the illuminating beam is assumed to be infinitely narrow and incident exactly upon the center of the scattering volume in the zero angle direction, and the scattered radiation is detected at different scattering angles in the far zone by a detector with a full angular aperture of $1°$ in the scattering plane and $5°$ in a plane that is perpendicular to the scattering one.

The angular distributions of the total scattering intensity for different scattering systems of spherical particles with a small radius, $a = 0.05\,\mu\text{m}$, or a large radius, $a = 0.3\,\mu\text{m}$, are presented in Fig. 4.2. The average multiplicity of the scattering of the detected radiation increases with increasing dimensions of the scattering system. For systems of small particles at illumination in the visible range, the Rayleigh approximation is applicable. For rather small dimensions of scattering volume, less than 1 mm in diameter, the contribution of single scattering is predominant. It follows from the intensity angular dependence, which is rather isotropic (compare Fig. 4.2a with Fig. 3.4 presenting a single scattering angular dependences for non-interacting particles). As the dimensions of the scattering system increase, the fraction of the contributions from higher multiplicity scattering grows as well. For a 20 mm diameter system, the detected light contains noticeable contributions of scattering from the 10th—20th multiplicity. With a further increase in system dimensions, most of the incident light is scattered in the backward direction and the scattering intensity in the forward half-plane vanishes. For this reason, after a certain value, the dimensions of the scattering system hardly affect the shape of the diagram of the scattering multiplicity distribution.

Systems comprised particles with a size on the order of the wavelength (Fig. 4.2b) also show an increase in the contributions from higher order

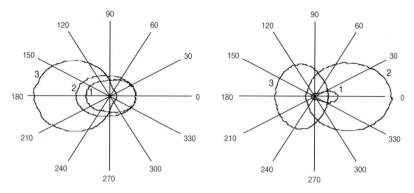

Fig. 4.2. Angular distributions of the total scattering intensity for multiply-scattering systems of spherical particles having a relative refractive index of $m = 1.2$ and uniformly distributed within a spherical volume at the volume fraction $f = 0.01$: (**a**) particles with small radius, $a = 0.05\,\mu$m, diameter of the system is equal to (1) 1, (2) 2, and (3) 20 mm; and (**b**) particles with large radius, $a = 0.3\,\mu$m, diameter of the system is equal to (1)2, (2)200, and (3)2,000 μm; the infinitely narrow unpolarized light beam incidents exactly upon the center of the scattering volume in the zero angle direction; the wavelength is 633 nm [35]

scattering with increasing dimensions of the scattering system. The system transforms from the forward to backward directed scattering mode at a rather small thickness, 2 mm in the diameter.

As can be seen, the intensity of unpolarized light at a higher scattering multiplicity weakly depends on the scattering angle and carries almost no information about the size of the scattering particles. Note that systems of small particles, at triple scattering, may already be considered nearly isotropic, while angular distributions for large particles, strongly elongated in the forward direction at single scattering (see Fig. 3.5), remain anisotropic for sufficiently high scattering multiplicity (four to six scattering events for a system of 0.2 mm in diameter, Fig. 4.2b).

The view of the LSM elements' angular dependences under the conditions of multiple scattering differs substantially from that for the LSM of a single-scattering system. It is seen from Figs. 4.3 and 4.4 that multiple scattering flattens the angular dependences of the LSM elements. The solid line shows the results of the calculation of a normalized LSM for an isolated spherical particle with a similar radius and relative index of refraction. All elements of the LSM are normalized to the M_{11} element (total scattering intensity) along the given direction, and the element M_{11} is presented in the plot as normalized to unity in the forward direction.

Since the single scattering angular distribution for particles with sizes substantially exceeding the Rayleigh limit is strongly asymmetric, the scattering intensity at large angles is very low. For this reason, one must trace the trajectories of a great number of photons to obtain good accuracy in this

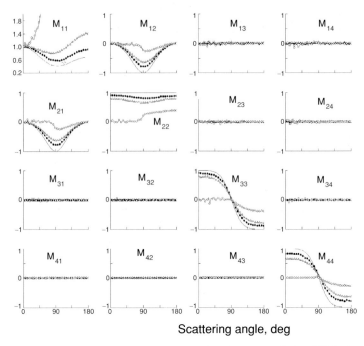

Scattering angle, deg

Fig. 4.3. The MC simulation: the angular distributions of the LSM elements for the multiple scattering systems of small spherical particles ($a = 0.05\,\mu m, m = 1.2$) uniformly distributed within a spherical volume ($f = 0.01$); diameter of the system is equal to 1 mm ($-\bullet-$), 2 mm ($-\triangle-$), and 20 mm ($-\circ-$); the solid line shows the results of calculations of the approximation of a single scattering; the infinitely narrow unpolarized light beam incidents exactly upon the center of the scattering volume in the zero angle direction; the wavelength is 633 nm [35]

angular range. Therefore, to demonstrate the fine structure of the angular dependence of the matrix elements, we need to use 10^7–10^8 photons in the simulation [33, 35].

When scattering by particle suspensions in a spherical cell of small diameter occurs, almost all of the detected photons are singly scattered. An increase in the optical thickness considerably enhances the contribution of multiple scattering. The angular dependences of the LSM elements have a form close to the single scattering LSM, provided that the optical thickness of the scattering system τ does not exceed unity for systems of large particles considered ten times or more larger than systems of small particles.

The multiple-scattering intensity (the element M_{11}) for a cell of large diameter decreases with an increasing scattering angle more slowly than the single-scattering intensity. As the cell diameter further increases, backward scattering becomes predominant (see Figs. 4.2–4.4). In systems of small particles (see Fig. 4.3), the growth of the multiple scattering contributions is accompanied by a gradual decrease in the magnitude of all the elements except

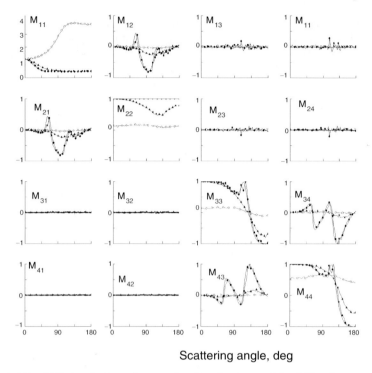

Scattering angle, deg

Fig. 4.4. The MC simulation: the angular distributions of the LSM elements for the multiple scattering systems of large spherical particles ($a = 0.3\,\mu\mathrm{m}$, $m = 1.2$) uniformly distributed within a spherical volume ($f = 0.01$); the diameter of the system is equal to 0.002 mm (–•–), 0.2 mm (–Δ–), and 2 mm (–○–); the solid line shows the results of the calculations of the approximation of single scattering; the infinitely narrow unpolarized light beam incidents exactly upon the center of the scattering volume in the zero angle direction; the wavelength is 633 nm [35]

for M_{11}; i.e., the form of the LSM approaches that of the ideal depolarizer. The magnitudes of the elements M_{12} and M_{21} decrease in nearly the same way while the elements M_{33} and M_{44} also decrease in magnitude but M_{44} decreases faster. As a result, multiple scattering gives rise to a difference in the detected values of the elements M_{33} and M_{44} even for systems of spherical particles. The values of the element M_{22} eventually become smaller than unity, this decrease being more substantial in the range of scattering angles close to 90°. Thus, the manifestation of the effect of multiple scattering in monodisperse systems of spherical particles, which is revealed in the appearance of non-zero values of the difference $|M_{33} - M_{44}|$ and $|1 - M_{22}|$, is similar to the manifestation of the effect of non-sphericity on the scatterers observed under conditions of single scattering [37].

For large particle systems, multiple scattering also decreases the magnitudes and smoothes out the angular dependences of the normalized elements of the LSM (see Fig. 4.4). The corresponding angular dependences, as compared to the LSM of small particles, show the following specific features: the minimum value of the element M_{22} is reached not at 90° but rather at large scattering angles; the fine structure of the angular dependences for all of the elements are smeared, even in the presence of a small fraction of multiply-scattered light; and finally, the very important result that the element M_{44}, unlike other elements, at the limit of high scattering multiplicity, tends to 0.5 rather than zero for all scattering angles. Such a form of the LSM means that the radiation scattered by the large particles retains circular polarization at higher scattering multiplicities. This result serves as a confirmation of the preferential survival of some types of polarization under conditions of multiple scattering for different sizes of scattering particles or tissue structures [28, 38].

The process of multiple scattering photons during their migration can be described as a series of successive rotations of their coordinate systems, determined by the scattering planes and directions. Since these rotations are random, the detected photons are randomly polarized and, hence, the detected light is partially depolarized. Depolarization increases with the increasing multiplicity of the scattering. An important integral characteristic of the depolarizing ability of the scattering object is the quantity $\|M\|^2$ (see (3.28)). For a non-depolarizing object, it is equal to 4; for a depolarizing object, it takes values between 4 and 1. For example, the depolarizing ability of systems of small particles ($a = 0.05\,\mu m$) increases monotonically with increasing optical thickness, i.e., $\|M\|^2$ calculated for $\theta = 90°$ changes from 3 to 2.2, and further to 1, as the diameter of the system goes from 1 to 2, and further to 20 mm at a particle volume fraction of $f = 0.01$. The depolarization weakly depends on the scattering angle and shows a weakly pronounced maximum in direction close to $\theta = 90°$ [35, 39]. These tendencies fit well the values of the squared norm of the experimental matrices of the human normal and cataract eye lenses, which are minimal for $\theta = 90°$, and correspondingly equal to 3.3 and 2.5 [40]. For systems of large particles ($a = 0.3\,\mu m$) at optical thicknesses where a single-scattering regime dominates, $\|M\|^2 \approx 4$ for all scattering angles. For optically thick systems (2 mm, $f = 0.01$), $\|M\|^2$ is close to unity for all scattering angles. For moderate optical thicknesses (0.2 mm, $f = 0.01$), the depolarizing ability is strongly different for different directions. The scattered light may be almost completely polarized in the region of the small scattering angles, completely depolarized at the large angles ($\theta = 120°$) and partly polarized in the backward direction. The angular range of the strongest depolarization corresponds to the angle at which the element M_{22} acquires a minimum value (see Fig. 4.4).

The simulated dependences allow one to estimate the limits of applicability of the single scattering approximation when interpreting the results of experimental studies of disperse scattering systems. It follows from these simulations that modifications of the LSM of monodisperse systems of spherical particles due to the effects of multiply scattering have much in common with modifications of the LSM of singly scattering systems due to deviation in the shape of the particles from spherical. This fact imposes serious limitations on the application of the measured LSM of biological objects for solving the inverse problem to determine particle nonsphericity. The appropriate criteria to distinguish the effects of multiple scattering and particle nonsphericity have yet to be developed.

It is important to note that the comparison of the MC simulation, which accounted for all of the orders of multiple scattering, with the analytical double-scattering model indicated no essential change in the backscattering polarization patterns [32,41]. This is due to the fact that the main contribution comes from the near-double-scattering trajectories in which light suffers two wide-angle scatterings and many near-forward scatterings among the multiple-scattering trajectories. The contributions of such multiple, but near-double, scattering trajectories are obviously well approximated by the contributions of the corresponding double-scattering trajectories.

The above MC technique of photon trajectory modeling is well suited to the simulation of multiple scattering effects in a system of randomly arranged particles. Furthermore, this scheme allows for an approximate approach to describe the interference effects caused by space particle ordering. To this end, one should include the interference of scattered fields into calculations of the single scattering Mueller matrix and integral cross-sections of a particle. In other words we account for the interference effects of the simulation of the *single scattering* properties at the first stage, and then use these properties in the MC simulation of *multiple scattering*. Such an approach is admissible if the size of the region of the local particle ordering is substantially smaller than the mean free photon pathlength.

4.4 Densely Packed Particle Systems

The spatial correlation of individual scatterers results in the necessity to consider the interference of multiply scattered waves. The particle reradiation in the densely packed disperse system induces the distinction of an effective optical field from the incident one in a medium. Under these conditions, the statistical theory of multiple wave scattering seems to be most promising for describing the collective interaction between an ensemble of particles and electromagnetic radiation [42, 43].

The rigorous theory of wave multiple scattering is constructed on the basis of fundamental differential equations for the fields, followed by using statistical considerations [42]. The total field $\vec{E}(\vec{r})$ at the point $\vec{r}$ is the sum of the incident field $\vec{E}_i(\vec{r})$ and the scattered fields from all particles with regard to their phases,

$$\vec{E}(\vec{r}) = \vec{E}_i(\vec{r}) + \sum_{j=1}^{N} \vec{E}_j^s(\vec{r}), \qquad (4.33)$$

where $\vec{E}_j^s(\vec{r})$ is the scattered field of the jth particle. The field scattered by the jth particle is defined by the parameters of this particle and by the effective field incident on the particle.

Twersky has derived a closed system of integral equations describing the processes of multiple scattering [44]. A rigorous solution in a general form has not been found yet for this problem. In actual calculations, various approximations are exploited in order to perform the averaging of (4.33) over statistical particle configurations. For example, the quasi-crystalline approximation proposed for densely packed media by Lax [45] is the most efficient for tissue optics.

The averaging of (4.33) over statistical particle configurations results in an infinite set of equations that is truncated at the second step by applying the quasi-crystalline approximation. The closed system of equations obtained for the effective field is reduced to a system of linear equations by expansion in terms of vector spherical or cylindrical harmonics. The explicit expressions [46, 47] for the expansion coefficients involve the radial distribution function as well as the Mie coefficients for a single particle. The equality to zero for the determinant of this system of linear equations yields the dispersion relation for the effective propagation constant k_{eff} of this medium [48]. For systems of particles whose sizes are small, as compared with wavelength, the expression for k_{eff}, obtained in this way, has the view [46]:

$$k_{\text{eff}}^2 = k^2 + \frac{3fy}{D}k^2\left[1 + i\frac{2}{3}\frac{k^2a^2y}{D}S_3(\theta = 0)\right], \qquad (4.34)$$

where

$$y = \frac{n_1^2 - n_0^2}{n_1^2 + 2n_0^2}, \ D = 1 - fy, \ S_3\left(\theta = 0\right) = \frac{1}{1 - H_3}, \ H_3 = -24f\left(\frac{\alpha}{3} + \frac{\beta}{4} + \frac{\delta}{6}\right); \qquad (4.35)$$

f is the volume fraction occupied by particles with the refractive index n_1, and the α, β, δ values are found according to the approximation of hard spheres (see (3.39)–(3.41)). The calculated effective index of refraction

$$n_{\text{eff}} = n'_{\text{eff}} + in''_{\text{eff}} \qquad (4.36)$$

is complex, even if the particles and a base substance surrounding them exhibit no intrinsic absorption. The imaginary part of the effective index of refraction n''_{eff} describes the energy diminishing for an incident plane wave due to scattering in all directions. The transmittance of this layer with thickness z is

$$T = \exp(-\frac{4\pi}{\lambda}n''_{\text{eff}}z). \qquad (4.37)$$

The quantity $\mu_e = (2\pi/\lambda)n''_{\text{eff}}$ is the extinction coefficient. The value of the imaginary part of the effective index of refraction grows for these systems with a higher radiation frequency and it nonmonotonously depends on the particle concentration in the layer. As a result, the transmittance of the disperse layer decreases for small particle concentrations with a greater concentration of particles, and starting at $f \approx 0.1$, the transmittance grows up, or the so-called clearing effect takes place. The real portion of the effective index of refraction in this approximation is essentially independent of the wavelength and alters monotonously with growing particle concentration to approach the refractive index of the particles. The near ordering in the scatterers' arrangement with their greater concentration not only provides conditions for the manifestation of secondary scattered wave interference but also changes the regime of the propagation of noncoherent multiply scattered light [49]. This may be accompanied by the so-called concentration effects of clearing and darkening.

The optical softness of tissues enables one to employ, during calculation, an expansion by scattering multiplicities while restricting by low orders. In [50], an expression was obtained for the effective index of refraction of the eye cornea modeled on a system of cylinder scatterers in the form of expansion by scattering multiplicities; the effects of polarization anisotropy were then analyzed with respect to the double scattering contributions.

Using the theory of multiple scattering, Twersky [51] succeeded in deriving approximate expressions for absorption μ_a and scattering μ_s coefficients that describe light propagation in blood. The blood hematocrit H is related to the erythrocyte concentration ρ and to the volume of an erythrocyte V_e by the following ratio [13]

$$\rho = H/V_e. \tag{4.38}$$

Thus, the absorption factor μ_a is

$$\mu_a = (H/V_e)\sigma_a. \tag{4.39}$$

For sufficiently small values of $H(H < 0.2)$, the scattering coefficient is given by the equation

$$\mu_s = (H/V_e)\sigma_s. \tag{4.40}$$

For $H > 0.5$, the particles become densely packed and the medium is almost homogeneous. In this case, blood may be considered a homogeneous medium containing hemoglobin in which the scattering particles formed by plasma surrounding the red blood cells are embedded. Within the limits of $H \to 1$, "plasma particles" disappear and the scattering coefficient should tend to zero. This results in the following approximate equation for μ_s [13,52]:

$$\mu_s \approx \frac{H(1-H)}{V_e}\sigma_s, \tag{4.41}$$

where coefficient $(1 - H)$ regards the scattering termination with $H \to 1$. However, the absolute dense packing $(H = 1)$ is not attainable in reality; for example, for the hard sphere approximation, H may not exceed 0.64.

Considering this fact and keeping in mind the physiological conditions, the affect of cell packing on light scattering might be described by a more complex function

$$\mu_s = (H/V_e)\sigma_s F(H), \tag{4.42}$$

where the packing function $F(H)$ accounts for the physiological condition of the red blood cells, particularly the cell deformability at high concentration.

Although the equations from Twersky's wave-scattering theory [44, 51] agree reasonably well with the measured optical density data for a whole blood layer [52], researchers have had to resort to curve-fitting techniques to evaluate the parameters in Twersky's equations. This theory also does not describe the spatial distribution of reflected and transmitted light and, therefore, does not accommodate light detectors and sources that do not share a common optical axis. By contrast, the RTT discussed above, particularly its more simple diffusion approximation, overcomes the mentioned limitations of wave-scattering theory. However, to be applied to densely packed tissues, this theory must account for particle interaction and size distribution effects. In combination with theories describing particle interactions, the use of empirical data can be considered a fruitful and practical approach for modeling the optical properties of tissues.

For example, by using diffusion theory, Steinke and Shepherd [52] have corrected the dependence (4.41) of the scattering coefficient μ_s for a thin blood layer on the hematocrit H, as follows:

$$\mu_s \approx (H/V_e)\sigma_s(1 - H)(1.4 - H). \tag{4.43}$$

Using the concept of the combination of photon-diffusion theory and particle representation of tissues, Schmitt and Kumar have developed a micro-optical model which explains most of the observed scattering properties of soft tissue [53]. The model treats tissue as a collection of scattering particles whose volume fractions are distributed according to a skewed log-normal distribution modified by a packing factor to account for correlated scattering among densely packed particles (see (2.12), (2.17), and (2.18)).

Assuming that the waves scattered by the individual particles in a thin slice of the modeled tissue volume add randomly, then the scattering coefficient of the volume can be approximated as the sum of the scattering coefficients of the particles of a given diameter,

$$\mu_s = \sum_{i=1}^{N_p} \mu_s(2a_i), \tag{4.44}$$

where

$$\mu_s(2a_i) = \frac{\eta(2a_i)}{v_i}\sigma_s(2a_i); \tag{4.45}$$

N_p is the number of particle diameters $\eta(2a_i)$ is the volume fraction of the particles of the diameter $2a_i$ (see (2.12), (2.17) and (2.18)), and $\sigma_s(2a_i)$ is the optical cross-section of an individual particle with diameter $2a_i$ and volume ν_i. The volume-averaged phase function $p(\theta)$ (and scattering anisotropy parameter g) of the tissue slice is the sum of the angular-scattering functions $p_i(\theta)$ (and anisotropy parameters, g_i) of the individual particles weighted by the product of their respective scattering coefficients:

$$p(\theta) = \frac{\sum\limits_{i=1}^{N_{\mathrm{p}}} \mu_{\mathrm{s}}(2a_i)p_i(\theta)}{\sum\limits_{i=1}^{N_{\mathrm{p}}} \mu_{\mathrm{s}}(2a_i)}; \tag{4.46}$$

$$g = \frac{\sum\limits_{i=1}^{N_{\mathrm{p}}} \mu_{\mathrm{s}}(2a_i)g_i(2a_i)}{\sum\limits_{i=1}^{N_p} \mu_{\mathrm{s}}(2a_i)}. \tag{4.47}$$

The reduced scattering coefficient is usually defined as $\mu' = \mu(1 - g)$. The volume-averaged backscattering coefficient can be defined as the sum of the particle cross-sections weighted by their angular-scattering functions evaluated at $180°$,

$$\mu_{\mathrm{b}} = \sum\limits_{i=1}^{N_{\mathrm{p}}} \frac{\eta(2a_i)}{v_i} \sigma_{\mathrm{s}}(2a_i)p_i(180°). \tag{4.48}$$

The product of $\mu_{\mathrm{b}}(\mathrm{cm}^{-1}\,\mathrm{sr}^{-1})$ and the thickness of the tissue slice yield the fraction of the incident irradiance backscattered per unit solid angle in the direction opposite the incident light.

An evaluation of the model conducted by applying Mie theory to a collection of spheres with a wide range of sizes gave a set of parameters for the distribution and packing of the particles $[D_{\mathrm{f}} = 3.7, \bar{n}_0 = 1.352, \bar{n}_{\mathrm{s}} = 1.420, F_\nu = 0.2, 2a_{\mathrm{m}} = 1.13\,\mu\mathrm{m}, \sigma = 2\,\mu\mathrm{m}, p = 3$ (see (2.5), (2.12) and (2.17))] that yield credible estimates of the scattering coefficients and scattering anisotropy parameters of representative soft tissues. Table 4.1 summarizes the optical properties predicted by the model at three wavelengths (633, 800, and 1,300 nm) for a soft tissue containing different dry-weight fractions of connective tissue fibers ($f_{\mathrm{f}} = 0.03, 0.3$, and 0.7). The coefficients $\mu_{\mathrm{s}}, \mu'_{\mathrm{s}}, \mu_{\mathrm{b}}$, and g were computed for determined parameters of the particle system. In general, these calculations fit well with the experimental data for in vitro and even in vivo measurements of optical parameters of soft tissues.

It follows from the model [53] that: (1) as an optical medium, tissue is represented best by a volume of scatterers with a wide distribution of sizes; (2) fixing the total volume fraction of particles and their refractive indices places upper and lower bounds on the magnitude of the scattering coefficient; (3) the scattering coefficient decreases with wavelength approximately as $\mu_{\mathrm{s}} \sim \lambda^{2-D_{\mathrm{f}}}$ for $600 \leq \lambda \leq 1{,}400\,\mathrm{nm}$, where D_{f} is the limiting fractal dimension, and

Table 4.1. Wavelength dependent optical coefficients of model tissues with three different dry-weight fiber fraction (f_f), for $D_f = 3.7$ [53]

optical coefficients	633 nm f_f			800 nm f_f			1,300 nm f_f		
	0.03	0.3	0.7	0.03	0.3	0.7	0.03	0.3	0.7
$\mu_s(\mathrm{cm}^{-1})$	105	224	402	69	146	274	29	63	119
$\mu_s'(\mathrm{cm}^{-1})$	8.0	20	45	5.7	14	32	3.0	7.5	16.5
$\mu_b(\mathrm{cm}^{-1}\mathrm{sr}^{-1})$	0.8	2.2	5.0	0.5	1.3	3.1	0.3	0.9	2.0
g	0.92	0.91	0.89	0.92	0.90	0.88	0.90	0.88	0.86

(4) scatterers in tissue with diameters between $\lambda/4$ and $\lambda/2$ are the dominant backscatterers, and the scatterers that cause the greatest extinction of forward-scattered light have diameters between 3λ and 4λ.

A reduced scattering coefficient, decreasing with wavelength in accordance with a power law, was experimentally demonstrated for normal, dehydrated and coagulated human aorta and rat skin in an in vitro study [54–56]:

$$\mu_s' \propto \lambda^{-h}. \tag{4.49}$$

For the human aorta under conditions of direct heating (100° C), h was reduced from 1.38 for the normal tissue sample to 1.06 for the heated one. For the rat skin impregnated by glycerol (mostly dehydration effect) in the wavelength range $500 - 1,200$ nm, h was 1.12 for the normal skin, decreased subsequently with increasing time in glycerol($h = 1.09$ for 5 min in glycerol, 0.85 for 10 min, 0.52 for 20 min), and went back to 0.9 for the rehydrated sample [56].

In vivo backscattering measurements for the human skin and underlying tissues also demonstrated the power law for wavelength dependence on the reduced scattering coefficient [57]:

$$\mu_s' = q\lambda^{-h}(\lambda\ \mathrm{in}\ \mu\mathrm{m}). \tag{4.50}$$

In particular, for reflectance spectra from the human forearm in the wavelength range 700–900 nm, constants q and h were determined to be 550 ± 11 and 1.11 ± 0.08, respectively. From Mie theory, it follows that the power constant h is related to an averaged size of the scatterers – the so-called Mie-equivalent radius a_M. Once h is determined, this radius can be derived from [57]

$$h = -1109.5a_M^3 + 341.67a_M^2 - 9.36961a_M - 3.9359(a_M < 0.23\ \mu\mathrm{m}), \tag{4.51}$$
$$h = 23.909a_M^3 - 37.218a_M^2 + 19.534a_M - 3.965(0.23 < a_M < 0.60\ \mu\mathrm{m}). \tag{4.52}$$

These relations were determined for a relative refractive index between the spheres and surrounding medium, $m = 1.037$. The in vivo measured constant $h = 1.11$ leads to a a_M value of $0.30\,\mu\mathrm{m}$, which is about two times less than the mean radius $(0.57\,\mu\mathrm{m})$ used in the above discussed model of a collection of packed spheres with a wide range of sizes [53].

4.5 Summary

In general, for polarized light propagated in a strongly scattering medium, multiple scattering decreases the magnitudes and smoothes out the angular dependencies of the normalized LSM elements characterizing the polarized light interaction with the medium. For media composed of large particles that are specified by a high degree of single scattering anisotropy or considerable photon transport length, the scattered radiation retains preferential circular polarization at higher scattering multiplicities. This and other theoretical results of this chapter serve as a confirmation of the preferential survival of certain types of polarization under conditions of multiple scattering for different sizes of scattering particles or tissue structures (see Chap. 5).

The simulated LSM elements, and their angular dependences, allow one to estimate the limits of the applicability of single scattering approximations to interpret the results of experimental studies of tissues and tissue-like phantoms. In particular, it was shown that the modification of the LSM of a monodisperse system of spherical particles associated with multiple scattering has much in common with the modification of single scattering LSM associated with particle nonsphericity.

The MC technique of photon trajectory modeling described in the chapter is well suited to the simulation of multiple scattering effects in a system of randomly arranged particles. Furthermore, on this basis, an approximate approach for describing the interference effects caused by space particle ordering can be suggested. For that, one must include the interference of scattered fields in the calculations of the single scattering Mueller matrix and integral cross-sections of the particles. Initially we account for the interference effects of the simulation of the *single scattering* properties (see Chap. 3) and then use these properties in a MC simulation of the *multiple scattering* effects. Such an approach is quite admissible because the size of the region of the local particle ordering (a few particle diameters) is substantially smaller than the mean free photon pathlength.

Polarization Properties of Tissues and Phantoms

5.1 Introduction

It has been shown that polarized light propagation in tissues depends on their scattering, absorption, and anisotropic properties. The anisotropic properties of a tissue in turn depend on the tissue morphology, i.e., the scatterers' size and shape, refractive index, internal structure, and the birefringence and optical activity of the materials of the tissue components [1–9]. The polarization properties of elastically scattered light are described by a 16-element light scattering matrix (LSM), each element being dependent on the incident and scattering angles, wavelength, size, shape, and material of the scatterers (see (3.18), (3.48), (3.63), (4.32) and Figs. 4.3, 4.4).

The LSM elements and their dependence on the scattering angles provide information on the structure and properties of a tissue under study. The solution of the appropriate inverse problem provides values for the size distribution function of the scattering particles, their index of refraction, shape, and orientation. Most tissues and biological liquids have quite complicated structures and the complexity of their LSMs depends on the object's parameters; hence, often the solution to the inverse problem is quite difficult or even impossible. Along with this, in a number of cases, a qualitative estimation of the object's properties is sufficient, and an exact solution of the inverse problem is not required. The general view of experimental LSM, its norm values and the symmetry relations of the elements (see (3.11)–(3.21) and (3.23)–(3.26)) allow one to compare an object under study with a certain class of scattering systems [10]. An estimate of the state of the studied objects using experimental LSM can be obtained from the relationships among the matrix elements. These relations and the LSM norm values can also be criteria for estimating the correctness of experimental results.

In this chapter, we will consider the principles of scattering matrix measurements and describe the actual measuring instruments that have been described in the literature. The peculiarities and distinctive properties of the experimental LSM of thin tissue slices or cell monolayers, as well as of thick strongly scattering tissues and tissue-like phantoms, will be analyzed.

5.2 Light Scattering Matrix Meters

The simplest and, probably, the most pictorial way to measure the LSM elements' angular dependencies is to use a conventional scatterometer (nephelometer) with various polarization optical elements placed ahead of, and after, the scattering medium. Usually linear polarizers and quarter-wave plates are employed as such elements [8, 9, 11–13]. A possible result of the measurements is combinations of the LSM elements. Actually, these combinations are obtained by multiplying the matrices of optical elements placed ahead of the scattering object, the matrices of the scattering object itself, and those of optical elements placed after the scattering object (Fig. 5.1). In general, four measurements are necessary to obtain one LSM element. Despite the fact that this technique is reasonable, if rather cumbersome, its application may be followed by significant relative errors that are associated with small matrix elements obtained as differences in the great signals [8]. These errors can be partially avoided by modulating the polarization state in the incident or the scattered beam.

The operation principle of the LSM meter [14–19] (Fig. 5.1), based on the modulation of the polarization of the incident laser beam followed by the scattered light demodulation (transformation of polarization modulation to intensity modulation), is described by the following matrix equation:

$$\mathbf{S} = \mathbf{AF'MFPS_0},\tag{5.1}$$

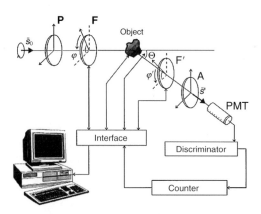

Fig. 5.1. Scheme of laser scattering matrix meter based on the principle of mechanically rotated phase plates (see text for details) [14, 19]

where $\mathbf{S}$ and $\mathbf{S}_0$ are the Stokes vectors of the recorded and source radiation, respectively; $\mathbf{P}$, $\mathbf{A}$, and $\mathbf{F}'$, are the Mueller matrices for the linear polarizers and the phase plates placed, respectively, ahead of and after the scattering medium. As the phase plates are rotated, the intensity recorded by a photodetector, i.e., the first element of the Stokes vector $\mathbf{S}$, depends on time. By multiplying the matrices in (5.1) and performing the appropriate trigonometric transformations, one can show that the output intensity can be represented as a Fourier series, namely [20],

$$I = a_0 + \sum_{k=1}^{K} (a_{2k} \cos 2k\varphi + b_{2k} \sin 2k\varphi), \tag{5.2}$$

where

$$a_{2k} = \sum_{i=1}^{N} I(\varphi_i) \cos 2k\varphi_i; \quad b_{2k} = \sum_{i=1}^{N} I(\varphi_i) \sin 2k\varphi_i; \tag{5.3}$$

$I(\varphi_i)$ is the intensity of the scattered light detected by the photoreceiver for a certain orientation of the fast axis of the first retarder, φ_i; and N is the number of measurements per rotation cycle of the first phase plate F.

The coefficients of the series described by (5.2) are defined by the values of the matrix $\mathbf{M}$ elements of the object under study, and their measurement ensures a system of linear equations to determine the matrix $\mathbf{M}$. The number of equations and the degree of stipulation for this system of equations are dependent on the choice of the ratio between the rotation rates of the phase plates (retarders). An optimal choice of the rotation rates relationship at 1:5 allows an optimally stipulated system of linear equations ($K = 12$ to be derived to find the full matrix $\mathbf{M}$ of the object under study [20].

The LSM meter presented schematically in Fig. 5.1 is designed with rotating retarders ($\lambda/4$-phase plates). Due to its comparatively simple measuring procedures and software, it avoids many of the experimental artifacts peculiar to DC measurements and to systems utilizing electrooptic modulators [8]. The LSM meter has a fixed polarizer P and analyzer A and two rotating-phase plates, F and F', before and after the sample. The polarizer and analyzer are aligned in parallel with each other and their transmission planes are orthogonal to the scattering plane; the fast axis of each phase plate, F and F', forms an angle with the scattering plane φ and φ', as a result, the respective phase differences, δ and δ', are induced. The ratio of the rotation rates of the phase plates is set equal to 1:5, i.e., $\varphi' = 5\varphi$, because all of the 16 matrix elements are uniquely determined in this case. The computer-controlled LSM meter provides automatic scattering angle scanning in the range $0 \pm 175°$ with a step of $4'$ and an accuracy of $5''$. A single-mode stable He:Ne laser (633 nm) is used as a light source. Computer-driven retarders provide $N = 256$ indications per one rotation cycle of the first phase plate F. A photon counting system is used with the photomultiplier tube (PMT), amplitude discriminator (clipping amplifier), and counter. The fast Fourier transform analysis allows one to measure and calculate all 16 S-matrix elements for the fixed scattering angle during about 1 s with an accuracy of 3–5%.

A number of LSM meters (Mueller matrix scatterometers) and instruments for polarization measurements designed for in vitro and in vivo studies of biological olbjects are also available (see, for example, [6,8,9,12,13,21–42]). Some of them provide fast and automatic measurements of all 16 LSM elements over a wide range of scattering angles [30] or for a certain scattering angle, 90° [25]. For example, the system described in [30] has a 0–360° range of scattering angles; it uses two electro-optical modulators for polarization modulation at two different frequencies 2,000 and 251 Hz of the incident He:Ne (633 nm) laser beam; it provides simultaneous four-detector polarization sensitive detection and matrix elements calculation with an accuracy of 1–3%, depending on the element, and a measuring time of 1 s for the whole matrix at one incident angle and one scattering angle.

Many other systems also work in the backscattering or transmittance mode and are used mostly for imaging tissue or scattering media. Examples of such schemes include the Stokes-vector imaging system [28, 29, 31–33], a scheme for mapping degree of polarization [36, 37, 42, 43], a method of polarization-difference imaging [38], and a polarization-multispectral imaging system [34]. Both scanning polarimetry, which incorporates ellipsometry into a confocal scanning laser opthalmoscope [28, 29], and full-field CCD-based single wavelength or multispectral polarization imaging systems [31,32,34,36,42] are also in use.

Various instrumental errors or imperfections characteristic of LSM meters (Mueller matrix meters), such as the dual-rotating-retarder LSM [19], the Mueller matrix polarimeter [31], rotating-retarder Stokes polarimeters [40], and the liquid crystal based Stokes polarimeter [41], are widely discussed as various means for their optimization.

5.3 LSM of Thin Tissue and Cell Layers

A number of brief reviews of experimental results from studies on the polarization-scattering characteristics of biological objects are available in the literature [4,6,8,9,14,44–48]. The following regularities can be cited that allow the parameters of scattering particles (tissue or other bio-object components) to be classified by analyzing their LSM. The distinction between elements M_{22} and M_{11} serves as the measure for scattering particles to be nonspherical. These peculiarities have been studied in different kinds of pollen [49] and marine organisms [50, 51]. However, as follows from Monte Carlo modeling (see Figs. 4.3 and 4.4), with a certain turbidity in a system of spherical particles, an analogous distinction between elements M_{22} and M_{11} is caused by multiple scattering.

As noted in [48, 49] and [52–56], element M_{34} is most specific for various biological microorganisms. This element is sensitive to small morphological alterations in scatterers. It has been shown that the element M_{34} is affected by a small surface roughness on a sphere [53]. It has also been proven that M_{34} measurements may be a basis for determining the diameters of rod-shaped bacteria (*Escherichia coli* cells) that are difficult to measure using other techniques [54]. The angular dependences of the normalized element M_{34}/M_{11} for different bacteria turn out to be oscillating functions whose maxima positions are very sensitive to varying sizes of the bacteria [54–56]. This allows bacterial growth to be followed [56].

In [49] and [52], the measuring results are presented for the whole LSM of some biological particles. A high specificity of the normalized element M_{34}/M_{11} is shown for every type of biological scatterer. Stable distinctions are revealed in the values of M_{34}/M_{11} for spores of two mutant varieties of bacteria, which are distinguished by variations in their specific structures which are invisible with traditional scattering techniques. The distinctions of other matrix elements, however, are seen less clearly for these two types of similar scatterers.

When scattering is well described by the Rayleigh–Gans theory, then $M_{34} = 0$. Thus, a nonzero value for this element can be associated with deviation of the particle characteristics from those that satisfy the Rayleigh–Gans theory. This is possibly the reason why M_{34} is so sensitive to the characteristics of biological scatterers [11].

The polarization characteristics for suspensions of biological particles have been described in [44], where the sensitivity of different matrix elements to variations in scatterer shape and size is analyzed. It is noted that the values of elements M_{33} and M_{44} in the backward scattering direction may serve as indicators of particle nonsphericity.

LSM measurement is also used to examine the optical parameters of blood cells. For example, the determination of the real part of the relative index of refraction of blood cells, m, is based on the study of the angular structures of the LSM nonzero elements [57]. This method is applicable to normal or gamma distributions of polydisperse particle systems and does not require data on particle concentration; only the conditions of single scattering must be obeyed. The technique is reduced to finding a scattering angle at which the LSM element is zero within a range of scattering angles of 80–120°. Further, the relative index of refraction m is derived from the nomograms valid for $m = 1.02–1.07$. If the element is nonzero within the angular range 80–120°, thus $m > 1.07$, one needs to determine a scattering angle at which element M_{34} is zero.

The measurement of angular dependences for the total LSM of blood erythrocytes enables one to distinguish between disc-like and spherulated cells (spherocytes) in relation to their packing density in a monolayer [18] (Fig. 5.2). The angular dependence of the matrix element M_{11} in both cell types turns out to be influenced by the packing density in the angular range of $\theta = 15–16°$. The angular dependences of the elements M_{11}, M_{22}, M_{33}, and M_{21} at $\theta = 110–170°$ are found to be far more affected by the shape of the scatterers than by their

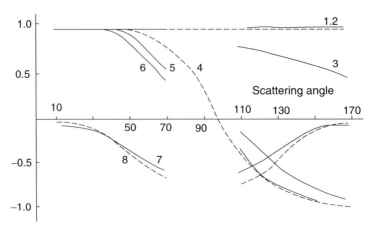

Fig. 5.2. Angular distributions of the normalized (to M_{11}) LSM elements of the monolayer of erythrocytes: experimental: M_{22} (1, 3), M_{33} (5, 6), M_{12} (7), M_{21} (8); theoretical (Mie theory): M_{22} (2), M_{33} (4); for disc-like erythrocytes (3, 5, 7, 8) and spherocytes (1, 2, 4, 6) (from [18] with corrections)

concentration. It is also possible to derive the refractive indices of erythrocytes from measurements of the M_{12} magnitude at scattering angles $\theta \approx 140$–$160°$. A study [57] revealed the high susceptibility of the angular dependences of LSM elements (M_{11} and M_{12}) to the degree of erythrocyte aggregation in blood plasma.

LSM measurements have also been used to examine the formation of liposome complexes with plague capsular antigens [14] and various particle suspensions, e.g., those of spermatozoid spiral heads [8, 9].

Determination of the LSM elements is equally promising for more effective differentiation between blood cells by time-of-flight cytometry [9, 48, 58, 59]. It is noted in [48] that the comparison between measured signals for all types of human white blood cells allows one to distinguish between two types of granulocytes.

LSM elements measurement in the backscattering mode for thin tissue samples, 25–250 μm of the human skin dermis, bone and muscle, shows a high level of birefringence and random orientation of the local structure of these tissues [60, 61].

The measurement of angular dependences of LSM elements in a human lens reveals a significant difference for clear and opaque (cataractous) eyes (Fig. 5.3). This distinction is due to large nonspherical scattering particles appearing in the medium of the turbid lens (because of the formation of high-molecular proteins). A transparent lens contains a monodisperse system of small-diameter scatterers. A turbid lens contains a reasonable fraction of larger scatterers. The high sensitivity of the LSM element angular dependences to the variations in the medium structure makes it possible to employ LSM

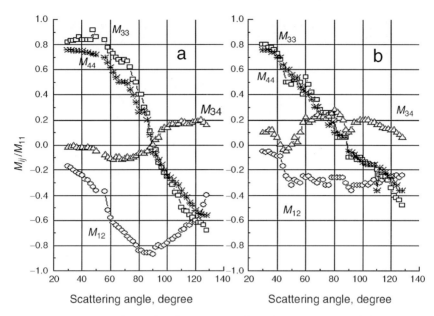

Fig. 5.3. Experimental angular dependences for LSM elements of (**a**) normal lens 5 h after the death of a 56-year-old subject and (**b**) cataractous lens 5 h after the death of an 88-year-old subject. Measurements were performed at a wavelength of 633 nm [62]

measurements for early diagnosis of alterations in the tissue structure which are related to the appearance of cataracts.

This inference can be illustrated by the results of the direct model experiments presented in Fig. 5.4 [6, 63, 64]. The measurements are performed in an α-crystalline solution (quasi-monodisperse particle fraction of $\approx 0.02\,\mu$m in diameter) from a freshly isolated calf lens and in solutions of high molecular-weight proteins (mean diameter $\approx 0.8\,\mu$m) from opaque lenses. The figure shows that measurements of the angular dependences of the LSM elements permits the identification of a coarsely dispersed fraction of scatterers that is difficult to achieve by spectrophotometry.

The method of polarizing light biomicroscopy has been shown to yield qualitative characteristics that permit detection of images of the eye lens that correlate with visual acuity; with standard biomicroscopy, there is no correlation [65].

Laser scattering matrix measurements are also employed for the in vitro examination of various eye tissues, from cornea to retina. In vivo measurements in the intact eye are equally feasible, provided a fast LSM meter is used to exclude a sensorimotor eye globe response. In this case, structural information about selected eye tissues can be obtained to diagnose cataract and other ophthalmologic disorders.

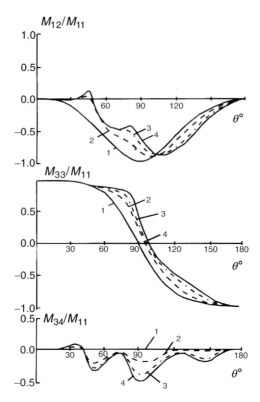

Fig. 5.4. Angular dependences for LSM elements of α-crystallin solutions and a fraction of large-size scatterers isolated from the cataractous lens. The volume fractions of α-crystallin $f_1 = 0.3$ and the large-particle fraction: (1) $f_2 = 0$ $(T = 99\%)$; (2) $f_2 = 5 \times 10^{-5}$ $(T = 98\%)$; (3) $f_2 = 1.4 \times 10^{-4}$ $(T = 94\%)$; and (4) $f_2 = 2.5 \times 10^{-4}$ $(T = 90\%)$, T is the transmittance of the 5 mm thick solution at $\lambda = 633$ nm [14]

A survey of rabbit eye LSM has demonstrated that the aqueous humor in the anterior eye chamber is actually a transparent isotropic substance exhibiting weak light scattering properties (the intensity of scattered light does not exceed 1.5–2% of the incident light intensity) owing to the presence of dissolved organic components. The results of an LSM study in the vitreous humor indicate that its amorphous tissue does not affect the polarization of straight-transmitting light, offering a possibility for ocular fundus image and optic nerve structure examination, which is important for early diagnosis of glaucoma [28, 29, 34, 66]. On the other hand, certain pathological changes in the vitreous humor may be responsible for the alteration of the LSM elements. Specifically, a minor intraocular hemorrhage is easy to identify by virtue of conspicuous light scattering from erythrocytes.

Scanning laser polarimetry [28, 29] and multispectral imaging micro-polarimeter [34, 66] assess the retinal nerve fiber layer (RNFL) for glaucoma diagnosis by detecting the birefringence of the peripapillary RNFL. It has been found that RNFL behaves as a linear retarder. The retardance is constant at a wavelength range between 440 and 830 nm and persists after tissue fixation [34]. The average birefringence measured in the transmittance mode by the multispectral imaging micropolarimeter [34, 66] for a few rat RNFLs with an average thickness of $13.9 \pm 0.4\,\mu\mathrm{m}$ is $0.23 \pm 0.01\,(\mathrm{nm}\,\mu\mathrm{m}^{-1})$ (or 2.3×10^{-4}) before and $0.19 \pm 0.01\,(\mathrm{nm}\,\mu\mathrm{m}^{-1})$ (or 1.9×10^{-4}) after tissue fixation.

Many tissues demonstrate effects of optical activity that are manifested in circular dichroism and circular birefringence. The optical activity of tissues may be conditioned by the optical activity of the substance they are formed from and by their structural peculiarities. Circular intensity differential scattering (CIDS) is the difference between scattered intensities for left and right circularly polarized incident light. CIDS effects can be investigated by measuring the LSM element M_{14} [11]. The so-called "form-CIDS" is an anisotropy caused by the helical structure of a particle [48]. The CIDS interrelation with the scatterer structure has been considered [67]. Measurements of CIDS are used to study the secondary and tertiary structures of macromolecules [68], the polymerization of hemoglobin in sickle red blood cells [69].

To account for the fibrilar-lamellar structure of eye cornea (see Figs. 2.3, 2.5, and 2.6), theoretical spectra of linear dichroism and dependences of circular dichroism and birefringence on the angle between lamellas have been obtained [70]. A system of plane anisotropic layers, each of which is represented by a densely-packed system of long cylinders (fibrils) whose optical axes were oriented along a spiral, has also been considered as a corneal model. The theoretical results correspond qualitatively to the in vitro experimental data for the LMS of the rabbit cornea.

The polarimetric quantification of glucose is based on the phenomenon of optical rotatory dispersion (ORD) whereby a chiral molecule in an aqueous solution rotates a plane of linearly polarized light passing through the solution [71, 72]. The angle of rotation depends linearly on the concentration of the chiral species. At physiological concentrations and pathlengths of about 1 cm, the optical rotation due to glucose is on the order of $0.005°$. The anterior chamber of the eye (the fluid-filled space directly below the cornea) has been suggested as a sight well suited to polarimetric noninvasive measurements [72, 73], since scattering in the eye is generally very small compared to other tissues.

5.4 Strongly Scattering Tissues and Phantoms

As follows from Chap. 4, given the known character of the Stokes vector transformation for each scattering act, the state of polarization following multiple light scattering in a highly scattering medium can be found using various approximations of multiple scattering theory or the MC method. For small particles, the effects of multiple scattering are apparent as a broken symmetry relationship between the LSM elements (see (3.23)–(3.26)), $M_{12}(\theta) \neq M_{21}(\theta)$, $M_{33}(\theta) \neq M_{44}(\theta)$, and a significant reduction of linear polarization of the light scattered at angles close to $\pi/2$ [74].

For a system of small spatially uncorrelated particles, the degree of linear $(i = L)$ and circular $(i = C)$ polarization in the far region of the initially polarized (linearly or circularly) light transmitted through a layer of thickness d is defined by the relation [75].

$$P_i \cong \frac{2d}{l_S} \sin h(l_S/\xi_i) \cdot \exp(-d/\xi_i), \qquad (5.4)$$

where $l_s = 1/\mu_s$ is the scattering length,

$$\xi_i = (\zeta_i \cdot l_s/3)^{0.5} \qquad (5.5)$$

is the characteristic depolarization length for a layer of scatterers, $d \gg \xi_i$, $\zeta_L = l_s/[\ln(10/7)]$, $\zeta_C = l_s/(\ln 2)$.

As can be seen from (5.4), the characteristic depolarization length for linearly polarized light in tissues that can be represented as ensembles of Rayleigh particles is approximately 1.4 times greater than the corresponding depolarization length for circularly polarized light. One can employ (5.4) to assess the depolarization of light propagating through an ensemble of large-scale spherical particles whose sizes are comparable to the wavelength of incident light (Mie scattering). For this purpose, one should replace l_s by the transport length $l_t \cong 1/\mu'_s$ (see. (4.17) and (4.18)) and take into account the dependence on the size of the scatterers in ζ_L and ζ_C. With growth in the size of the scatterers, the ratio ζ_L/ζ_C changes. It decreases from $\sim$1.4 down to 0.5 as $2\pi a/\lambda$ increases from 0 up to $\sim$4, where a is the radius of the scatterers and λ is the wavelength of the light in the medium; it remains virtually constant at the level of 0.5 when $2\pi a/\lambda$ grows from $\sim$4 to 15.

Monte Carlo numerical simulations and model experiments in aqueous latex suspensions with particles of various diameters demonstrate that there are three regimes of dependence of the ratio of the degree of linear polarization to circular polarization for transmitted light, P_L/P_C, on the ratio d/l_t (Fig. 5.5) [75]. In the Rayleigh range, P_L/P_C grows linearly with an increase of d/l_t. In the intermediate range, this ratio remains constant. In the range of Mie scattering, this quantity linearly decreases. The behavior of this quantity is associated with the transition of the system under study from one of isotropic scattering to one of anisotropic scattering.

Qualitatively, the physical mechanism behind the change in the depolarization character for isotropic single-scattering is associated with the fact that a considerable probability of backward scattering in each event of light–medium interaction does not distort linear polarization, whereas backward scattering

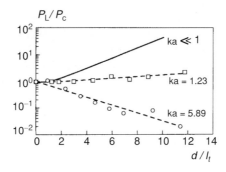

Fig. 5.5. Semilogarithmic dependences of the degree of polarization ratio P_L/P_C on d/l_t for three ka values, $k = 2\pi/\lambda$. The *solid line* corresponds to Rayleigh scattering ($ka \ll 1$) and the *dashed lines* indicate a correspondence between experimental findings and Eq. (5.4) at $l_s = l_t$. The experimental points are measurements for aqueous suspensions of polystyrol latex spherical particles having diameters 0.22 (□) and 1.05 (○) μm, $\lambda_0 = 670$ nm [75]

for circular polarization is equivalent to the reversal of polarization direction (similar to reflection from a mirror), i.e., it is equivalent to depolarization. For the same reason, in the case of a strongly elongated scattering phase function, the degree of circular polarization in an individual scattering event (anisotropic single-scattering) for light propagating in a layer remains nonzero for lengths greater than the degree of linear polarization.

These arguments also follow from the above MC simulation of polarized light interaction with multiply scattering systems [76] and from experimental works [25, 27]. For example, with high scattering multiplicities, the radiation scattered by large particles holds preferential circular polarization (LSM element M_{44} is far from zero for all scattering angles) (see Fig. 4.4). With multiple scattering, the LSM for a monodisperse system of randomly distributed spherical particles is modified to be approximately identical to the single-scattering LSM of a system containing nonspherical particles, or optically active spheres [76, 77].

Thus, different tissues, or the same tissues in various pathological or functional states, should display different responses to probes with linearly or circularly polarized light. This effect can be employed both in optical medical tomography and in determining the optical and spectroscopic parameters of tissues. As follows from (5.4), the depolarization length in tissues should be close to the mean transport path length l_t of a photon (see (4.17) and (4.18)), because this length characterizes the distance within which the direction of light propagation and, consequently, the polarization plane of the linearly polarized light, becomes totally random after many sequential scattering events.

Since the length l_t is determined by the parameter g characterizing the anisotropy of the scattering, the depolarization length also substantially depends on this parameter. Indeed, the experimental data of [78] demonstrate

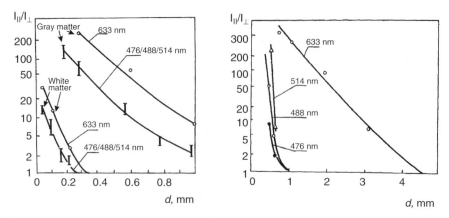

Fig. 5.6. Dependence of the depolarization degree ($I_{\parallel} I_{\perp}$) of laser radiation (He:Ne laser, $\lambda = 633$ nm; Ar laser, $\lambda = 476/488/514$ nm) on the penetration depth for (**a**) brain tissue (gray and white matter) and (**b**) whole blood (low hematocrit) [78]. Measurements were performed within a small solid angle (10^{-4} sr) along the axis of a laser beam 1 mm in diameter

that the depolarization length l_{p} of linearly polarized light, which is defined as the length within which the ratio $I_{\parallel}/I_{\perp}$ decreases down to 2, displays such a dependence. The ratio mentioned above varies from 300 to 1, depending on the thickness of the sample and the type of tissue (Fig. 5.6). These measurements were taken within a narrow solid angle ($\sim 10^{-4}$ sr) in the direction of the incident laser beam. The values of l_{p} differed considerably for the white matter of brain and tissue from the cerebral cortex: 0.19 and 1.0 mm for $\lambda = 476$–514 nm and 0.23 and 1.3 mm for $\lambda = 633$ nm, respectively. Human skin dermis (bloodless) has a depolarization length of 0.43 mm ($\lambda = 476$–514 nm) and 0.46 mm ($\lambda = 633$ nm). The depolarization length at $\lambda = 476$–514 nm decreases in response to a pathological change in the tissue of aorta wall: 0.54 mm for a normal tissue, 0.39 mm for the stage of tissue calcification, and 0.33 mm for the stage of necrotic ulcer. Whole blood with a low hematocrit is characterized by a considerable depolarization length (about 4 mm) at $\lambda = 633$ nm, which is indicative of dependence on parameter g, whose value for blood exceeds the values of this parameter for tissues of many other types and can be estimated as 0.982–0.999 [4, 5, 79].

In contrast to depolarization, the attenuation of collimated light is determined by the total attenuation coefficient μ_{e} (see (4.21)). For many tissues, μ_{e} is much greater than μ'_{s}. Therefore, in certain situations, it is impossible to detect pure ballistic photons (photons that do not experience scattering), but forward scattered photons retain their initial polarization and can be used for imaging [80–82]. This is illustrated by Figs. 5.7 and 5.8, which present the experimental data for the decay of degree of linear polarization P_{L} (see (3.8)), obtained for a gelatin gel–milk phantom (a model of bloodless dermis) within a broad wavelength range [46, 83], and for various tissues and blood as a func-

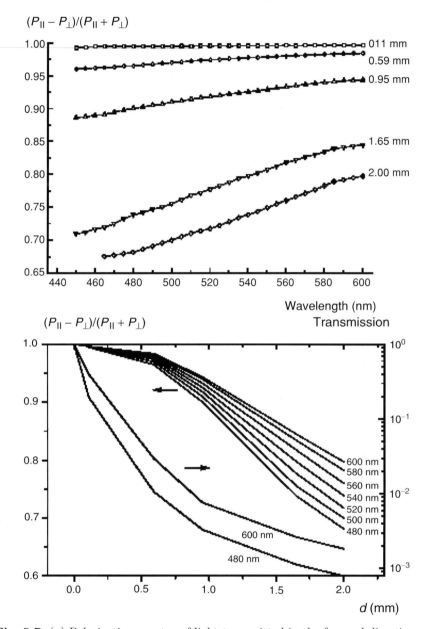

Fig. 5.7. (a) Polarization spectra of light transmitted in the forward direction and (b) the relevant dependencies on the layer thickness d for a gelatin gel–milk (20%) phantom [83]

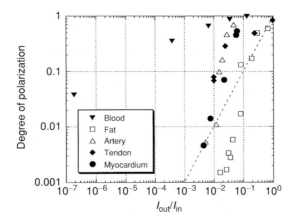

Fig. 5.8. Degree of linear polarization in different tissues as a function of the sample optical transmittance, $I_{out}/I_{in} \equiv T$, on 633 nm. Each point is an average of three measurements [27]. The error bars representing the standard deviation of the measurements are smaller than the symbols used

tion of light transmission [27]. The kink in the characteristics of polarization decay (Fig. 5.7b), which is observed for a small thickness of 0.6 mm, can be attributed to the transition of a medium to the regime of multiple scattering.

The authors of [84] experimentally demonstrated that laser radiation retains linear polarization at the level of $P_L \leq 0.1$ within $2.5 l_t$. Specifically, for skin irradiated in the red and NIR ranges, we have $\mu_a \cong 0.4 \, \text{cm}^{-1}$, $\mu'_s \cong 20 \, \text{cm}^{-1}$, and $l_t \cong 0.48 \, \text{mm}$. Consequently, light propagating in skin can retain linear polarization within a length of about 1.2 mm. Such an optical path in tissue corresponds to a delay time of 5.3 ps, which provides an opportunity to produce polarization images of macro-inhomogeneities in a tissue with spatial resolution equivalent to the spatial resolution that can be achieved with the selection of photons using more sophisticated time-resolved techniques. In addition to the selection of diffuse-scattered photons, polarization imaging makes it possible to eliminate specular reflection from the surface of a tissue, which allows one to apply this technique to the imaging of microvessels in facile skin and to the detection of birefringence and optical activity in superficial tissue layers [36, 84–86].

Polarization imaging is a new direction in tissue optics [4–6, 24, 27, 29, 31–36, 60, 75, 77, 81–106]. The most promising approaches for polarization tissue imaging – linear polarization degree mapping, two-dimensional backscattering Mueller matrix measurements, polarization-sensitive optical coherence tomography (OCT), and full-field polarization-speckle technique – are discussed in detail in the following chapters.

It should be noted that in media containing large-scale scatterers (a common tissue model), depolarization is a higher order effect ($\sim \theta^4, \theta < 1$) than polarization ($\sim \theta^2$) [107]. It should be emphasized that analysis of the polarization state in cases involving small-angular multiple scattering is important for many problems pertaining to the optical diagnosis of biological media which can be represented as random systems with long-range correlations of fluctuations of dielectric permittivity. Such systems display coherent scattering effects [108, 109] or may be expected to show fluctuations of polarization similar to those in disordered media with large-scale inhomogeneities [110].

In weakly absorbing media showing small-angular multiple scattering, the degree of linear polarization for a Henyey–Greenstein phase function (see (4.12)) is described by the following formula [107]:

$$P_{\rm L} = - \left[(\mu'_s z)^4/2\theta^2\right] \left[\sqrt{1 + (\theta/\mu'_s z)^2} - 1\right]^2 \left[1 + (\theta/\mu'_s z)^2\right]. \qquad (5.6)$$

This means that, in a very small angle range $\theta \ll \mu'_s z$), the degree of polarization does not depend on the depth (z)

$$P_{\rm L} = -\theta^2/8. \qquad (5.7)$$

At the wings of the scattering angle dependence $(\theta \gg \mu'_s z)$, it tends to

$$P_{\rm L} = -\theta^2/2, \qquad (5.8)$$

which equals the degree of polarization of singly scattered light.

5.5 Summary

A large variety of polarization-sensitive optical instruments have been used in the in vitro and in vivo studies of biological objects that are described in the literature. Light scattering matrix meters (or Mueller matrix scatterometers) that provide fast and automatic measurements of all 16 LSM elements for a wide range of scattering angles are the most appropriate measuring systems for in vivo medical diagnostics. The most advanced systems have a 0–360° range of scattering angles, electro-optical modulation of polarization, parallel polarization-sensitive detection, a matrix elements calculation accuracy of 1–3%, and a measuring time of 1 s for the whole matrix at one incident angle and one scattering angle.

LSM measurements are widely exploited for diagnostic purposes to examine the optical parameters of thin tissue layers and blood cells. For example, the angular measurements of the total LSM of blood erythrocytes enable us to distinguish between disc-like and spherulated cells, to account for their packing density in a monolayer, and to estimate the refractive indices of erythrocytes and their degree of aggregation. Such measurements also show promise for the examination of liposome complexes, other biological particle suspensions, and r cell differentiation in time-of-flight cytometry.

LSM measurements can be employed for in vivo studies of various human eye tissues, from cornea to retina. Polarization studies of corneal, eye lens, vitreous humor, and optic nerve structures are important for early diagnosis of cataracts, glaucoma, intraocular hemorrhages, etc.

CIDS measurements are used to study secondary and tertiary structures of macromolecules, the polymerization of hemoglobin in sickle red blood cells.

The anterior chamber of the eye has been suggested as a potential sight for noninvasive polarimetric rotational glucose sensing with a high degree of accuracy, since scattering in the eye is generally very small compared to that in other tissues.

For strongly scattering tissues, there are three different regimes for linear and circular polarization light transportation. For tissues predominantly composed of small (Rayleigh) particles, the ratio of linear to circular polarization components grows linearly with an increase in tissue thickness. In the range of Mie scattering, this quantity linearly decreases, and in the intermediate range, remains constant. Such behavior of different scattering systems is associated with the isotropic, anisotropic, or intermediate mode of single scattering attributed to the system under study. All of these regimes are important for designing polarization-sensitive technologies for medical diagnosis.

Polarization imaging is a new direction in tissue optics. The most promising approaches for polarization tissue imaging include the following: linear polarization degree mapping, the polarization-spectral method, two-dimensional backscattering Mueller matrix measurements, polarization-sensitive OCT, and the full-field polarization-speckle technique. These topics are discussed in detail in the following chapters.

6

Polarization-Dependent Interference of Multiply Scattered Light

6.1 Introduction

With increasing turbidity in a scattering system, the stochasticity of light propagation in random media causes a number of dissipation phenomena [1] related to the increasing entropy in the distributions of the local parameters of the scattered optical fields. Among these phenomena, the decay of polarization of the multiply scattered light is one of the most important features of radiative transfer in random media. This phenomenon is related to the vector nature of electromagnetic waves running through a scattering system. From physical observation, we expect that the specific relaxation scale characterizing the rate of suppression of the initial polarization of light propagating in a multiply scattering medium will be closely related to other relaxation scales which characterize an increase in the uncertainty of the other fundamental parameters of electromagnetic radiation. An obvious goal, therefore, is to determine the relationship between the polarization relaxation parameters, which can be introduced as the characteristic spatial scales of the decay of the polarization characteristics chosen to describe the scattered field [2–4], and the relaxation parameter that characterizes the spatial scale in which the almost total loss of information about the initial direction of the light propagation occurs. In terms of radiative transfer theory, the latter parameter is defined as the mean transport free path (MTFP, see Chap. 4, (4.18); see also, e.g., [5]). The relationship between the MTFP and the polarization decay parameters is controlled by the individual properties of each scattering medium, and, consequently, a given scattering system can be specified with adequate reliability using measurements of the polarization decay rate for the given scattering and detection conditions. Thus, the introduction of additional polarization measurement channels into those systems traditionally used for optical diagnostics and visualization of optically dense scattering media provides a novel quality and expands the functional ability of these systems.

An instance of particular interest is the appearance of polarization effects in the case of stochastic interference [6] of electromagnetic waves traversing random media. The most familiar examples of such an appearance are the polarization-dependent effect of coherent backscattering and the polarization dependence of temporal correlations of electric field fluctuations induced by the multiple scattering of coherent light by nonstationary media. These phenomena indicate the vector character of electromagnetic radiation propagating in random media.

In this chapter, a variety of fundamental effects, related to the manifestation of the vector nature of light in the case of multiple scattering by disordered media, are considered from the viewpoint of their applications in the optical diagnostics of scattering systems with complex structures such as biological tissue.

6.2 Coherent Backscattering

Research on the coherent backscattering effect for the case of bulk scattering of coherent light by dense random media was pioneered by Van Albada and Lagendijk [7] and Wolf and Maret [8]. Their qualitative explanation of the coherent backscattering phenomenon obviously follows from analysis of the formation of a backscattered field induced by the multiple scattering of a plane scalar monochromatic wave in a random medium with half-space geometry Fig. (6.1). For a given scattering angle θ, considered to be the superposition of partial multiply scattered waves with equal wave vectors $\overline{k}_s$, the scattered field features an increase in the resulting amplitude for the exact backscattering geometry $(\overline{k}_s = -\overline{k}_i$, where $\overline{k}_i$ is the wave vector of the incident wave, and, respectively, $\overline{k}_s$ is the wave vector of the outgoing multiply scattered wave). This feature is obviously caused by the constructive interference of partial waves associated with "direct" and "reverse" sequences of elementary scattering events (see, e.g., the sequences 1, 2, ...,$m-1, m$ and $m, m-1,$....,2,1 shown in Fig. 6.1).

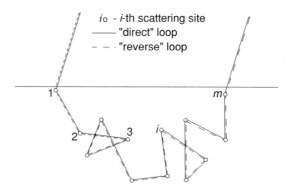

Fig. 6.1. Interpretation of the coherent backscattering effect in the case of multiple scattered scalar waves. The scattered field is detected exactly in the backward direction with respect to the incident wave

Both sequences are characterized by the same values of the propagation paths in a random medium, and, therefore, the phase delay between the "direct" and "reverse" partial waves for each pair of these waves is equal to zero. The constructive interference of all partial contributions in the case of $\theta = 180°$ causes a twofold increase in the scattered field intensity in comparison with the noncoherent summation of scattered partial waves. Any small divergence between the directions of $\overline{k}_i$ and $\overline{k}_s$ causes an abrupt decrease in the resulting intensity for all of the partially scattered waves with the same wave-vector $\overline{k}_s$ down to the value determined by the noncoherent summation of the partial waves.

A theoretical consideration of the coherent backscattering effect based on the scalar theory of wave propagation in disordered media [9–11] provides the triangular form of the backscattering peak with the half width depending on the ratio l^*/λ (l^* is the MTFP value of the scattering medium and λ is the wavelength of the probe light in the scattering medium). Using the scheme presented in Fig. 6.1, we conclude that, if $\overline{k}_m$ differs from $\overline{k}_0$, the phase shift between the partial contributions characterized by "direct" and "inverse" propagation through the scattering medium can be estimated as $(\overline{k}_0 + \overline{k}_m) \cdot (\overline{r}_1 - \overline{r}_m)$, where $\overline{r}_1$ and $\overline{r}_m$ are the positions of the first and last scattering sites in the considered sequence of scattering events. In this case, the average interference term, due to all contributions with m scattering events, will be nonzero and positive for $\left|\overline{k}_m + \overline{k}_0\right| < L_m^{-1}$, where L_m is the average diameter of the loops corresponding to the trajectories of scattered waves in the scattering medium (these loops are not necessarily closed). It is obvious that the minimum value of L_m is the average distance between two successive scattering events; that is, the elastic mean free path l (in the case of isotropic scattering, when $l \approx l^*$). Hence, an increase in the scattered intensity, from an incoherent background value up to a factor of 2 inside a cone of angular width on the order of λ/l centered at the backscattering direction, should be expected. More rigidly, an increase can also be described by using diagrammatic calculations of the intensity of the scalar wave backscattered from a semi-infinite medium [11]. These calculations lead to the following form for the angular dependence of the backscattered light:

$$I\left(\theta\right) \sim \left(l^{-1} + k\theta\right)^{-2} \left\{1 + [1 - \exp\left(-4k\,z_{\text{ext}}\theta/3\right)]/lk\theta\right\}, \qquad (6.1)$$

where z_{ext} (of the order of l) is the extrapolation length which can be obtained from the boundary conditions for the diffuse light propagation in the semi-infinite medium. In the case of anisotropic scattering, the mean scattering length l should be replaced by the transport mean free path l^*. For small scattering angles, the angular dependence of the intensity of the coherently backscattered light in the case of anisotropic scattering has the asymptotic form:

$$I\left(\theta\right) \sim 5k\theta\,l^*/3, \qquad (6.2)$$

which predicts the triangular form of the coherent backscattering peak observed in the experiments.

The vector nature of optical fields contributes an additional feature to the coherent backscattering phenomenon which is manifested as the dependence of the angular distribution of the intensity inside the backscattering cone on the polarization state of the detected backscattered radiation. Typically, in early studies of coherent backscattering, a random medium was illuminated by a linearly polarized collimated laser beam and the angular dependencies of the intensity of the two linearly polarized components of backscattered light were analyzed. One of them is a co-polarized component with the same polarization azimuth as the incident light, and the other is a crosspolarized component with the electric vector orthogonally directed with respect to the electric vector of the incident wave. A typical scheme of an experimental setup used to analyze the shape of the polarization-dependent coherent backscattering peak is shown in Fig. 6.2.

The normal to flat surface of a cell filled by an optically thick scattering medium (such as, an aqueous suspension of polystyrene beads) is slightly inclined with respect to the axis of the illuminating laser beam. This is necessary to avoid the influence of specular reflections from the cell surface on the coherence backscattering measurements. A manually rotated polarizer P2 is

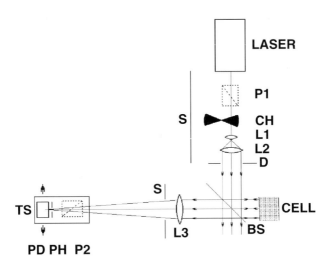

Fig. 6.2. An optical scheme used to study the coherent backscattering effect for orthogonally polarized components of backscattered light (see, e.g., [8]). P1 and P2 are polarizers; CH is a chopper; lenses L1 and L2 form the telescopic system – beam expander and collimator; D is a diaphragm; BS is a beam splitter; lens L3 is used to transform the angular distribution of the backscattered light intensity to the corresponding spatial distribution in its focal plane; S is a screen; the detection unit used to analyze the angular dependence of the backscattered light intensity consists of a pinhole diaphragm PH, a photodetector PD, and a translation stage TS

used to select the co-polarized or crosspolarized component of the backward scattered light. Typical forms of the angular distributions of the intensity of the "polarized" (co-polarized) and "depolarized" (crosspolarized) parts of the backscattered radiation are illustrated by Fig. 6.3.

A qualitative explanation, based on transfer matrix formalism, can be provided to interpret the difference between the values of the backscattering enhancement factors (i.e., the ratio of the intensity of the light scattered exactly in the backward direction to the backscattered intensity outside the coherent backscattering cone) for co-polarized and crosspolarized detected light [7]. By introducing the two-dimensional polarization vector of propagating light as p, we can establish the relationship between the incoming and outgoing light as $p_{\text{out}} = Mp_{\text{in}}$, where M is a 2×2 matrix, defined by a certain scattering diagram (light path). The resulting intensity of the scattered light can be obtained from the square of the sum of all possible light paths. The scattered light intensity outside the coherent backscattering cone can be obtained by an incoherent summation, and this multiple-scattered background will be almost totally depolarized. This leads to the following relationship for the elements of a transfer matrix:

$$\langle M_{11}^2 \rangle = \langle M_{22}^2 \rangle = \langle M_{12}^2 \rangle = \langle M_{21}^2 \rangle, \qquad (6.3)$$

where the ensemble averaging is provided by a summation over all possible light paths. Inside the coherent backscattering cone, the reverse light path should be added coherently to each light path. The reverse path is obtained from a given path by reversing all of its momenta; its transfer matrix $\widetilde{M}$ can be obtained by using the following rule: $\widetilde{M}_{ij} = M_{ji}$. By taking into account the symmetry requirements, we can conclude that all of the crossproducts should

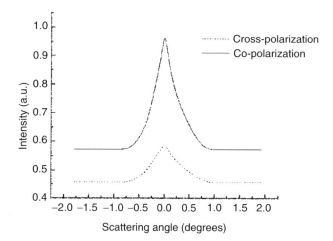

Fig. 6.3. Typical shapes of the coherent backscattering peaks for co-polarized and crosspolarized components in the case of bulk scattering by random dielectric media (for specified scattering systems, see, e.g., [7] and [8])

vanish: $\langle M_{ij}M_{kl}\rangle \sim \delta_{ik}\delta_{jl}$. When this result is used to add two backscatter diagrams coherently, a polarization-dependent enhancement factor is obtained, and its value is equal to 2 for the backscattered light polarized parallel to the incident light and 1 for the perpendicular component. Note that in real scattering systems, the experimentally measured values of the enhancement factor differ from those obtained on the basis of the above simple qualitative considerations. In particular, in the case of aqueous suspensions of 1.091 μm polystyrene beads, the values of the enhancement factor obtained by Van Albada and Lagendijk [7] are close to 1.6 for the co-polarized component (instead of the predicted value of 2) and close to 1.3 for the crosspolarized component (instead of the predicted unity value). Similarly, in an experiment with 0.46 μm polystyrene spheres in water, Wolf and Maret [8] obtained values on the order of 1.75 for the co-polarized component and 1.25 for the crosspolarized component. On the other hand, an experimental study of coherent backscattering from a disordered sample consisting of $BaSO_4$ particles (Kodak reflection filter, [12]) showed an absence of an expressed peak for angular dependence in the intensity of the crosspolarized component while the dependence of the co-polarized component was characterized by an enhancement factor approximately equal to 1.9.

More rigid theoretical considerations provided by Stephen and Zwilich [10] allowed them to obtain values of an enhancement factor equal to 1.9 and 1.2, respectively. An expanded theoretical study of the coherent backscattering of light by disordered media based on diagrammatic calculations was carried out by Akkermans et al. [11]. Other variables, including the time-dependent effects, the effects of absorption and the fractal nature of the scattering media were considered with particular attention devoted to anisotropic scattering and the influence of the polarization state of the backscattered light. They found that the angular dependencies of the intensity of the co-polarized and crosspolarized components of backscattered light for the scattering system, which is characterized by isotropic scattering, can be described by the following relationship:

$$I_{\mathrm{II}}(\theta) \sim \sum_{n} d_{n\mathrm{II}}I(n)\,\exp\left(-\frac{n}{3}\left(\frac{2\pi l}{\lambda}\right)^2 \theta^2\right),$$

$$I_{\perp}(\theta) \sim \sum_{n}(1-d_{n\mathrm{II}})C(n)I(n)\,\exp\left(-\frac{n}{3}\left(\frac{2\pi l}{\lambda}\right)^2 \theta^2\right), \qquad (6.4)$$

where summation is carried out over the number of scattering events which correspond to various contributions to the scattered optical field. $I(n)$ is the incoherent contribution obtained for scalar waves within the diffusion approximation ($I(n) \sim n^{-1.5}$); and the values of d_{nII} are the depolarization ratios, which determine the intensity transfer from the initial polarization to the perpendicular component. In the case of isotropic scattering, $d_{n\mathrm{II}} = \left(1+2(0.7)^{n-1}\right)/\left(2+(0.7)^{n-1}\right)$. The values of $C(n)$ are the average coherence ratios between pairs of time-reversed sequences of scattering events, which vary from 1 for $n = 2$ to zero for $n \to \infty$. It should be noted, however, that this approach breaks down in the case of short paths because of the inaccuracy of the expressions for $d_{n\parallel}$, $I(n)$, and $C(n)$ under this condition. Nevertheless, the obtained expressions are in good agreement with the experimental results [5]. These authors noted that the main results obtained for Rayleigh scattering systems can be easily expanded to the case of nonpoint-like scatterers characterized by expressed scattering anisotropy. Anisotropic scattering can be described as the replacement of the mean elastic free path l by the MTEP l^* in the above expressions.

Of particular interest is the manifestation of coherent backscattering of circularly or linearly polarized light for random dielectric media whose optical properties are altered by a Faraday rotation, or for scattering media with natural optical activity. Such specific scattering systems, which were considered by MacKintosh and John [13], can be defined as time-reversal-noninvariant and parity-nonconserving media. It has been established that the effect of breakdown of the time-reversal and parity symmetries is similar to that of the case of backscattering from systems with confined geometry or from an absorptive medium. In these cases, the suppression of the long scattering paths reduces the intensity of the coherent backscattering peak. However, the influence of a Faraday rotation or the natural optical activity of a scattering medium on the coherent backscattering of circularly polarized light is characterized by the opposite tendencies: the first suppresses only backscattered light of the same helicity as the incident light, whereas the latter suppresses coherence in the opposite helicity channel while the helicity-preserving channel remains unaffected. This study shows that in the case of circularly polarized light the helicity-preserving component of the backscattered peak is quantitatively similar to the peak calculated for scalar waves. The enhancement factor for these conditions is exactly equal to 2. This similarity is also manifested in the dependence of the backscattered peak parameters on the thickness and absorption properties of the scattering medium.

It should be noted, however, that observable changes in the backscattered peak shape can be obtained in the case of a very strong Faraday rotation. For instance, estimates of the required rotation for a scattering medium characterized by an elastic mean free path equal to $20\,\mu m$ provide a value on the order of $500°$ per mm. In real materials, such a strong rotation is accompanied by absorption, and it is difficult to distinguish between these effects.

6.3 Polarization-Dependent Temporal Correlations of the Scattered Light

A difference in propagation conditions for partial waves with different polarization states running through disordered media is also manifested as the influence of the polarization discrimination of detected multiply scattered light on the decay of the temporal correlation $G_1(\tau) = \langle E(t) E^*(t+\tau) \rangle$ of scattered field fluctuations, which are induced by random motions of scattering sites in the probed medium. This influence results from differences in the pathlength statistics of partial components of the multiply scattered optical field. These partial components are characterized by different polarization states. In particular, if the multiple scattering nonstationary medium is probed by a linearly polarized coherent light, the difference in the pathlength statistics for the co-polarized ($I_\parallel$) and crosspolarized ($I_\perp$) components of the detected scattered light manifests itself as the difference in the values of the correlation time of the intensity fluctuations of the scattered light. Consequently, the correlation functions of the detected intensity fluctuations $g_2(\tau) = \langle I(t) I(t+\tau) \rangle / \langle I^2(t) \rangle - 1$ for the co-polarized ($g_{2\parallel}(\tau)$) and the crosspolarized ($g_{2\perp}(\tau)$) components are characterized by different values of the decay rate, estimated as the asymptotic value of the slope of curves $g_{2\parallel}(\tau)$ and $g_{2\perp}(\tau)$ in the vicinity of $\tau = 0$. Such divergence of the decaying polarization-sensitive correlation functions of the scattered light intensity fluctuations is especially pronounced for the specific case of backscattered light detection, in which the difference between pathlength distributions of the $I_\parallel$ and $I_\perp$ components becomes dramatic. Experimental study of the influence of the detected light polarization state on the decay of the temporal correlations of coherent radiation backscattered by random nonstationary media was pioneered by MacKintosh et al. [14]. The most fundamental results obtained in their work are listed as follows:

(1) The decay rates, as well as the ratio of the intensities of the backscattered components with opposite polarization states in the case of scattering media illuminated by linearly polarized light (i.e., the co-polarized and crosspolarized components), are strongly influenced by the scatter size parameter ka (k is the wave number of the probe light and a is the characteristic size of the scattering particle, e.g., the radius for spherical particles). In the case of Rayleigh scattering systems consisting of small-sized particles and characterized by an isotropic phase function, the difference between the slopes of $g_{2\parallel}(\tau)$ and $g_{2\perp}(\tau)$ is maximal and the backscattered light is characterized by a significant degree of residual polarization. On the other hand, in media with an expressed scattering anisotropy (Mie scattering regime), the degree of residual polarization of the backscattered radiation approaches 0, and the difference in the slopes of $g_{2\parallel}(\tau)$ and $g_{2\perp}(\tau)$ is small;

(2) In the case of scattering media probed by circularly polarized light, the behavior of the detected components with opposite helicity (i.e., with right circularly polarized and left circularly polarized light) differs from that of linearly polarized light. In particular, with small-sized scattering particles, the correlation function for the helicity-preserving channel (i.e, for scattered light with the same helicity as the incident light) decays faster than that in a polarization channel with the same helicity. This feature can be explained by the fact that with circularly polarized light, the low-order

scattering sequences yield backscattered light that is primarily of the opposite helicity and the incident and reflected photons are related by mirror symmetry.

An important issue in studying the multiple scattering of coherent light by disordered nonstationary systems is the asymptotic behavior of the temporal autocorrelation functions of the scattered field fluctuations $G_1(\tau) = \langle E(t) E^*(t+\tau)\rangle$. Based on a diffusion approximation, a theoretical prediction for a case involving backscattering from a disordered Brownian medium has $G_1(\tau)$ decaying as $\sim 1 - \gamma\sqrt{6\tau/\tau_0}$ with an increasing time delay, where τ_0 is the single-scattering correlation time for a Brownian medium equal to $(k^2 D)^{-1}$. D is the self-diffusion coefficient of the Brownian particles. It should be noted that theoretical predictions of the shape of the coherent backscattering cone show the linear asymptotic, which is also dependent on the slope parameter γ. The value of γ is controlled by the boundary reflectivity and the mean scattering pathlength. Estimating on the basis of the scalar diffusion theory of light propagation in disordered media, the slope equals ≈ 2.4. For realistic scattering systems, the experimentally obtained values of γ are strongly influenced by the scattering anisotropy of the probed medium as well as the polarization state of the detected light. In particular, the experimental data obtained by MacKintosh et al., leads to the following relations between the scattering anisotropy, the polarization state of the probe light and the autocorrelation function slope:

- If a scattering medium consisting of small-sized particles is illuminated by a linearly polarized light and the linearly polarized backscattered component with the same polarization state as of an incident light (the co-polarized component) is detected, then the slope is approximately equal to $\gamma_\parallel \approx 1.45$; the "crosspolarized" autocorrelation function is characterized by $\gamma_\perp \approx 3.06$;
- The expressed anisotropic scattering (Mie scattering regime) is characterized by the close values of $\gamma_\parallel$ and $\gamma_\perp$: $\gamma_\parallel \approx 1.96$, $\gamma_\perp \approx 2.17$;
- If a Rayleigh scattering medium is illuminated by a circularly polarized light and the values of γ are estimated for circularly polarized components of the backscattered light with the same (γ_+) or opposite (γ_-) helicity as an incident light, then $\gamma_+ \approx 2.68$, $\gamma_- \approx 1.59$;
- Similar estimates for scattering systems consisting of large-sized scattering particles lead to the following relationships: $\gamma_+ \approx 1.72$, $\gamma_- \approx 2.62$. In this case, the circular polarization channels exhibit a high degree of polarization memory and their relative behavior appears reversed with respect to the case of Rayleigh scattering systems. This means that the helicity-preserving channel decays more slowly than the opposite-helicity channel. This effect is also reflected in the intensity ratio: $\langle I_+ \rangle / \langle I_- \rangle = 1.40$.

A theoretical consideration of these effects was carried out by MacKintosh and John [15]. In a case involving the probing of multiply scattering media

by linearly polarized light and backscattered light detection, the evolution
of the polarization state for the various partial components of the scattered
optical field allows for the following geometric interpretation (Fig. 6.4). Each
partial component propagating in the scattering medium and undergoing the
sequence of scattering events on the randomly distributed scattering sites,
can be characterized by the current orientations of its wave vector $\overline{k}_i$ and
unit vector $\hat{e}_i$ which characterize the current direction of the electric field
vector. Correspondingly, each $\overline{k}_i$ state can be represented by the position of
the imaging point on the surface of a sphere of radius k. The upper pole
of the sphere corresponds to the wave vector of the incoming probe light and
the bottom pole corresponds to the wave vector of the outgoing backscattered light. If a scattering system consists of large-sized scattering sites (Mie
scattering regime) and is characterized by an anisotropic phase function, then
each backscattered component is generated as a result of a sequence of a great
number of scattering events. This process can be interpreted as the gradual
movement of the imaging point from the upper pole of the sphere to the bottom pole. The following rule for the transformation of $\hat{e}_i$ to $\hat{e}_{i+1}$ due to the
ith scattering event, can be considered [14]:

$$\hat{e}_{i+1} \sim \hat{e}_i - \left(\overline{k}_i \hat{e}_i\right) \overline{k}_i \Big/ \left|\overline{k}_i\right|^2, \tag{6.5}$$

where $\overline{k}_i$ is the intermediate wave vector for the ith scattering event.

Each backscattered partial contribution can be associated with a certain
trajectory of the imaging point which connects both poles of the sphere.
Considering the various trajectories along the meridians, M_1, M_2 and M_3,

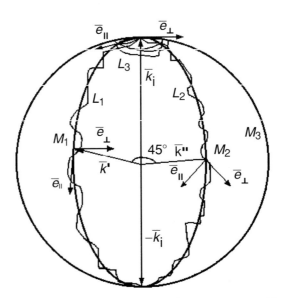

Fig. 6.4. Evolution of the polarization state of linearly polarized light propagating
in multiply scattering medium

it is easy to see that if the electric field vector of the incoming linearly polarized light is oriented in the meridian plane (M_1), or directed orthogonally to this plane (M_3), then the sequences of the scattering events, which correspond to the movement of the imaging point along the trajectories L_1 and L_3, will cause the direction of the electric field vector for the outgoing backscattered component to be the same as for incident light. But the sequence of scattering events presented by the L_2 trajectory will cause the outgoing component with the electric vector to be orthogonally oriented with respect to the electric vector of the incident light. Thus, for scattering media with strong scattering anisotropy, the contributions of the partial components that preserve the initial polarization state and the components that cause a 90° rotation of the electric vector, will be approximately the same, and the resulting outgoing light will be almost totally depolarized.

On the other hand, with scattering media characterized by an isotropic phase function (Rayleigh scattering systems), the relative contributions of the low-step-scattered partial components (single-scattered, twice-scattered, etc.) are sufficient and results in a noticeable degree of residual linear polarization.

By considering a nonstationary scattering medium and analyzing the correlation properties of the backscattered light propagating in two polarization channels which are characterized by opposite polarization states (such as channels with parallel and perpendicular polarization in the case of incident linearly polarized light), we can introduce two temporal correlation functions of scattered field fluctuations:

$$G_{1\text{II}}(\tau) = \langle E_\text{II}(t)\, E_\text{II}^*(t+\tau) \rangle \,;\ \ G_{1\perp}(\tau) = \langle E_\perp(t)\, E_\perp^*(t+\tau) \rangle. \tag{6.6}$$

MacKintosh and John obtained the following relationship for $G_{1\text{II}}(\tau)$, $G_{1\perp}(\tau)$ in a case involving backscattering from a Brownian medium consisting of uncorrelated point-like scatterers (the so-called white-noise model, in which the incident and reflected wave vectors do not influence the light propagation inside the scattering medium):

$$G_{1\text{II}}(\tau) \propto G(\tau,0) + \frac{20}{7} G\left(\tau, \sqrt{9/7}\right) \tag{6.7}$$

$$G_{1\text{II}}(\tau) \propto G(\tau,0) - G\left(\tau, \sqrt{9/7}\right),$$

where the parameter-dependent function $G(\tau, \zeta)$ is calculated as:

$$
\begin{aligned}
G(\tau,\zeta) = \int_0^\infty &\,dz\,\exp\left(-z/l\right) \int_0^\infty dz'\,\exp\left(-z'/l\right) \int d^2\rho \\
&\times \left[\frac{\exp\left\{ -\left(6\tau/\tau_0 + \zeta^2\right)^{0.5} \left[\rho^2 + (z - z')^2\right]^{0.5}/l \right\}}{\left[\rho^2 + (z - z')^2\right]^{0.5}} \right. \\
&\left. \quad - \frac{\exp\left\{ -\left(6\tau/\tau_0 + \zeta^2\right)^{0.5} \left[\rho^2 + (z + z' + 2z_b)^2\right]^{0.5}/l \right\}}{\left[\rho^2 + (z + z' + 2z_b)^2\right]^{0.5}} \right],
\end{aligned}
\tag{6.8}
$$

and the z axis is directed normally to the interface between the free space and the scattering medium, $\rho^2 = x^2 + y^2$. l is the mean scattering free path length, and z_b is the so-called extrapolation length depending on the boundary reflectivity of the scattering medium. The parameter ζ can be considered the "effective absorption" of the propagating light causing a cut-off of the long-path components due to the polarization discrimination of the detected backscattered light. With $\zeta \neq 0$, only short-path components with $s \leq l^*/\zeta$ contribute to the detected backscattered light. Thus, it is easy to see that the co-polarized correlation function $G_{1II}(\tau)$ is enhanced because of the cut-off of the long-path components and the predominating contribution of the short paths, while the correlation function for crosspolarized light is mainly influenced by the long-path components which obviously cause the depolarization of the propagating wave. Estimates of the asymptotic values of the autocorrelation function slope γ for co-polarized and crosspolarized light backscattered from a Rayleigh scattering medium were achieved by using (6.7) and (6.8) which yielded magnitudes of $\gamma_{II} \approx 1.6$ and $\gamma_{\perp} \approx 2.7$ which are very close to the experimental values obtained in the case of scattering 488 nm light from uncorrelated polystyrene latex spheres with a diameter of 0.091 µm [14].

A similar analysis can be used to consider a scattering media illumination with circularly polarized light. Calculations of the correlation functions for the two circularly polarized channels (one referred to as the helicity-preserving channel and the other as the opposite helicity channel) give the polarization-dependent correlation functions for light backscattered from a disordered medium consisting of point-like scatterers as:

$$G_{1+}(\tau) = G_1(\tau, 0) + \frac{1}{2}G_1\left(\tau, \sqrt{9/7}\right) - \frac{5}{6}G_1\left(\frac{5}{9}\tau, \sqrt{5/3}\right);$$

$$G_{1-}(\tau) = G_1(\tau, 0) + \frac{5}{7}G_1\left(\tau, \sqrt{9/7}\right) + \frac{5}{3}G_1\left(\frac{5}{9}\tau, \sqrt{5/3}\right). \tag{6.9}$$

These relations give the slope values for the helicity preserving and opposite helicity channel as $\gamma_+ \approx 2.4$ and $\gamma_- \approx 1.7$. The obtained relations are valid in the case of short delay times τ and for isotropic scattering by small-sized particles.

In comparison with linearly polarized light, circularly polarized light, which propagates in a disordered medium with expressed scattering anisotropy, demonstrates the existence of "polarization memory." This effect is related to a much lower decay rate for the wave helicity in comparison with the randomization rate for the propagation direction and results in a dramatic difference between γ_+ and γ_- for scattering anisotropy $g \geq 0.7$ as was mentioned in [15]. Following MacKintosh and John, the superior survival of circular polarization compared to that of linear polarization can be explained by the fact that the probability of scattering with, and without, a spin flip depends only on the scattering angles which are independent of the azimuthal rotations. Considered in terms of the Born approximation for a case of scattering from small particles, the amplitudes for scattering through an angle θ with, and without, a spin flip can be obtained as proportional to $(1 - \cos \theta)/2$ and $(1 + \cos \theta)/2$, respectively. This gives the probability of a spin flip for small scattering angles decreasing as θ^4, while the degree of randomization of the wave's direction grows as $n \langle \theta^2 \rangle$ after n scattering events.

In the case of large scattering particles, a similar analysis can be carried out based on Mie theory. With highly anisotropic scattering, the temporal correlation function of the field fluctuations in the helicity preserving channel is enhanced in comparison with the opposite helicity channel by the more probable paths without a spin flip. It has been shown [15] that the probabilities of scattering events without, and with, a spin flip can be expressed for anisotropic scattering as follows:

$$p_{\pm} \equiv \frac{1}{2} \left(1 \pm A \right), \tag{6.10}$$

where A is the asymmetry factor which depends on the scattering anisotropy (i.e., on the particle size). For Mie scattering systems, $A \approx 1$. Thus, the probability of n scattering events without a spin slip is equal to $[(1 + A)/2]^n$. After calculation, the following expression for the pathlength probability density distributions corresponding to the helicity preserving and the opposite helicity channels were obtained:

$$\rho^{\pm}(s) = \frac{\rho(s)}{2} \left(1 \pm e^{-s/n'l} \right), \tag{6.11}$$

where the asymmetry factor is related to n' as $A \equiv \exp(-1/n')$ and n' is determined by the number of scattering events which are required to randomize the wave's helicity. In particular, for scattering systems with $g = 0.9$, the value of n' was found to be equal to ~ 50. Such differences in the pathlength distributions for the helicity preserving and the opposite helicity channels resulted in the above mentioned divergence of the slope values γ_+ and γ_-.

Also, the polarization-dependent temporal correlations of light multiply scattered by nonstationary random media have been studied under a wide variety of scattering conditions by many research groups. In particular, Kuz'min and Romanov [16] studied the angular dependencies of correlation functions theoretically to determine the copolarized and cross-polarized components of light backscattered by Brownian scattering systems. The slope parameters for the field correlation functions $g_{1\text{II}}(\tau)$ and $g_{1\perp}(\tau)$ were calculated inside, and outside, of the coherent backscattering cone. It was found that the initial slope for $g_{1\text{II}}(\tau)$ decreases with an increasing scattering angle (i.e., the angle between the directions of the incident light propagation and scattered light collection). On the other hand, the initial slope for $g_{1\perp}(\tau)$ does not depend on the scattering angle.

A specific case of polarization-dependent behavior of temporal correlation functions of scattered light fluctuations involves multiple scattering from dense crystallizing colloidal systems which can be considered scattering media with partial correlations between separate scattering sites. It was discovered by Sanyal et al. [17] that the appearance of a crystalline phase in dense colloidal suspensions causes specific behaviors of the copolarized and crosspolarized temporal correlation functions of scattered light, namely, the decay to zero of the time correlation for depolarized light (crosspolarized component) as in a liquid, and the existence of a noticeable residual correlation at large decay times for co-polarized scattered light (as expected for the solid phase). It was also found that with colloidal samples that were aged for several weeks, the temporal correlation function for the crosspolarized signal also became non-decaying. This effect was observed for $0.115\,\mu m$ diameter charged polystyrene spheres in water with a volume fraction $f = 0.03$. A mixed bed of ion-exchange resins in suspension at the bottom of the cell was used to reduce the ionic impurities. The sample cell was probed by light from a Kr laser ($\lambda = 647.1\,nm$), and light scattered at $\theta = 165°$ was detected. The normalized autocorrelation functions of the intensity fluctuations of the co-polarized and crosspolarized components of the scattered light were measured using a Malvern correlator. The autocorrelation functions of the field fluctuations were obtained from corresponding intensity correlation functions based on the Siegert relationship [18]: $g_2(\tau) = 1 + |g_1(\tau)|^2$.

At the initial stage of the experiment (before adding resins), both correlation functions $g_{1II}(\tau)$ and $g_{1\perp}(\tau)$ demonstrate behavior which is typical for the above described noncorrelated random media; their decay fits well with the form: $\sim \exp\left[-\gamma(6\tau/\tau_0)^{0.5}\right]$ with $\tau_0 = 1.8086\,ms$ and $\gamma_{II} \approx 1.89$ and $\gamma_\perp \approx 2.87$. The appearance of a micro-crystalline phase after adding resins manifests itself as the decay of $g_{1II}(\tau)$ to a nonzero constant and the decay of $g_{1\perp}(\tau)$ to zero at essentially the same rate as in the liquid. Additional transmission measurements carried out on the same state of the sample also show similar behavior indicating that the sample is in a state of arrested translation motion throughout, which is typical for the solid state. Because of the ergodicity of the system being considered, the experimental autocorrelation functions of the co-polarized light, obtained by using ensemble and time averaging, exhibit different behavior (in particular, the ensemble averaged intensity correlation function is characterized by a slower decay rate and a larger residual value than the time-averaged correlation function).

This behavior of polarization-dependent temporal correlations of scattered light has been explained based on the consideration of wave propagation in densely packed random media. Because the studied system is characterized by the following ratio: $\lambda > af^{-1/3}$ (a is the particle radius), several particles are contained inside an arbitrarily chosen local volume of the probed medium, which is defined by a characteristic size on the order of the wavelength of the probe light. In this case the spatial fluctuations of the dielectric tensor components can be described in terms of a smoothly varying field. By considering the wave equation for the electric field of a monochromatic wave propagating in a densely packed random fluctuating medium, the following system of equations for the amplitudes of the co-polarized and crosspolarized components of the propagating waves can be written:

$$-\frac{\partial^2}{\partial s^2}\begin{pmatrix} E_{\text{II}}(s) \\ E_\perp(s) \end{pmatrix} = k_0^2 \vec{A}\left(\Re(s), t\right)\begin{pmatrix} E_{\text{II}}(s) \\ E_\perp(s) \end{pmatrix} + k_0^2\left[\aleph\left(\Re(s), t\right) + 1\right]\begin{pmatrix} E_{\text{II}}(s) \\ E_\perp(s) \end{pmatrix},$$
$$(6.12)$$

where $\Re(s)$ is the random-walk path of the propagation of the partial wave in the scattering medium; $\varepsilon_0\aleph$ and $\varepsilon_0\vec{A}$ are the local fluctuations of the isotropic and anisotropic parts of the dielectric tensor components; and ε_0 is the average dielectric constant of the medium which is isotropic at large spatial scales. With some simplifying assumptions about the structural properties of the scattering medium, the system, described by (6.12), may be solved analytically. After ensemble averaging, the following expressions can be obtained for the "single-path" polarization-dependent correlation functions of the field fluctuations:

$$G_{1\text{II}}(\tau) = \frac{E_0^2}{2}\exp\left[-Cs + Cse^{-\tau/\tau_I} - Ss\right]\left[\exp\left(Sse^{-\tau/\tau_A}\right) + \exp\left(-Sse^{-\tau/\tau_A}\right)\right]$$

$$G_{1\perp}(\tau) = \frac{E_0^2}{2}\exp\left[-Cs + Cse^{-\tau/\tau_I} - Ss\right]\left[\exp\left(Sse^{-\tau/\tau_A}\right) - \exp\left(-Sse^{-\tau/\tau_A}\right)\right],$$

where the constants C and S are defined by the covariance terms for the fluctuating anisotropic part of the dielectric tensor which are assumed to be delta-correlated in space and exponentially decaying in time. The values of τ_I and τ_A characterize the temporal decay of the covariance terms.

It is easy to see that at a large-time limit $(\tau \gg \tau_I, \tau_A)$, the "single-path" correlation function for the co-polarized component is characterized by a non-zero asymptotic value of $(E_0^2/2\exp[-(S+C)s]$, while the crosspolarized correlation function asymptotically approaches zero. In other words, the effect of the residual nonzero temporal correlation of the co-polarized component of the multiply scattered light is associated with the presence of some anisotropic entities caused by the imperfectly crystallized regime.

Similar behavior of the temporal correlation functions of the co-polarized and cross-polarized components of laser light multiply scattered by collagen-containing tissue was observed in a speckle-correlation experiment with human sclera [19]. In the terms of the theoretical interpretation presented here, it can be concluded that the presence of anisotropic components, such as collagen fibers in the tissue structure, cause noticeable diversity in $g_{2\parallel}(\tau)$ and $g_{2\perp}(\tau)$ when the time scale is large.

6.4 Polarization Microstatistics of Speckles

One of the fundamental manifestations of the polarization phenomena in the multiple scattering of coherent light by disordered media is the vector nature of speckle-modulated multiply scattered optical fields. When manifested as random spatial fluctuations of amplitudes and phases of the orthogonally directed components of the scattered field in the observation plane, the vector nature of the multiply scattered speckles usually appears at spatial scales that are associated with the characteristic size of the "coherence area" of the observed speckle field (i.e., the average speckle size). In the case of multiple scattering, the vector statistics of the speckle fields are related to the probability of a change in the polarization state of the partial waves propagating in a random medium along the different Feynman paths which are associated with the random sequences of many scattering events. Each scattering event rotates the incident polarized field randomly. Consequently, the resulting observed field at the arbitrarily chosen detection point is constructed as the superposition of the partial waves incoming to the detection point from the scattering medium. This process changes the incident polarization state in a statistical fashion.

In the particular case of scalar speckle fields, which are usually induced by the propagation of a coherent light in single scattering systems with no birefringence, the statistical properties of the spatial fluctuations of the speckle amplitude, phase and intensity are totally described by the corresponding probability density functions $\rho(E), \rho(\varphi)$ and $\rho(I)$. In the general case of a vector speckle field, the statistical description of the random fluctuations of scattered light requires a more sophisticated approach. A convenient method for arriving at such a description is to use the covariance, or the coherency matrix of the scattered light, which depends on both the properties of the scattering medium and the polarization state of the incident light. The basic variables for describing the in-observation-plane amplitude distribution of the scattered field are the real and the imaginary parts of the field components $E_x = E_x^r + iE_x^i$ and $E_y = E_y^r + iE_y^i$, where $-\infty < E_{x,y}^{r,i} < \infty$ and the coordinate system $0xy$ is usually introduced by taking into account the polarization state of the incident light. The Gaussian statistics of the real and imaginary parts of the scattered field components E_x, E_y are usually assumed in cases of the multiple scattering of coherent light by random media, and the following form of the joint probability density function for the above considered four variables can be introduced, as shown by Goodman [20] and used by Barakat [21]:

$$\rho\left(E_x^r, E_x^i, E_y^r, E_y^i\right) = \frac{1}{\pi^2 d} \exp\left(-\frac{1}{d}\left[j_{22}|E_x|^2 + j_{11}|E_y|^2 -\right.\right.$$
$$\left.\left. - 2Re\left(j_{12}E_x^*E_y\right)\right]\right),$$

(6.13)

where j_{km} are the elements of the 2×2 covariant Hermitian matrix $\Im$ introduced as:

$$\Im = \begin{pmatrix} \langle E_x^* E_x \rangle & \langle E_x^* E_y \rangle \\ \langle E_y^* E_x \rangle & \langle E_y^* E_y \rangle \end{pmatrix}$$

(6.14)

the brackets $\langle\ \rangle$ denote the ensemble averaging, and $d = \det \Im$.

In scattering systems with statistically independent Feynman paths, the following relations are valid for the elements of the coherency matrix:

$$\langle E_x^r E_y^r \rangle = \langle E_x^i E_y^i \rangle = \frac{1}{2}j_{12}^r; \langle E_x^r E_y^i \rangle = -\langle E_x^i E_y^r \rangle = \frac{1}{2}j_{12}^i.$$

(6.15)

In addition, the following relation takes place for $\Im$ as the Hermitian matrix:

$$j_{12} = j_{21}^* = j_{12}^r + i j_{12}^i. \tag{6.16}$$

The matrix elements j_{km} are simply related with the averaged values of the Stokes parameters of the scattered light:

$$\begin{aligned}
\langle I \rangle &= j_{11} + j_{22}; \\
\langle Q \rangle &= j_{11} - j_{22}; \\
\langle U \rangle &= 2j_{12}^r; \\
\langle V \rangle &= 2j_{12}^i,
\end{aligned} \tag{6.17}$$

and, correspondingly,

$$d = \frac{1}{4}(\langle I \rangle^2 - \langle Q \rangle^2 - \langle U \rangle^2 - \langle V \rangle^2).$$

The total degree of polarization can be easily calculated as:

$$P = \frac{\sqrt{\langle Q \rangle^2 + \langle U \rangle^2 + \langle V \rangle^2}}{\langle I \rangle} = \sqrt{1 - \frac{4\det^2 \Im}{|\mathrm{tr}\Im|^2}}. \tag{6.18}$$

Thus, by measuring the "global" polarization parameters of multiply scattered light, such as $\langle I \rangle, \langle Q \rangle, t\langle U \rangle, \langle V \rangle$, one can reconstruct the statistical properties of the Gaussian speckle fields of the expressed vector nature.

Another type of variables can be introduced to describe the polarization microstatistics of multiple scattered speckle fields. Here, the object of specific interest is the probability density functions of the parameters of the polarization ellipse reconstructed at each point of the observation plane. Figure 6.5 shows the variety of such local polarization ellipses where each

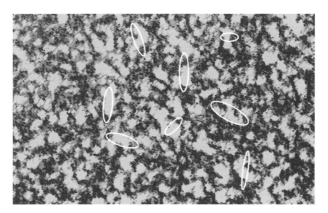

Fig. 6.5. Variety of shapes and orientations of the polarization ellipse for the vector speckle field

is associated with a certain speckle spot in the observation plane. A set of polarization ellipse parameters (Fig. 6.6). can be introduced [22] – the amplitudes of the major (A_a) and the minor (A_b) axes, and the corresponding intensities $I_a = (A_a)^2, I_b = (A_b)^2$, which are determined by the following relations:

$$I_a = (A_a)^2 = \frac{1}{2} \left\{ (A_x)^2 + (A_y)^2 \right.$$
$$\left. + \sqrt{\left[(A_x)^2 - (A_y)^2\right]^2 + 4(A_y)^2 (A_y)^2 \cos^2 \delta} \right\},$$

$$I_b = (A_b)^2 = \frac{1}{2} \left\{ (A_x)^2 + (A_y)^2 \right.$$
$$\left. - \sqrt{\left[(A_x)^2 - (A_y)^2\right]^2 + 4(A_y)^2 (A_y)^2 \cos^2 \delta} \right\},$$

$$A_x = \sqrt{(E_x^r)^2 + (E_x^i)^2},$$
$$A_y = \sqrt{(E_y^r)^2 + (E_y^i)^2},$$
$$\delta = \arg(E_y) - \arg(E_x).$$

In addition, the azimuth of the polarization ellipse is characterized by the angles $\psi_\pm$ between the major (+) and minor (−) axes relative to the x axis. For this set of parameters, the joint probability density function can be obtained in the case of Gaussian speckle fields [21] by taking into account (6.13):

$$\rho(I_a, I_b, \psi_\pm) = \frac{1}{\pi d} \frac{I_a - I_b}{\sqrt{I_a I_b}} \exp\left(-\frac{(j_{11} + j_{22})(I_a + I_b)}{2d} \right)$$
$$\times \cosh\left(\frac{2|j_{12}|}{d} \sin \beta \sqrt{I_a I_b} \right) \times \cosh\left[Z(\psi_\pm)(I_a - I_b) \right], \quad (6.19)$$

where $Z(\psi_\pm) = \frac{j_{12}}{d} \cos \beta \sin 2\psi_\pm + \frac{(j_{11} - j_{22})}{2d} \cos 2\psi_\pm$ and the value of β is determined as $\beta = \arg(j_{12})$.

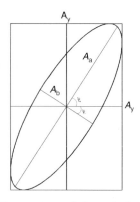

Fig. 6.6. Parameters of the polarization ellipse

We can see that $\rho(I_a, I_b, \psi_\pm)$ is always equal to zero if $I_a = I_b$. This means that the probability of finding the speckle spot in which the scattered light is circularly polarized is equal to zero for any scattering system except the trivial case of $d \equiv 0$, which corresponds to circular input polarization and a unit degree of polarization.

6.5 The Concept of Polarization-Correlation Universality

The concept of polarization-correlation universality was considered by Freund, Kaveh, Berkovits, and Rosenbluh [23] in relation to the microstatistics of optical waves in random media. Microstatistics is associated with the vector nature of speckle-modulated optical fields induced by the multiple scattering of coherent light in disordered media. Following this approach, we can denote the scattered field components by $E_{\mathrm{in,out}}$, so that E_{xy} represents a y-polarized output produced by a x-polarized input, and we can introduce the correlation matrix of the scattered fields as:

$$C = \begin{pmatrix} 1 & 0 & 0 & \Gamma \\ 0 & \rho & \delta & 0 \\ 0 & \delta^* & \rho & 0 \\ \Gamma^* & 0 & 0 & 1 \end{pmatrix},$$

(6.20)

where $C_{ij} = \langle E_i E_j^* \rangle$; the brackets $\langle \ \rangle$ imply ensemble averaging and the following condensed notation is used: $xx = 1, xy = 2, yx = 3$, and $yy = 4$. This form of the correlation matrix follows from the symmetry considerations of a statistically isotropic medium and is not model specific. All of the matrix elements are normalized by the value of the co-polarized intensity $C_{11} = \langle |E_{xx}|^2 \rangle$. With this formalism, the crosspolarized intensity $C_{22} = \langle |E_{xy}|^2 \rangle$ can be considered as the measure of the depolarization of the scattered light; Γ and δ describe the partial correlation of the scattered fields.

The relationship between the elements of the correlation matrix C and the elements of the coherency matrix $\Im$ can be obtained by consideration of the scattering matrix F, which establishes a connection between the components of the input and scattered fields. Following Eliyahu and Eliyahu et al. [22, 24, 25], we can write:

$$\Im = \left\langle \overline{E}_{\mathrm{s}} \overline{E}_{\mathrm{s}}^\oplus \right\rangle = \langle F^{\mathrm{T}} \Im_{\mathrm{i}} \left(F^{\mathrm{T}}\right)^\oplus \rangle,$$

$$\overline{E}_{\mathrm{s}} = F^{\mathrm{T}} \overline{E}_{\mathrm{i}}; \ \Im_{\mathrm{i}} = \overline{E}_{\mathrm{i}} \overline{E}_{\mathrm{i}}^\oplus,$$

where $\overline{E}_{\mathrm{i}}, \overline{E}_{\mathrm{s}}$ are the incident and scattered fields and $\Im_{\mathrm{i}}$ is the coherency matrix of the incident light. In this way, the elements of the correlation matrix ρ, Γ, and Δ can be obtained from elements of the scattering matrix F as:

$$\rho = \left\langle |F_{xy}|^2 \right\rangle,$$

$$\Gamma = \langle F_{xx}^* F_{yy} \rangle,$$

(6.21)

$$\Delta = \left\langle F_{xy}^* F_{yx} \right\rangle.$$

Analysis of polarization transfer for model scattering systems consisting of point-like noncorrelated scattering sites leads to the existence of the so-called sum rule: $\Gamma + \delta = 1 - \rho$. In the case of a predominating contribution of the long-path diffuse partial components to the formation of multiply scattered fields, the depolarization parameter ρ approaches 1, and the values of Γ and δ approach zero, i.e., the scattered light becomes almost totally depolarized and the fields are almost completely uncorrelated. Such scattering conditions are typical in the case of light transmission through an optically thick disordered slab. On the other hand, the diagrammatic calculations for the case of diffuse reflection lead to the following relation $\Gamma \gg \delta$ that gives $\Gamma \sim 1 - \rho$. These simple relations allow the statistical moments of multiply scattered fields to be expressed in terms of relatively simple functions of depolarization ρ and thus provide a simple universal description of the polarization correlations in random media.

Depending on the illumination and detection conditions, the polarization correlation of speckle-modulated multiply scattered fields will manifest itself in different forms. Consider that a linearly polarized beam is used to illuminate the sample which is characterized by arbitrary depolarization ρ, and the scattered light is detected without a polarizer. In this case, the normalized value of the intensity crosscorrelation function for two speckle patterns recorded for two values of the polarization azimuth angle of the incident light ($I(\theta_0)$ is the reference speckle pattern, and $I(\theta)$ is the second speckle pattern) can be presented in the following simple form:

$$C_{\mathrm{u}}(\theta_0, \theta) = (1 - \beta) \cos^2(\theta - \theta_0) + \beta; \qquad (6.22)$$
$$\beta = (1 - \rho)^2 / (1 + \rho)^2.$$

The subscript u means that no polarizer is placed in the output beam.

An experimental study of the crosscorrelation functions of speckle-modulated intensity distributions obtained from surface as well as bulk scatterers of a different nature, which are characterized by a wide range of ρ values (from $\rho = 0.002$ for bright stainless-steel surface scatterer to $\rho = 0.94$ for BaSO$_4$ diffuse reflectance coating) have shown gratifying agreement between the experimental data and the theoretical predictions expressed by (6.22). In this case, the existence of a simple universal form for the correlation of unpolarized speckle patterns, which allows description in terms of a single, easily determined depolarization parameter ρ, is evident.

The more specific case is the polarization discrimination of multiply scattered light outgoing from the sample. If a fixed polarizer is inserted into the output beam and the input polarization direction θ_0 is varied, then the observed form of the correlation function, $C(\theta_0, \theta)$, is strongly influenced by the choice of the reference speckle pattern $I(\theta_0)$. If both values, θ_0 and θ, are measured from the direction defined by the fixed output polarizer, then the crosscorrelation function $C(\theta_0, \theta)$ has the following form [23]:

$$C_{\mathrm{p}}(\theta_0, \theta) = \frac{[\cos \theta_0 \cos \theta + \rho \sin \theta_0 \sin \theta]}{(\cos^2 \theta_0 + \rho \sin^2 \theta_0)(\cos^2 \theta + \rho \sin^2 \theta)}, \qquad (6.23)$$

where the subscript p indicates the presence of a polarizer in the output beam. Experimental studies of crosscorrelation functions of speckle patterns, recorded with the use of a fixed polarizer, have also shown excellent agreement with the theoretical predictions given by (6.23).

When examined for an arbitrarily chosen coherence area (speckle spot), the intensity oscillations caused by rotation of the input polarization direction obey the generalized Malus law:

$$I(\theta) = a \cos^2(\theta - \alpha) + b, \tag{6.24}$$

where a (the "modulation depth") and b (the "baseline") are positively defined random values. Probability distributions of a and b are determined by the microstatistics of speckle field fluctuations. In the case of expressed multiple scattering, the statistical distribution of the initial phase α of the speckle intensity oscillations must be uniformly distributed between 0 and π, and it is necessarily statistically independent of a and b. The analytical description of the a and b distributions is a rather difficult problem but the specific case of totally depolarized outgoing light and without an output polarizer is characterized by the following probability distribution of "modulation depth" values [23]:

$$P_a(a/\langle a \rangle) = (3\pi/8\langle a \rangle)x^2 K_1(x), \quad x = 3\pi a/4\langle a \rangle, \tag{6.25}$$

where $\langle a \rangle = (3\pi/8)\langle I \rangle$, $\langle I \rangle$ is the mean speckle intensity, and K_m is a modified Bessel function of order m. The nth order statistical moment of a is expressed as follows:

$$\langle a^n \rangle / \langle a \rangle^n = (8/3\pi)^n \Gamma(1 + n/2)\Gamma(2 + n/2). \tag{6.26}$$

where Γ is the gamma function.

The probability distribution of the "baseline" can be approximated in the case of a large b by the negative exponential distribution:

$$P_b(b) \sim \exp\left(-4b/\langle I \rangle\right). \tag{6.27}$$

Another specific case of strong scattering with $\rho = 1$, in which the input polarization direction is kept fixed and a polarizer is rotated in the output beam, is described by:

$$\begin{aligned} P_a(a/\langle a \rangle) &= (\pi/2 \langle a \rangle) y K_0(y), \\ y &= \pi a/2 \langle a \rangle, \quad \langle a \rangle = (\pi/2) \langle I \rangle. \end{aligned} \tag{6.28}$$

The hypothesis of polarization-correlation universality was verified by examining the polarization microstatistics of speckle patterns induced by the multiple scattering of laser light by disordered media of a different nature. In particular, the suggestion that speckle microstatistics is a universal function of depolarization ρ was examined with the use of thin alumina membranes as the scattering samples [26]. These membranes, known as Anopore membranes, were stacked to obtain higher ρ values and submerged in water for lower ρ values. Thus, a set of scattering samples characterized by the wide range of ρ was obtained. The thickness of the Anopore membranes is equal to 60 µm and the pores are 0.1 µm (type I) or 0.2 µm (type II) in diameter. The surface density of pores is about 10^9 cm^{-2}. These parameters give values of depolarization ρ (in air) equal to 0.75 and 0.90, respectively. Estimates of the extinction mean-free-path lengths for these samples submerged in water give values of 33 and 22 ± 1 µm for type-I and type-II, respectively.

Light scattering experiments were carried out for transmission mode. In order to verify (6.24), the polarization microstatistics of speckle patterns were examined for the case of a varying polarization angle θ. An Ar-ion laser (20 mW output at 514.5 nm, linear polarization) was used as an illumination source and the laser beam was expanded to obtain a light spot of 6 mm in diameter on the sample surface. Changes in θ were provided by a half-wave plate which was placed between the laser and the beam expander and rotated by a step motor. Oscillations of the scattered light intensity were detected at different points of the detection plane (0.28 m from the sample) with a light-collecting optical fiber with a core of 3.7 µm in diameter. These detection conditions provided an average speckle size in the detection zone that was about 80 times larger than the detector aperture, which corresponded to the fiber core. Two-dimensional scanning of the distal end of the optical fiber allowed for the recording of the scattered light intensity oscillations for different coherence areas, or speckles in the detection plane. The signal from the output of the optical fiber was detected with the use of a cooled photomultiplier tube in the photon-counting mode.

In accordance with [26], the parameters a and b (see (6.24)) were calculated from an experimentally obtained intensity time series as:

$$a = [8\{\langle I^2 \rangle - \langle I \rangle^2\}]^{0.5}; \quad b = \langle I \rangle - a/2, \tag{6.29}$$

as follows from the theoretical predictions. Then both values of the modulation parameters were normalized by the average frequency for convenience. The empirical probability density distribution of the modulation depth parameter a obtained for the sample characterized by the unit value of ρ, demonstrates excellent agreement with the theoretical predictions. Because of the absence of a compact analytical form of the theoretical prediction for the probability distribution $\rho(a)$ for the intermediate values of $0 \le \rho < 1$, the following empirical model was used to fit the obtained experimental data:

$$\langle \tilde{a} \rangle P_{\tilde{a}}(\tilde{a}) = \beta \tilde{a}^\nu \exp(-\xi \tilde{a})$$

with β, ν and ξ as the fitting parameters. Note that this fitting relation asymptotically reduces to (6.28) with an increasing value of $\tilde{a}$ if $\beta = 3\pi^{1.5}/4$, $\nu = 1.5$, and $\xi = 2$.

The empirical statistics of the normalized baseline value are characterized by asymptotic decay $P_b(\bar{b}) \sim \exp(-\alpha \bar{b})$. In the case of a single-scattering regime ($\rho \to 0$), the concept of polarization-correlation universality predicts the negative exponential statistics with $\alpha \rightarrow 1$ for the baseline value. On the contrary, for a sufficient multiple scattering regime characterized by a unit value of depolarization, the value of the asymptotic decay parameter α reported in [26] is equal to 4 for $0.5 \leq \bar{b} \leq 1.5$. For comparison, a similar value for the Amopore membranes in air was found to be equal to $\approx 3 \pm 0.3$ for the same interval of $\bar{b}$. Analysis of the experimentally obtained values of α, depending on the depolarization value ρ, shows that the asymptotic decay parameter becomes relatively constant at a value of 3 down to $\rho \sim 0.5$. This means that theoretical predictions for the asymptotic decay of baseline statistics might need some revision. It should be noted that for empirically obtained baseline statistics, the exponential asymptotic falloff was observed at $\bar{b} \cong 1$, while the theoretical prediction of $\alpha = 4$ assumed $\bar{b} \gg 1$. Tarhan and Watson noted that because of the exceedingly low probability of observing large values of $\bar{b}$, it was not possible to establish with certainty whether $\alpha \cong 3$ will evolve into $\alpha = 4$ at the limit, or not. No indication of such a crossover was found in this work.

Comparative analysis of the polarization microstatistics of speckles for cases of speckle pattern formation by laser light multiply scattered in the forward as well as in the backward direction (the value of ρ for both scattering regimes is equal to 1) has shown "modulation depth" statistics to be in excellent agreement with the theoretical predictions given by Freund et al. [23].

Thus, the experimental verification of the hypothesis of polarization-correlation universality, carried out by Tarhan and Watson with high resolution for a wide range of values of the depolarization parameter, has demonstrated excellent agreement with the theoretical predictions of "modulation depth" statistics that have empirical distributions in the case of unity ρ. On the contrary, the asymptotic behavior of the baseline statistics differ from the predictions made on the basis of the concept of polarization-correlation universality. Also, the differences in the baseline statistics for different samples characterized by the same values of the depolarization parameter indicate that scattering parameters in addition to ρ may be required in order to completely characterize baseline distributions.

In addition, the major argument against polarization-correlation universality in multiple scattering is that the form of the temporal autocorrelation functions for multiply scattered light in reflection is not universal but instead depends on both the diffraction parameter of the scattering site and the polarization state of the incident light [26]. In their "Comment on 'Polarization memory of multiply scattered light,'" Freund and Kaveh [27] have presented the argument below, which, from their viewpoint, supports the hypothesis of polarization-correlation universality in multiple scattering.

Previously, the claim of universality was based on a scalar-wave treatment of the vector optical fields and the agreement of this treatment with some experimental results. It was noted by Freund and Kaveh that for the temporal correlation problem, the major differences between the scalar and the vector fields can be considered in terms of the different relative weights of the Feynman paths associated with transport of multiply scattered light. Different polarization channels of the vector field weigh these paths differently. For a linearly polarized output, for example, the parallel, co-polarized output channel weighs short, polarization preserving paths more heavily than does the scalar approach, while the perpendicular, crosspolarized channel places greater relative weight on long, depolarizing paths. Following Freund et al., it is possible to provide an equivalent to the scalar approach by the simultaneous collection of both channels, i.e., to simply collect all of the multiply scattered light without any polarization bias. For instance, in the case of incident linearly polarized light, the unbiased total intensity can be simply presented as the sum of partial intensities corresponding to co-polarized and crosspolarized output channels:

$$I(t) = I_{\mathrm{II}}(t) + I_\perp(t).$$

Following from an assumption about the absence of correlations of waves propagating in the opposite polarization channels, the unbiased temporal correlation function of the scattered field can be written as:

$$G_1(\tau) = [\langle I_{\mathrm{II}} \rangle \, G_1(\tau, \gamma_{\mathrm{II}}) + \langle I_\perp \rangle \, G_1(\tau, \gamma_\perp)]/[\langle I_{\mathrm{II}} \rangle + \langle I_\perp \rangle], \qquad (6.30)$$

where γ parametrizes the differences between the two channels. A possible way to obtain the unbiased temporal correlation function of the scattered field fluctuations is to use the heterodyne measurements in combination with polarization discrimination of the scattered light with the use of a polarizer at $45°$ with respect to $\mathrm{II}, \perp$ directions in order to combine the optical fields from both channels with equal weights.

In the frames of the scalar wave approach, the field autocorrelation function of light backscattered from the random Brownian medium can be presented for short times in the following approximate analytical form:

$$G_1(\tau, \gamma) \approx \exp\left(-\gamma \sqrt{6\tau/\tau_0}\right),$$

where τ_0 is the single scattering correlation time. Similarly, the normalized unbiased temporal correlation function can also be considered for intensity fluctuations of the scattered light:

$$G_2(\tau) \approx \exp\left(-2\langle \gamma_2 \rangle \sqrt{6\tau/\tau_0}\right). \qquad (6.31)$$

Considering the asymptotic behavior of both $g_1(\tau)$ and $g_2(\tau)$ at $\tau \to 0$, the parameters $\langle \gamma_1 \rangle$ and $\langle \gamma_2 \rangle$ can be expressed as:

$$\langle \gamma_1 \rangle = [\langle I_{\mathrm{II}} \rangle \, \gamma_{\mathrm{II}} + \langle I_\perp \rangle \, \gamma_\perp]/[\langle I_{\mathrm{II}} \rangle + \langle I_\perp \rangle];$$
$$\langle \gamma_2 \rangle = \left[\langle I_{\mathrm{II}} \rangle^2 \, \gamma_{\mathrm{II}} + \langle I_\perp \rangle^2 \, \gamma_\perp\right] \Big/ \left[\langle I_{\mathrm{II}} \rangle^2 + \langle I_\perp \rangle^2\right]. \qquad (6.32)$$

The calculations of the "unbiased" parameter $\langle \gamma_1 \rangle$ carried out by Freund and Kaveh with the use of experimental data for the backscattered light presented in, [9] yielded a value of $\langle \gamma_1 \rangle$ equal to 2.06 ± 0.03 for widely different particle sizes and for both circularly and linearly polarized input fields. Similarly, a "universal" value of 2.0 ± 0.1 was obtained for $\langle \gamma_2 \rangle$. They also mentioned that the average 5% dispersion matches well the stated 5% error limits of [9].

Following Freund and Kaveh, the simple relations between the values of the slope parameter γ for the "unbiased" or "scalar" autocorrelation function, the autocorrelation functions for polarization channels with the opposite polarization states, and the depolarization parameter ρ as the central point of the concept of universality could be established:

$$\gamma_{\text{opposite}} = \left[\frac{1 + \rho}{\rho + \rho^a} \right] \gamma_0;$$

$$\gamma_{\text{same}} = \rho^a \gamma_{\text{opposite}}, \qquad (6.33)$$

where $a = 4/3$ and γ_0 is the scalar value of γ. A comparison of the measured values of γ and the values calculated with the use of (6.33) has shown that the suggested simple relations "provide a nearly perfect description of the data for widely different particle sizes for both linear and circular polarizations."

Freund and Kaveh have noted, however, that although the comparison of theoretical predictions based on the universality concept with the results of correlation experiments supports the suggestion about system-independent behavior, the obtained value of γ_0 (2.06) is somewhat larger than the expected value predicted by scalar theory:

$$\gamma_0 = 1 + \Delta = 1.7104\ldots, \qquad (6.34)$$

where Δ, which is determined by the boundary conditions for the diffusion problem in the case of semi-infinite geometry, can be obtained from Milne theory.

This discrepancy manifests itself as a too large decay rate in the experimentally obtained temporal correlation of intensity fluctuations in comparison with the theoretical prediction (the measured $g_2(\tau)$ appears too narrow). The explanation of such behavior suggested by Freund and Kaveh is related to the problem of internal surface reflections which act to reinject a portion of the scattered light into the scattering medium and thus elongate the Feynman paths traversed by the photons.

The concept of polarization-correlation universality (especially in the part of the system-independent microstatistics of speckle-modulated multiply scattered fields) has met with criticism based on the conclusion that the diffusion approximation cannot be used to predict the value of the correlation decay parameter γ; nor can it be used to make any statement about the universality of the value of γ [28]. It was mentioned that "given the very broad variety of media that exhibit diffusive-light propagation, this apparent system independence may be purely fortuitous, and it would seem premature to claim that the value of γ is universal. Moreover, given the tenuous state of current theoretical work on this point, any claims of strong universality are clearly premature and misleading."

6.6 Summary

A specific property of optical fields induced by the multiple scattering of coherent radiation in disordered media is the wide variety of local polarization states associated with the speckle structure of the scattered field. Multiply scattered speckles as the product of the stochastic interference of coherent light propagating through random media are characterized by elliptical polarization with the parameters of a polarization ellipse varying from one speckle to another in a stochastic manner.

The parameters of polarization microstatistics of multiply scattered speckles are controlled by the optical properties of the scattering medium and thus an analysis of the polarization-dependent stochastic interference of multiply scattered waves can be applied to characterize scattering systems. Among other interference phenomena accompanying the multiple scattering of coherent light by random media, polarization-dependent coherent backscattering and polarization-sensitive temporal fluctuations of coherent light scattered by nonstationary systems can be considered as the physical basis for the development of novel diagnostic techniques applicable to the analysis of the structure of weakly ordered media such as, biological tissues. In particular, our specific interest is the study of the different behaviors of polarization-dependent temporal correlations of light multiply scattered by collagen-containing tissues. The presence of anisotropic (collagen-containing) components in tissue structures causes the apparent differences between the correlation characteristics of multiply scattered light with different polarization states and thus can be used for tissue structure characterization.

7

Decay of Light Polarization in Random Multiple Scattering Media

7.1 The Similarity in Multiple Scattering of Coherent Light

The relationship between the statistical properties of Feynman path distributions, which characterize the pathlength statistics of partial waves propagated in random media, and the statistical properties of multiply scattered vector optical fields manifest themselves in a number of theoretically predictable and experimentally observable effects [1–9]. One of these is the appearance of similarities in multiple scattering. A group of relaxation phenomena in the case of coherent light propagation in a disordered system can be considered a manifestation of this similarity, or its likelihood, in multiple scattering. This similarity is related to the same forms of dependencies of certain statistical moments of scattered optical fields on the specific spatial scales which characterize the decay of the corresponding moments in the course of coherent light propagation in disordered media. The following relaxation effects should be considered [10–12]:

- The existence of temporal correlations of amplitude and intensity fluctuations in scattered optical fields at a fixed detection point for nonstationary systems of scattering particles
- The decay of the polarization of light propagated in disordered systems
- The manifestation of Bougier's law in the case of multiple scattering with noticeable absorption.

Relaxation of the statistical moments of the scattered optical fields can be considered in terms of Feynman path distributions, i.e., by the statistical analysis of ensembles of optical paths for partial waves, which propagate in the scattering medium and from which the observed scattered field can be constructed. In the diffusion scattering mode, each partial component of the multiply scattered optical field is associated with a sequence of a great number, N, of statistically independent scattering events and is characterized by path s. The statistical moments of the scattered field can be considered as the integral transforms of the probability density function, $\rho(s)$. Within a weak scattering limit, when $l, l^* \gg \lambda$ (l and l^* are the mean elastic free path and the mean transport free path for the scattering medium [see [5] in Chap. 6] and λ is the wavelength of the probe light) such second-order statistical moments as the average intensity of scattered light, the temporal correlation function of the field fluctuations and the degree of polarization of the multiply scattered light at the arbitrarily chosen detection point can be expressed for $N = s/l \gg 1$ as the Laplace transforms of $\rho(s)$. In particular, the average intensity of the scattered light for a multiply scattering medium with nonzero absorption can be written using the modified Bougier's law:

$$\langle I \rangle \cong \int_0^\infty \exp(-\mu_a s)\rho(s)\mathrm{d}s = \int_0^\infty \exp(-s/l_a)\rho(s)\mathrm{d}s, \tag{7.1}$$

where averaging is carried out over all possible configurations of the scattering sites. The normalization condition can be written in the following form:

$$\int_0^\infty \rho(s)\mathrm{d}s = \langle I \rangle_0, \tag{7.2}$$

where $\langle I \rangle_0$ is the average intensity in the absence of absorption.

For nonstationary disordered media consisting of moving scattering particles, the normalized temporal autocorrelation function of the scattered field fluctuations is expressed as [13–16]:

$$g_1(\tau) \approx \frac{\int_0^\infty \exp[-B(\tau)s/l]\rho(s)\mathrm{d}s}{\int_0^\infty \rho(s)\mathrm{d}s}, \tag{7.3}$$

where $B(\tau)$ is determined by the variance of the displacements of the scattering sites for the time delay τ. In the particular case of Brownian systems, the exponential kernel of the integral transform (7.3) is equal to $\exp\left(-2\tau\, s/\, \tau_0 l^*\right)$.

The relaxation of the initial polarization state of coherent light propagating in a disordered multiply scattering medium is caused by an energy flux interchange between partial waves with different polarization states. In particular, if propagated light has initial linear polarization, the linearly "co-polarized" and "cross-polarized" partial components of the scattered field can be considered. The first of these is characterized by the the electric field vector in the same direction as the incident illuminating beam and the other in an orthogonal direction to it. In a similar way, the interrelationship between the left circularly polarized component and the right circularly polarized component can be analyzed if an illuminating light with initial circular polarization is used. Propagation of linearly polarized light in a strongly scattering disordered medium can be considered by solution of the Bethe–Salpeter equation for the case of the transfer of a linearly polarized partial, "single-path" contribution, which undergoes n scattering events in a disordered medium with isotropic scattering [5]. This consideration yields the following expressions for the intensities of the "single-path" cross-polarized and co-polarized components [17]:

$$I_\parallel^s = f_\parallel^s(n) I^s(n),$$
$$I_\perp^s = f_\perp^s(n) I^s(n), \tag{7.4}$$

where the single-path "scalar" intensity $I^s(n)$ can be obtained by evaluating the photon density for a scalar wave propagating at a distance corresponding to n scattering events, and the weighting functions $f_\parallel^s(n)$, $f_\perp^s(n)$ can be determined by their dependencies on the number of scattering events [5,17]:

$$f_\parallel^s(n) = \left[\left(\tfrac{10}{15} \right)^{n-1} + 2 \left(\tfrac{7}{15} \right)^{n-1} \right] \Big/ 3,$$
$$f_\parallel^s(n) = \left[\left(\tfrac{10}{15} \right)^{n-1} - \left(\tfrac{7}{15} \right)^{n-1} \right] \Big/ 3. \tag{7.5}$$

Thus, by introducing a value for the polarization degree of the arbitrary single-path contribution of a scattered optical field with a propagation path equal to $s \approx nl$ as $P^s(n) = \left[I_\parallel^s(n) - I_\perp^s(n) \right] \Big/ \left[I_\parallel^s(n) + I_\perp^s(n) \right]$, we can obtain the following:

$$P^s(n) = \frac{f_\parallel(n) - f_\perp(n)}{f_\parallel(n) + f_\perp(n)} = \frac{3 \left(\tfrac{7}{15} \right)^{n-1}}{2 \left(\tfrac{10}{15} \right)^{n-1} + \left(\tfrac{7}{15} \right)^{n-1}} = \tag{7.6}$$
$$= 3 \left[2 \exp \left\{ (n-1) \ln \frac{10}{7} \right\} + 1 \right]^{-1}.$$

Correspondingly, the single-path polarization degree $P^s(n)$ for a linearly polarized light obeys the exponential decay $P^s(n) \cong 1.5 \exp \left\{ (n-1)/n' \right\}$ with the decay parameter equal to $n_L' \approx 2.804$ for long propagation distances with a great number of scattering events $n \gg 1$.

If a multiple scattering disordered medium is illuminated by circularly polarized light, then the single-path degree of the circular polarization of the multiple scattered light can be introduced as the ratio $P_C^s(n) = [I_+(n) - I_-(n)]/[I_+(n) + I_-(n)]$, where $I_+(n), I_-(n)$ are the intensities of the circularly polarized partial contributions which undergo n scattering events and have the same helicity as an incident circularly polarized light $(+)$ or the opposite helicity $(-)$. Similar consideration of the multiple scattering of circularly polarized light also leads to exponential decay of the single-path degree of the circular polarization with the value of the decay parameter equal to $n_C' \approx \ln 2 \approx n_L'/2$ [17].

If polarized light propagates in a disordered medium characterized by a sufficiently nonzero value of the anisotropy parameter g (the case of anisotropic scattering), then the decay parameter $n_{L,C}'$ should be replaced by the effective value $\tilde{n}'_{L,C}$, which is determined by the optical properties of the scattering particles that form the scattering system. Introducing the depolarization length, $\xi_{L,C} = \tilde{n}'_{L,C}l$, as one of the dimension scales which characterize the scattering system, we find the relationship between $\xi_{L,C}$ and another important scale – the mean transport free path l^*. This relationship is strongly influenced by the optical properties of the scattering medium as well as by the illumination and detection conditions.

The degree of residual polarization in a scattered optical field at an arbitrarily chosen detection point can be determined by averaging the single-path polarization degree over the ensemble of partial components of a scattered optical field characterized by the path length density distribution $\rho(s)$:

$$P_{L,C} = \int_0^\infty P_{L,C}^s(s)\rho(s)ds \approx \frac{3}{2}\int_0^\infty \exp\left(-\frac{s}{\xi_{L,C}}\right)\rho(s)ds, \qquad (7.7)$$

where the probability density function $\rho(s)$ is determined by the conditions of the light propagation in the scattering medium between a source of polarized light and a detection system which is used for polarization discrimination of the scattered light.

The theoretically predicted exponential decay of the single-path polarization degree with an increasing path length s was directly observed in the experiments using time-resolved intensity measurements of the co-polarized and cross-polarized components of a backscattered light in a case involving optically dense media illumination by a short pulse of linearly polarized laser light [18]. In these experiments, the colloidal systems consisted of aqueous suspensions of 1-μm-diameter silica spheres with an ionic strength of 0.03 m L^{-1} NaNO$_3$, pH = 9.5, and the volume fractions ranged from 5% to 54%. The scattering samples were probed by laser pulses with a duration of 150 fs emitted by a dispersion-compensated, self-mode-locked Ti:sapphire laser pumped by a frequency-doubled Nd:YAG laser. The backscattered light pulses were analyzed with a background-free cross-correlation technique. The Ti:sapphire laser, which had a repetition frequency of 76 MHz, was tuned to a wavelength of 800 nm, and its output was split into two beams by a 50:50 beam splitter. One beam passed through a delay stage and served as a gating pulse in the crosscorrelator. Data runs were typically recorded with a 3-mm (20 fs) step size. The other beam passed through a mechanical chopper, a second beam splitter, and a 15 cm focal length converging lens to reach a sample placed at the focus of the beam. The estimated value of the photon density corresponding to a single pulse of probe light was found to be equal to 5.3×10^{13} cm^{-2} per pulse.

The degree of polarization of the backscattered light was determined by using a half-wave plate and a Glan-Tompson polarizer. Typical shapes of the detected pulses of the co-polarized and cross-polarized components of the backscattered light from two scattering samples with strongly differing scattering coefficient values are illustrated in Fig. 7.1. The inset illustrates the

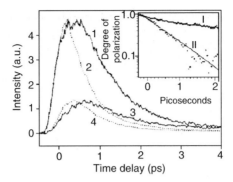

Fig. 7.1. The pulse shapes for co-polarized and cross-polarized components of backscattered light. *Solid lines*, the scattering sample with a 5% volume concentration of silica spheres; *dotted line*, the scattering sample with a 25% volume concentration of silica spheres. (1, 2) Intensity of the co-polarized component; (3, 4) intensity of the cross-polarized component. Inset shows the evolution of the time-dependent degree of linear polarization of backscattered light for both samples (I, 5% volume concentration of the scattering sites; II, 25% concentration of the scattering sites) [18]

tendency of the time-dependent degree of linear polarization of the backscattered light to decay.

The single-path degree of the linear polarization can be expressed in simple exponential form as $P_L^s(n) \approx \exp(-n/n_L')$ where n_L' is regarded as the average number of scattering events needed to depolarize the optical wave. For an effective speed of light, v_{eff} and the mean elastic scattering free path l, the time scale of the depolarization process can be estimated to be on the order of $\tau = n_L' l / v_{\mathrm{eff}}$.

The validity of the exponential decay model for describing the dissipation of the initial polarization state of light propagating in multiple scattering random media was confirmed by experimental studies (using slab geometry) of the depolarizing properties of optically thick random media which were probed in the transmittance mode [10,11,17]. By applying a diffusion approximation, we can find the path-length density distributions $\rho(s)$ for optically thick slabs in the transmittance mode and, after this, evaluate the single-sided Laplace transformation $L_\rho(m) = \int_0^\infty \exp(-s/ml^*)\rho(s)\mathrm{d}s$, which, when analyzed for the fixed value of m, exponentially decays with an increase in the dimensionless slab thickness L/l^*. This tendency is illustrated in Fig. 7.2. The above discussed exponential decay of the "single-path" degree of polarization causes the approximately exponential decay of the degree of polarization of light transmitted through an optically thick slab ($L/l^* \geq 5$) with the increasing ratio L/l^*. Indeed, the dependencies of the degree of polarization of linearly or circularly polarized light transmitted through optically dense scattering slabs on a dimensionless slab thickness (obtained in experiments

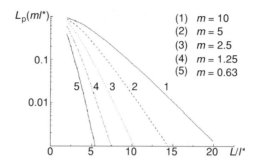

Fig. 7.2. The Laplace transformations of the pathlength density distributions for probe light, which is transmitted through a scattering slab, that depends on the normalized slab thickness [10,12]. The probability density functions $\rho(s)$ were calculated with the use of a diffusion approximation

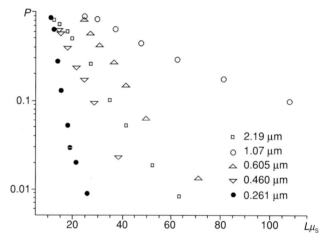

Fig. 7.3. The measured values of the degree of linear polarization of light transmitted through the scattering slabs [10]. Scattering systems are aqueous suspensions of polystyrene beads of various sizes. The values of the degree of polarization are plotted against the dimensionless scattering coefficient of the corresponding scattering system. Wavelength and cuvette thickness: 514 nm (Ar-ion laser) and 10 mm – for 0.460 μm and 1.07 μm particles; 532 nm (diode-pumped Nd-laser) and 20 mm – for 0.261 μm, 0.605 μm, and 2.19μm particles

with monodisperse aqueous suspensions of polystyrene beads of various sizes) show that $P_{\mathrm{L,C}}$ falls as: $P_{\mathrm{L,C}} \sim \exp(-KL/l^*)$ with K depending on the size of the scattering particles and the type of polarization of the incident light (Fig. 7.3).

The principle of similarity in multiple scattering that follows from the exponential form of the "single-path" parameters of multiply scattered optical fields, such as the "single-path" degree of polarization and the "single-path" temporal correlation function of scattered field fluctuations for nonstationary scattering media, is manifested as the equality of spatial scales which characterize the decay rate for the corresponding parameter.

In particular, such equality allows us to introduce a specific parameter for nonstationary scattering media such as a characteristic correlation time [19,20]. This parameter establishes the relationship between the characteristic spatial scale of the dissipation of the optical field correlation that is due to multiple scattering in a fluctuating random medium, the depolarization length, and the dynamic properties of a Brownian scattering medium. It can be written as follows:

$$\tau_{\mathrm{cd}} \approx \frac{l^*}{2\xi_{\mathrm{L}}Dk_0^2}, \tag{7.8}$$

where ξ_{L} is the depolarization length for linearly polarized radiation in the scattering medium, D is the translation diffusion coefficient of the scattering particles, and k_0 is the wave-number of the probe light.

It is easy to conclude that the characteristic correlation time is independent of the concentration of the scattering sites but is determined by their optical and dynamic properties. Thus, the characteristic correlation time can be considered the universal parameter of multiple scattering dynamic media. Figure 7.4 illustrates the principle of evaluation of τ_{cd} by using results taken from simultaneous measurements of the temporal correlation function and the degree of polarization of the multiply scattered light.

Experiments with aqueous suspensions of polystyrene sphere irradiated by linearly polarized light from an Ar-ion laser demonstrate the independence of the characteristic correlation time from the volume fraction of the scattering particles (Fig. 7.5).

The values of τ_{cd} were determined by the method illustrated in Fig. 7.4. Normalized values of the module of the field correlation functions were obtained from experimentally measured intensity correlation functions by using

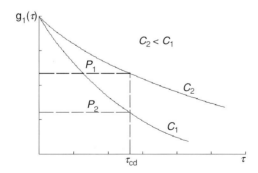

Fig. 7.4. Method for determining the characteristic correlation time for multiply scattering Brownian medium

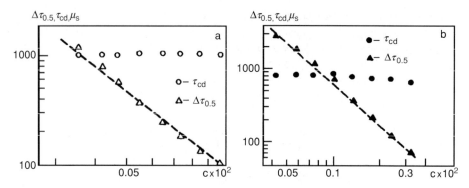

Fig. 7.5. Concentration dependencies of the characteristic correlation time and the half-width of the autocorrelation function of the intensity fluctuations obtained in experiments with aqueous suspensions of polystyrene beads (left, bead diameter $0.46\,\mu m$; right, bead diameter $1.07\,\mu m$) [19]

the Siegert relation (see Chap. 6, [18]). Moreover, measurements of the "conventional" correlation time, reported as the half-width of normalized field correlation functions, were performed. Figure 7.5 shows a logarithmic plot of the experimentally measured concentration dependencies of τ_{cd} and $\Delta\tau_{0.5}$ (the "conventional" correlation time estimated as the halfwidth of the correlation peak).

Analysis of the experimental data shows that in the experimental range of the concentrations of the aqueous suspensions of polystyrene beads, the concentration dependencies $\Delta\tau_{0.5} = \varphi(c)$ are close to power-law functions $\Delta\tau_{0.5} \sim c^{-\alpha}$. The exponents α in the power-law functions approximating the experimental values of $\Delta\tau_{0.5}$ in Fig. 7.5 are ≈ 2.21 and ≈ 1.96 for polystyrene beads of diameters 0.46 and $1.07\,\mu m$, respectively. These values are in satisfactory agreement with the value $\alpha = 2$ given by the diffusion approach. Specifically, as was mentioned in [6], for an optically thick layer of thickness L consisting of Brownian scattering particles, the normalized autocorrelation function of the amplitude fluctuations of the scattered coherent radiation allows the following approximation:

$$g_1(\tau) \approx \exp\left(-\sqrt{6\tau/\tau_0}\,\frac{L}{l^*}\right).$$

Thus, analysis of polarized light transfer based on the principle of similarity provides additional possibilities for describing the scattering properties of probed media. In particular, the influence of the size parameter of the scattering sites on the decay of polarization of propagated light can be studied with this approach, as was shown in [19].

7.2 Influence of Scattering Anisotropy and Scattering Regime

Study of the influence of the size parameter of scattering centers on the decay of the initial polarization state of coherent light backscattered by random media was pioneered by MacKintosh et al. [8]. On the basis of measurements of the intensity of backscattered light that corresponded to the opposite polarization channels (co-polarized and cross-polarized backscattered light for a linearly polarized probe light and components of scattered light with the opposite helicity for a circularly polarized probe light), they concluded that the backscattering of linearly polarized light from a random medium consisting of large-sized dielectric particles (Mie scattering regime) is accompanied by significant suppression of the polarization of the outgoing multiple scattered light (i.e., the backscattered light is almost totally depolarized). On the contrary, backscattering by random media, consisting of small-sized dielectric particles (Rayleigh scattering regime), is characterized by a significant degree of polarization of the backscattered light. If circularly polarized light is used to probe scattering media in the backscattering mode, then scattering ensembles consisting of small-sized particles are characterized by similar values of intensity of backscattered light in the polarization channels with the opposite helicity. By contrast, the backscattering of circularly polarized light by media with expressed scattering anisotropy exhibits a high degree of polarization memory, which is manifested as a noticeable difference between the values of intensity for a helicity-preserving polarization channel and a polarization channel with the opposite helicity: $\langle I_+ \rangle / \langle I_- \rangle = 1.40$ for a scattering system with $l^* = 10l$ [8].

A Monte Carlo simulation was used to analyze the influence of the size parameter of the scattering dielectric spheres on the decay of linear polarization in a backscattering mode [21]. In the procedure that followed, a transformation of the complex amplitude of partial waves, which form a backscattered optical field due to random sequences of scattering events, was simulated (Fig. 7.6). Each partial wave was induced by an incident linearly polarized monochromatic plane wave propagating along the z-axis of the "fundamental" coordinate system (x, y, z). The electric field of an incident wave was directed along the x-axis. A scattering medium was considered as the disordered ensemble of nonabsorbing dielectric particles with a given value of size parameter. The relative refractive index of the spheres was taken to be 1.2, which is approximately equal to the refractive index of polystyrene beads in water. The direction of the propagation of the incident linearly polarized plane monochromatic wave relative to the "fundamental" coordinate system was characterized by a normalized wave-vector $\bar{k}_0/|\bar{k}_0| = (0, 0, 1)$ where the z-axis was aligned with a normal to the scattering medium surface.

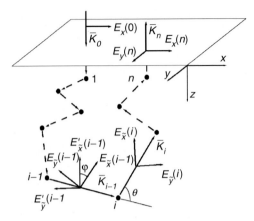

Fig. 7.6. The scheme of the transformation of the polarization state of a partial wave due to a random sequence of scattering events (Monte-Carlo simulation) [21]

The transformation of the electric field of the propagating partial wave was analysed for a sequence of n scattering events. For each ith step, the transformation of the complex amplitude for both of the orthogonally polarized components of the propagating wave was described by a (2×2) scattering matrix $(S_{km}(\theta, \phi, i)) = (S''_{km}(\theta, i)) \times (S'_{km}(\phi, i))$. The complex elements of the scattering matrix were calculated for simulated random values of the scattering angle θ and azimuth angle ϕ by using the current coordinates $(\tilde{x}_i, \tilde{y}_i, \tilde{z}_i)$ that were related to the ith scattering event.

The $\tilde{z}_i$-axis is directed along the wavevector of the partial wave that propagates after the ith scattering event, and the $\tilde{x}_i$-axis is perpendicular to the scattering plane. The scattering angle distribution that corresponds to the Mie phase function for a single scatter with a given value of size parameter was used for simulation of the random value of θ for each scattering event. Random values of the azimuth angle ϕ were considered to be uniformly distributed within the range $(0, 2\pi)$. The $(S'_{km}(\phi, i))$ matrix characterizes the transformation of the $\tilde{x}$ and $\tilde{y}$ components of the electric field of the partial wave, which propagates after the $(i-1)$th scattering event, due to rotation by the angle ϕ during conversion of the $(\tilde{x}_{i-1}, \tilde{y}_{i-1}, \tilde{z}_{i-1})$ current coordinates to the $(\tilde{x}_i, \tilde{y}_i, \tilde{z}_i)$ coordinates:

$(E_{\tilde{x}}(i-1), E_{\tilde{y}}(i-1) \Rightarrow E'_{\tilde{x}}(i-1), E'_{\tilde{y}}(i-1))$, see Fig. 7.6.

The $(S''_{km}(\theta, i))$ matrix elements, which are calculated using Mie theory, characterize the transformation of the electric field components that is due to the ith scattering event:

$(E'_{\tilde{x}}(i-1), E'_{\tilde{y}}(i-1) \Rightarrow E_{\tilde{x}}(i), E_{\tilde{y}}(i))$, see Fig. 7.6.

During the simulation, only the n-times scattered partial waves, which were characterized by a z component of the normalized wavevector with values between -0.985 and -1 (relative to the "fundamental" coordinates), were selected for further analysis. The magnitudes $I_x = |E_x|^2$ and $I_y = |E_y|^2$

were evaluated by calculating the x and y components of the electric field in the "fundamental" coordinates for each selected n-times scattered outgoing partial wave. After this, values $\langle I_\| \rangle = \langle I_x \rangle$ and $\langle I_\perp \rangle = \langle I_y \rangle$ were calculated by averaging over the whole ensemble of selected partial waves with $-1 \leq k_{zn}/|\hat{k}_n| \leq -0.985$ and a single-path value of P_L^s for a given number of scattering events was obtained as $P_L^s = (\langle I_\| \rangle - \langle I_\perp \rangle)/(\langle I_\| \rangle + \langle I_\perp \rangle)$.

Figure 7.7 illustrates the typical dependencies of the degree of single-path linear polarization on the number of scattering events as a result of the simulation procedure described above for two different scattering regimes (the Rayleigh scattering regime for small values of the anisotropy parameter [Fig. 7.7a] and the Mie scattering regime for large values of g [Fig. 7.7b]).

For a given number of scattering events, the values of $\langle I_\| \rangle$, $\langle I_\perp \rangle$ and $P_L^s(n)$ were calculated for a simulated scattering system, which was characterized by a given value of the size parameter by averaging over the ensemble of 10,000 outgoing partial waves. After this, the obtained values of the single-path residual polarization were plotted in semilogarithmic coordinates against the number of scattering events n. The bars show an increase in the deviation of the obtained $P_L^s(n)$ values (with respect to the mean value of the single-path residual polarization) with an increase in the number of scattering events. The value of the anisotropy parameter for each simulated scattering system was calculated as the mean cosine of the scattering angle by using Mie theory.

Typically, all of the curves obtained by the simulation procedure are characterized by the presence of two specific regions: a relatively small "low-step scattering" region with values of the single-path polarization degree which are close to 1, and a "diffusion scattering" region characterized by an approximately exponential decay of the single-path polarization degree $P_L^s(n)$. The

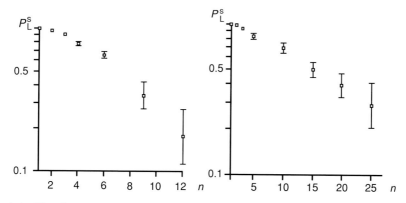

Fig. 7.7. The dependencies of the "single-path" degree of residual linear polarization in the backscattering mode on the number of scattering events (results of monte-carlo simulation): (**a**) isotropic scattering ($ka = 1, g \approx 0.178$); (**b**) anisotropic scattering ($ka = 6.5, g \approx 0.915$)

location of the overlap between these regions, as well as the polarization decay rate for the diffusion scattering region, strongly depends on the anisotropy parameter of the scattering particles.

Values of the normalized depolarization length m_{L}, which were estimated as $m_{\mathrm{L}} = (1 - g)\bar{n}'_{\mathrm{L}}$ by using the exponential approximation $P_{\mathrm{L}}^{\mathrm{s}} \propto \exp(-n/\bar{n}'_{\mathrm{L}})$, are presented in Fig. 7.8 (represented by full circles) as dependent on the anisotropy parameter.

In order to obtain the dependence of the normalized depolarization length on g, the dependencies of single-path residual polarization on n (which are similar to those presented in Fig. 7.8) were obtained by using the above described Monte Carlo procedure for scattering systems that are characterized by given values of the size parameter and, correspondingly, the anisotropy parameter. After this, the values of $\bar{n}'_{\mathrm{L}}$ were determined versus g by evaluation of the slope of the corresponding dependencies of $\ln P_{\mathrm{L}}^{\mathrm{s}} = f(n)$ for the "diffusion scattering" region.

For small scatterers (the Rayleigh scattering regime), the value of m_{L} that was obtained was approximately equal to 4.2. This magnitude diverges from the above presented theoretical value $m_{\mathrm{L}} \approx \bar{n}'_{\mathrm{L}} \approx 2.8$ [17] by approximately 35%. With an increase in the anisotropy parameter up to values on the order of 0.6–0.8, m_{L} decreases insignificantly. With larger values of g, the decay rate becomes large and m_{L} falls to values on the order of 1.0–1.2 in the vicinity of the first Mie resonance ($ka \approx 8$, $g \approx 0.93$).

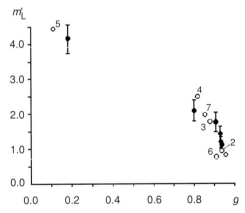

Fig. 7.8. The normalized depolarization length for linearly polarized light in the backscattering mode versus the parameter of scattering anisotropy. *Full circles*, results of Monte-Carlo simulation; *open circles*, experimental data; (1) 1.07 μm polystyrene beads in water, volume fraction is 10%, $L = 3$ mm, $\lambda = 488$ nm; (2) the same as 1, but $\lambda = 633$ nm; (3) Teflon, $L = 30$ mm, $\lambda = 488$ nm; (4) same as (3), but $\lambda = 633$ nm; (5) 0.091 μm polystyrene beads in water, volume fraction is 5%, $\lambda = 488$ nm [8]; (6) 0.605 μm polystyrene beads in water, volume fraction is 2%, $\lambda = 488$ nm [8]; (7) 0.46 μm polystyrene beads in water, volume fraction is 10%, $L = 3$ mm, $\lambda = 515$ nm [22]

In the "forward-scattering" mode (i.e., when the simulated partial waves are selected using the condition $0.985 \leq k_{zn}/|\hat{k}_n| \leq 1$), the dependencies of the single-path polarization degree on the number of scattering events obtained by the Monte-Carlo simulation for the Rayleigh scattering system are similar to those obtained for the backscattering mode (Fig. 7.9).

Thus, it can be concluded that the estimates of the depolarization length for linearly polarized light in scattering systems characterized by $ka \ll 1$ are insensitive to the regime of scattered light collection. On the contrary, the depolarization length for linearly polarized light, estimated under similar conditions for forward scattering systems consisting of large-sized particles, significantly exceeds the value of the mean transport free path (Fig. 7.9). Results of experimental studies on polarization decay in the forward scattering of linearly polarized light by optically thick disordered layers of dielectric spheres ([10, 17], see also Fig. 7.3) express a depolarization length ξ_L that increases with an increase in the size parameter ka of the scattering sites.

With multiple scattering systems consisting of optically soft dielectric spheres (e.g., aqueous suspensions of polystyrene spheres), the maximal value of ξ_L in the forward scattering mode was found in the vicinity of the first Mie resonance [17]. Theoretical analysis of the polarization decay of linearly polarized light multiply scattered in the forward direction by disordered media [24, 25] also shows better preservation of linear polarization of forward scattered light in random media with an expressed scattering anisotropy.

When compared with a phantom scattering media, the propagation of polarized light in tissue is characterized by some features related to the rate of polarization dissipation. These features were studied by Jacques et al. [26–29], Sankaran et al. [30–32], Wang, Schmitt, and many other researchers (see, e.g., [33–35]) in experiments with varoius in vivo and in vitro tissues such as human skin, porcine adipose tissue, and whole blood, etc.

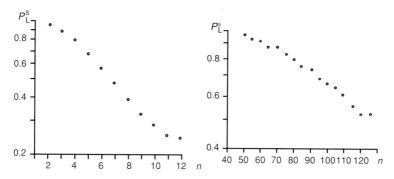

Fig. 7.9. The dependencies of the "single-path" degree of residual linear polarization in the forward scattering mode on the number of scattering events (results of Monte-Carlo simulation): (**a**) isotropic scattering ($ka = 1, g \approx 0.178$); (**b**) anisotropic scattering ($ka = 6.5, g \approx 0.915$) [23]

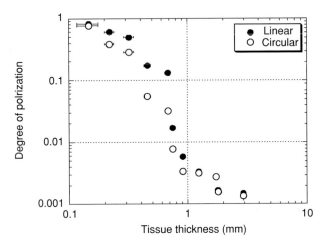

Fig. 7.10. The degree of linear and circular polarization in sections of porcine adipose tissue as a function of tissue thickness [32]

The difference between the values of the depolarization length of linearly or circularly polarized light estimated in experiments with tissue layers and corresponding parameters of phantom scattering media (for instance, aqueous suspensions of polystyrene beads) with the same optical properties (the mean transport free path and the parameter of scattering anisotropy) as the examined tissue samples can be considered the main peculiarity of polarization decay in biological tissue. Figure 7.10 shows the values of the degree of linear and circular polarization of light transmitted through a porcine adipose layer; these are dependent on the layer thickness. As in the case of phantom monodisperse scattering systems consisting of dielectric spheres of equal size, the dependence of P_L and P_C (the degree of circular polarization) on the thickness of the tissue layer demonstrates the presence of two characteristic regions: the region of nondiffuse scattering in optically thin $(L \leq l^*)$ tissue samples, which is characterized by the slow decay of the initial polarization, and the region in which an abrupt decrease of the degree of polarization takes place with the increasing thickness of the tissue layer. It is necessary to note that the decay rates for linear and circular polarization in the latter case are characterized by similar values and appear significantly smaller in comparison with scattering phantoms with similar optical properties. In any case, at the present time the peculiarities of polarized light transfer in real tissues at the cellular and subcellular levels of the spatial scales require further theoretical and experimental investigation.

7.3 Residual Polarization of Incoherently Backscattered Light

Specific conditions in the formation of multiply scattered optical fields in the backscattering of polarized light by optically thick disordered media (in

particular, the sigificant contribution of nondiffusing components characterized by short propagation paths $s \sim l^*$ in a scattering medium and the strong
influence of the boundary conditions of radiative transfer at the interface
"free space-scattering medium") cause the existence of a noticeable degree of
residual polarization in backward scattered light. This effect is not related
to the above discussed effect of polarization-dependent coherent backscattering (see Chap. 6). The residual polarization of backscattered light is caused
by the summation of intensities of partial waves propagating in the scattering medium at different distances and thus is characterized by different
values of the "single-path" degree of polarization $P_{L,C}^s$. Because of its incoherent nature, the residual polarization must be observed outside the coherence
backscattering cone. The approximate analytical expression for the degree of
residual polarization of incoherently backscattered light can be obtained by
using the above discussed principle of similarity for the particular conditions
(illumination of an optically thick scattering medium by a linearly polarized
plane wave and detection of the backward scattered plane wave component
of a multiply-scattered optical field). On this basis, by using the assumption
about the approximately exponential decay of P_L^s, the degree of residual linear
polarization of the detected light can be expressed as [21]:

$$P_L^r \approx \frac{3}{2} \exp(-\gamma \sqrt{3/m_L'}), \qquad (7.9)$$

where the above introduced slope parameter γ (see Chap. 6, (6.34) is controlled
by the total boundary reflectivity of the scattering medium (in particular,
the value of γ is strongly influenced by the existence of multiple reflections
of partial waves at the interface between the free space and the scattering
medium).

An important property of the degree of residual polarization measured under the above-mentioned conditions is that the value of P_L^r does not depend
on the concentration, c, of the scattering sites in the weak-scattering limit
($l \gg \lambda$) but is determined only by the optical properties of an individual
scatterer. This follows from the relationship between the values of the depolarization length and the mean transport free path for the scattering medium:
$\xi_{L,C} = m_{L,C}' l^* \sim c^{-1}$ where the factors $m_{L,C}'$ are controlled by the scattering
anisotropy. For scattering media such as disordered ensembles of spherical
dielectric particles, the relationship discussed above between the scattering anisotropy and the normalized depolarization length ξ_L/l^* (see Fig. 7.8)
leads to a significant degree of residual linear polarization in cases involving backscattering from an optically thick medium consisting of small-sized
particles (the Rayleigh scattering regime). On the other hand, backscattering
from disordered systems of large-sized dielectric particles (the Mie scattering regime) are characterized by an almost totally suppressed residual linear
polarization. It is easy to evaluate the extreme values of the degree of residual linear polarization of linearly polarized light backscattered from optically
thick media with expressed scattering anisotropy by substituting $m_L' \approx 1$ in
(7.9): $P_L^r \approx 0.043$.

Figure 7.11 displays the experimentally obtained values of P_L^r of light backscattered from the layers of aqueous suspensions of 1.07 μm polystyrene beads [21,36]. With this size of scattering particles, the Mie scattering regime is typical for a probe radiation with values of the wavelength of 632 nm and 488 nm, respectively.

The existence of three regions with strongly differing behavior in terms of the degree of residual polarization, which depends on the dimensionless thickness L/l^* of the layer, should be mentioned:

- The region where an abrupt decrease in P_L^r occurs with an increase in L/l^*, which is caused by an increasing concentration of scattering particles; in this scattering regime, the degree of residual polarization is strongly influenced by the optical thickness of the layer.
- The plateau-wise region which is reached only at larger values of scatter concentration; in this case, the scattering layer can be considered a semi-infinite disordered medium and the experimentally obtained P_L^r is close to the value defined by (7.9);
- The region where further increase in the scatter concentration is accompanied by gradual increase in the degree of residual polarization; such behavior can be interpreted as the manifestation of cooperative phenomena in multiple scattering where the scattering by each scatterer must be considered by taking into account the influence of closely neighboring scattering sites; typically, one of the concentration-dependent phenomena

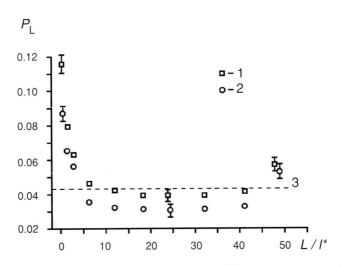

Fig. 7.11. The dependencies of the degree of residual linear polarization of backscattered light on the normalized thickness of a scattering slab (the parameter L/l^* was changed by changing the volume fraction of the scattering particles). The scattering medium is the aqueous suspension of 1.07 μm polystyrene beads; (1) 633 nm probe light; (2) 488 nm probe light; (3) the extreme value of the degree of residual linear polarization for anisotropic scattering

is the nonmonotonic behavior of the mean transport free path (a decrease in l^* with an increasing c, which is typical for disordered systems in a weak-scattering limit, is changed to the opposite behavior in a densely packed media due to the role of cooperative effects) [37].

The extreme value of P_L^r measured within the plateau-wise region at 488 nm is noticeably smaller than the corresponding value measured at 632 nm; this is an obvious manifestation of the influence of the increasing scattering anisotropy. The manifestation of the effect of the preservation of residual polarization in a backscattering experiment with biological tissue was discussed by Studinski and Vitkin [34].

7.4 Polarization Decay in Absorbing Media

Analysis of the dependencies of the degree of residual polarization of light backscattered from a medium on the wavelength of a probing linearly polarized radiation can be proposed as an optical method for analyzing multiply scattering media that exhibit selective absorption caused by the presence of certain chromophores. Such an approach is feasible because of the high level of sensitivity of the polarization state of the radiation, that is scattered and absorbed by the medium at the wavelength of the probe radiation, to the changes in the optical parameters of the medium [38]. This effect is connected with the cut-off of a part of the partial components of the scattered optical field that are characterized by a large value of the effective optical path in media with finite absorption. Thus, comparison of the degree of residual polarization of the radiation scattered from a medium measured inside the absorption band and also far from it, can be used, in particular, to estimate the concentration of a chromophore in the medium.

In the scattering medium, the effect of the probe light absorption on the degree of residual polarization P_L^r can be taken into account by introducing an additional Bouguer factor describing the cut-off of partial components characterized by large values of s and then by modifying (7.7) to the following form:

$$P_L^r \approx \frac{3}{2} \int_0^\infty \exp(-\{s/\xi_L + \mu_a s\})\rho(s)\mathrm{d}s \left/ \int_0^\infty \exp(-\mu_a s)\rho(s)\mathrm{d}s \right., \qquad (7.10)$$

in which a change in the condition for the normalization of the probability density of the optical paths due to additional absorption is taken into account. It should be noted that within the framework of the classical diffusion approximation of radiation-transfer theory, the effective value of the radiation diffusion coefficient is considered dependent on the absorption coefficient of the medium $\mu_a : D = c/3\{\mu_s(1-g) + \mu_a\}$ (see (4.16), (4.18), (4.19), and, for example, [39]). In its turn, this leads to the dependence of the probability density function for the optical paths on $\mu_a : \rho(s) = \rho(s, \mu_a)$. But, as has been shown in a number of papers [40–43], the calculation of the characteristics of scattered radiation within the framework of the diffusion approximation agrees considerably better with the experimental data and the simulation results if the radiation diffusion coefficient is taken to be independent of μ_a, i.e., $D = c/3\{\mu_s(1 - g)\}$. Thus, when analyzing the influence of absorption on the degree of residual polarization of backscattered light, we can choose $\rho(s)$ to correspond to the case of a nonabsorbing medium. Consequently, the effect of absorption on the distribution of the optical paths of partial components of a scattered field can be taken into account by introducing additional Bouguer factors into the integral transformations of $\rho(s)$ in (7.10).

By considering the detection of co-polarized and cross-polarized components of an optical field backscattered by an optically thick semi-infinite absorbant medium, we can conclude, on the basis of the principle of similarity, that the degree of residual polarization of the backscattered light can be expressed as follows:

$$P_L^r \approx \frac{3}{2} \exp(-\gamma \left\{ \sqrt{\frac{3l^*\{1 + \mu_a \xi_L\}}{\xi_L}} \right\} - \sqrt{3l^* \mu_a}). \tag{7.11}$$

The validity of this expression is related to the exponential approximation of the kernel of the integral transformation, obtained within the framework of the solution of the Bethe–Salpeter equation (see above). Thus, we conclude that the prediction error for P_L^r, with the use of (7.11), depends on the modal value of s_{mod} of the distribution of the optical paths of the backscattered partial components under backscattering and decreases with an increasing anisotropy factor of scattering when $s_{mod} \gg l$.

Figure 7.12 presents the theoretical dependencies of P_L^r on the dimensionless parameter $\mu_a l$ for the backscattering of a plane linearly polarized monochromatic wave from a semi-infinite random medium consisting of dielectric spherical particles with substantially differing values of the diffraction parameter ka and a relative refractive index equal to 1.2 (this value corresponds, for example, to such often used model scattering media as polystyrene particles in water when visible-range radiation is used).

According to the results of the Monte Carlo simulation and the experimental data presented in [21], the values of the normalized depolarization length ξ_L/l^* were taken to be equal to 1 for large-sized scattering particles (the Mie scattering regime, $ka \gg 1$) and 4 for Rayleigh particles with $ka \ll 1$.

When random scattering media are probed by a nonmonochromatic linearly polarized light, the spectral dependencies of the degree of residual polarization of the backscattered light are controlled, on the one hand, by the effect of the wavelength λ on ξ_L and l^* and, on the other hand, by the dependence of the absorption coefficient of the scattering medium on λ. The selective absorption of a scattering medium caused by the presence of certain chromophores

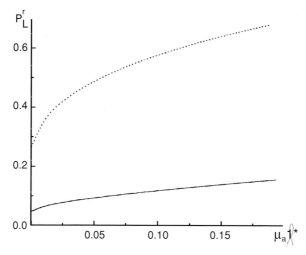

Fig. 7.12. The theoretical dependencies of the ultimate degree of residual polarization of backscattered light on the dimensionless parameter $\mu_a l^*$. *solid line*, Mie scattering system; *dotted line*, Rayleigh scattering system

leads to an increase in the degree of residual polarization within the spectral intervals that correspond to the absorption bands of the chromophores.

Figure 7.13 presents a typical dependence of the degree of residual polarization of backscattered light on the wavelength of the probe radiation, obtained in the experiment with whole milk as the phantom scattering medium. The inset illustrates the spectral dependence of the effective optical density $D(\lambda)$ of the sample under study as determined from its diffuse reflection spectrum $R(\lambda)$ according to the known relationship $D = -\lg[R(\lambda)]$.

In the 550–650-nm wavelength range, where the intrinsic absorption of whole milk is relatively small (inset in Fig. 7.13), the dependence of the degree of residual polarization on the wavelength demonstrates an increase in P_L^r with an increasing λ, which is caused by an increase in ξ_L with a decreasing effective value of the diffraction parameter for the scattering medium in the regime of detection of the backscattered radiation [21]. The increase in the intrinsic absorption of whole milk in the 450–530-nm spectral range results in a substantial increase in the degree of residual polarization, P_L^r, in the short-wavelength region of the visible spectrum. It should be noted that samples of whole milk from different lots that were studied in the experiment demonstrate considerable variation of spectra $P_L^r(\lambda)$ (in particular, the measured values of the degree of residual polarization at wavelength of 610 nm vary from 0.05 to 0.11), which, obviously, is caused by a spread of the statistical characteristics of the size distributions of the different components of milk from different lots.

The adding of chromophore (for instance, a food dye) to whole milk leads to a substantial increase in the degree of residual polarization for the intervals of the wavelengths that correspond to the absorption bands of the

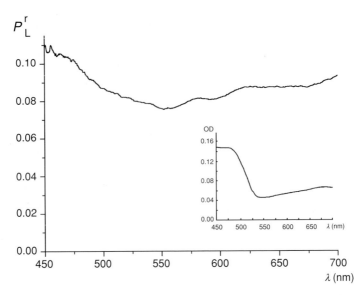

Fig. 7.13. The spectral dependence of the degree of residual polarization of backscattered light with an initial linear polarization for the whole milk as a scattering medium. The inset shows the spectral dependence of the optical density calculated from the spectrum of the diffuse reflectance

chromophore (Fig. 7.14). The dependencies of the degree of residual polarization on the absorption coefficient are approximated with satisfactory accuracy by an analytical expression (7.11) when ξ_L and l^* are used as the fitting parameters. In particular, Fig. 7.15 demonstrates the values of P_L^r obtained in the experiment and the corresponding approximated curve as functions of the values of the absorption coefficient of the aqueous solutions of the dyes (Fig. 7.14), measured at the maximum points of absorption (515 and 610 nm). In the process of approximation, we obtained the following fitting values: $\xi_L \approx 1.3l^*$ and $l^* \approx 0.4$ mm. In accordance with [21], such a relationship between ξ_L and l^* must correspond to the effective values of the anisotropy factor, $g \approx$ 0.85–0.90, for scattering systems consisting of optically "soft" dielectric particles with a relative refractive index on the order of 1.2. This result is in satisfactory agreement with the known data on milk microstructure [44] and the optical characteristics of similar scattering systems such as Intralipid [45].

Hemoglobin and melanin are the main chromophores that control the absorption spectra of biological tissues in visible/near infrared range probe radiation [46]. An increase in the content of blood in the surface layers of in vivo biological tissues results in an increase in the degree of residual polarization of backscattered light for those spectral intervals corresponding to the absorption bands of blood in the visible range. In particular, this effect is observed as the quantitative correlation between P_L^r at the wavelength corresponding

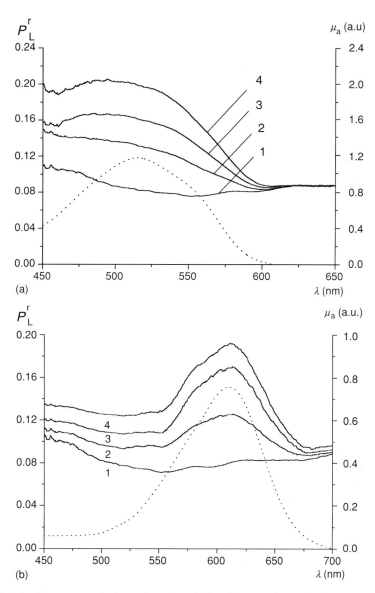

Fig. 7.14. The spectral dependencies of the degree of residual polarization of backscattered light with initial linear polarization obtained with different concentrations of absorbers for whole milk with an added absorber (food dye). (**a**) 1: $\mu_a = 0, 2: \mu_a = 0.141$ mm^{-1}, 3: $\mu_a = 0.188$ mm^{-1}, 4: $\mu_a = 0.327$ mm^{-1}; (**b**) 1: $\mu_a = 0, 2: \mu_a = 0.107$ mm^{-1}, 3: $\mu_a = 0.247$ mm^{-1}, 4: $\mu_a = 0.388$ mm^{-1}. Absorption spectra of both absorbers are shown by *dotted lines*

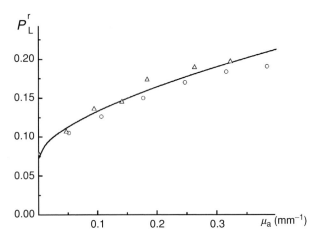

Fig. 7.15. The results of the fitting of experimental data for whole milk (Fig. 7.14) with the use of (7.11). (1) Theoretical dependence, (2, 3) experimental results corresponding to the maximal absorbance of the added absorbers (see Fig. 7.14)

to the maximum of an absorption band of blood (535–575 nm) and the value of the erythema index for human skin with induced erythema. The erythema index for treated skin was determined according to the procedure described in [47]. The erythema was produced in the skin of the forearm of a volunteer by removing the surface layers of epidermis with the use of skin strip technology [48] which involves the application of an adhesive film. In this case, the thickness of the removed layers of epidermis was approximately 4–8 μm. The layer-by-layer removal of epidermis produced by mechanical action produces an increase in the level of microcirculation and, consequently, an increase in the blood content of the subsurface layers of skin.

Figure 7.16 demonstrates the spectra of the degree of residual polarization, P_L^r, for samples of in vivo healthy skin in the initial state (1) and skin with different degrees of induced erythema (2, 3). The corresponding spectra of the optical density of the biological tissues are also presented for comparison. These spectra display increasing absorption in the 535–575-nm spectral region (the absorption band of the oxygenized form of hemoglobin) with the progress of erythema. It should be noted that beyond the absorption bands of blood, the dependencies, $P_L^r(\lambda)$, demonstrate a decrease in the degree of residual polarization with increasing wavelength, which can be caused by the effect of additional absorption of the probe radiation in the tissue volume by natural chromophores (in particular, by melanin). The spectral dependence of the absorption coefficient of melanin in the visible range is characterized by a rapid monotonic decrease with an increasing wavelength and can be approximated by the power-law dependence of the following form: $\mu_a \sim \lambda^{-3.5}$ [49].

Thus, if we disregard the absorption of the probe radiation caused by hemoglobin, we find that the presence of melanin in the subsurface layers

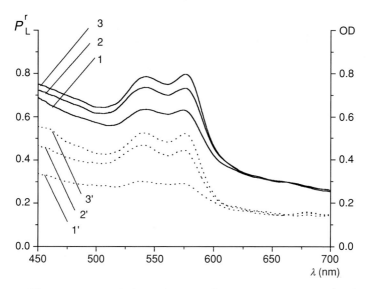

Fig. 7.16. The experimental dependencies of the optical density (1–3) and the degree of residual polarization (1'–3') for in vivo human skin. (1) normal tissue, (2, 3) tissue with erythema

Table 7.1. The relationship between the measured values of the erythema index and the degree of residual linear polarization of backscattered light in in vivo human skin

the index of erythema	the degree of residual polarization
40 ± 2	0.24 ± 0.02
60 ± 3	0.31 ± 0.03
96 ± 4.8	0.39 ± 0.04
128 ± 6.4	0.47 ± 0.04
174 ± 8.7	0.54 ± 0.05

The wavelength of the probe light corresponds to the maximal absorbance of hemoglobin

of human skin leads to an increase in the degree of residual polarization of backscattered light in the short-wavelength region in the visible range.

For comparison, the values of P_L^r are presented in Table 7.1 for the probe radiation wavelength that corresponds to the maximum absorption of the hemoglobin of blood (the 535–575-nm band) for different values of the erythema index for the skin region under study. These were measured according to the procedure described in [47]. Similar results were obtained recently for a case of erythema induced by the laser-mediated ablation of in vivo human epidermis as a basic part of the skin resurfacing procedure [50]. The unambiguous correlation between P_L^r and the values of the erythema index allows us to propose spectral measurements of the degree of residual polarization of backscattered linearly polarized probe radiation as a method for functionally diagnosing the state of the dermal layer of human skin.

Thus, the effect of light absorption by a multiply scattering medium on the degree of residual polarization of backscattered light can be interpreted within the framework of the concept of the optical path distribution for specified conditions of scattered field formations. Spectral analysis of the degree of residual polarization, including the technology of controlled selective absorbance of the media under study, can be proposed as a new method for investigating the optical properties of multiply scattering media, including biological tissues.

Highly absorbing random medium, in which the light penetration depth is comparable with the characteristic size of the scattering sites, manifests itself as a quasi-two-dimensional, rough-surface structure. On the contrary, a transition between the volume and surface scattering effects takes place with an increase in the light penetration depth. An experimental study of the manifestation of probe light absorption in random media on the formation of backscattered polarization patterns was carried out by Dogariu et al. [51]. Because of the strong localization of the backscattered light patterns in the vicinity of the point of incidence of the narrow probe beam, the effect of internal reflections on a dielectric interface was used in these experiments to enlarge the spatial extent of the observed polarization patterns and to, thereby, facilitate the quantitative data collection. The geometry of the polarization experiment used by Dogariu et al. is shown in Fig. 7.17. A cuvette with a 1-mm-thick window was filled with a highly absorbant suspension. A focused linearly polarized He:Ne laser beam was directed on the outer interface ("II") of the cell window (the diameter of a light spot was equal to 0.3 mm). The images of window surface II were captured with the use of a CCD camera arranged with a polarizer in orthogonal orientation to polarization direction of the incident light. The highly absorbant scattering media schematically shown above interface I were aqueous suspensions of carbonyl iron (polydisperse particles with an average size of 4.5 μm) and cerium oxide

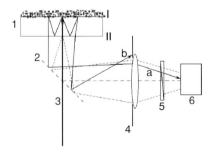

Fig. 7.17. The scheme of the experimental setup used to study the polarization patterns in the case of the backscattering of linearly polarized light from a highly absorbant medium [51]: (1) dielectric layer, (2) beam splitter, (3) narrow illuminating beam, (4) imaging system; (5) polarizer, (6) detector

particles ($3.5\,\mu$m average size) with various volume fractions. Under the simplifying assumption about the monodisperse character of these suspensions,
their optical parameters, such as the scattering and absorption coefficients,
were calculated using Mie theory. In the course of recording the cross-polarized
patterns, 50 frames were typically acquired and averaged.

The typical shapes of cross-polarized patterns (equal intensity contour
plots) obtained for different values of light penetration depth are illustrated
in Fig. 7.18. As can be seen, increasing absorbance causes the transformation
of cross-polarized patterns from axially symmetric equal intensity contours
to the fourfold intensity distributions that are typical of the cross-polarized
patterns obtained from dielectric layers with rough surfaces. The observed
evolution of the polarization patterns with a decrease in the light penetration
depth can be explained in terms of the ray-optics approach as was shown by
Dogariu et al. In the ray representation, the field emitted at one point after
a single-scattering event is completely determined by the incident field and
the local properties (geometry and complex refractive index) of the boundary.
This approach is applicable if the characteristic length scales of the scattering
system are significantly larger than the wavelength of the probe light. Formation of the cross-polarized patterns is explained by the use of the following
ray-propagation model schematically shown in Fig. 7.19.

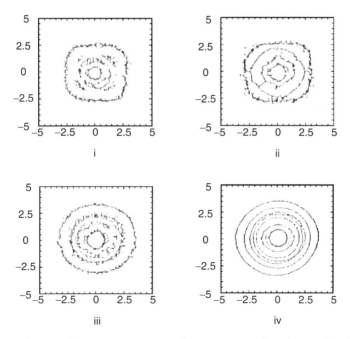

Fig. 7.18. The equal intensity contours for experimentally observed polarization
patterns obtained under different absorption conditions; (i–iv) the values of the
penetration depth of the probe light are equal to $7\,\mu$m, $12\,\mu$m, $20\,\mu$m, and $40\,\mu$m,
respectively [51]

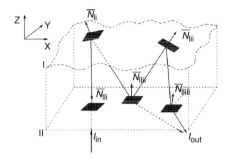

Fig. 7.19. Schematic of an in-plane, multiple scattering sequence developed inside a dielectric layer in contrast to a strongly absorbant medium [51]

A typical feature of the observed patterns is the existence of three distinct regions, such as a bright central spot, a dark surrounding ring, and a clearly expressed four-lobe structure with contrast which decreases rapidly with increasing radial distance. The appearance of the central depolarized spot is evidently caused by the direct retroreflection of the incident beam and imperfect depolarizers. Also, the existence of the dark ring is easily explained by the ray tracing inside the dielectric layer (cell window, see Fig. 7.19). In the case of the small aperture angle of the imaging system, the rays leaving the layer at points of type IIii will be eliminated, and only rays exiting at points of type IIiii, which suffered cornerlike scattering sequences of two reflections on the inner interface (I) and one reflection on the outer interface (II), will contribute to the formation of the observed pattern. As the incidence angles on the outer interface are in general smaller than the limit angle of the total internal reflection, most of the light is transmitted through the outer interface at large angles and, therefore, cannot be collected. Thus, a very small part is reflected toward the points of type Iii. At the limit angle, an abrupt increase in brightness is evident because all of the light incident at IIii is reflected toward points Iii.

The main features of the observed cross-polarized patterns, such as the four-lobe structure, can be considered as follows. The coplanar scattering sequences in the plane containing the incident polarization vector as well as in the orthogonal plane preserve the initial polarization azimuth of the propagating light, while the scattering sequences in the planes at azimuthal angle φ cause the rotation $\pi - 2\varphi$ of the polarization vector. Such scattering sequences cause the formation of similar polarization patterns in weak volume scattering of narrow beams [52]. The randomization of the polarization vector due to multiple scattering is the competing factor that diminishes the contrast of the observed patterns as it takes place with increasing radial distance.

It is easy to show that in cases of axial symmetry and reflections from the inner interface between the dielectric layer and "surface-like" medium, which is isotropic and homogeneous, the outgoing wave can be describedby the following relationship [51]:

$$\bar{E}_{\text{out}} \sim e_x \left| \bar{E}_{\text{out}} \right| \sin 2\varphi + e_y \left| \bar{E}_{\text{out}} \right| \cos 2\varphi, \tag{7.12}$$

where e_x is determined by the orientation of the polarization vector of the incident beam, e_y corresponds to orthogonal direction, and φ is the azimuthal angle of the corner-like sequence of scattering events. The scattering event on the inner interface can be described in terms of the ray-optics approach by introducing the effective surface with the local slope specified by a unit random vector $\overline{N}(\theta, \varphi)$. In this case, the outgoing field at point IIii will be determined by deterministic reflection events governed by Snell's law and by a random process described as stochastically changing $\overline{N}(\theta, \varphi)$.

The influence of polarization discrimination on the outgoing light from the crossed polarizer is simply defined by the following relationship that describes the azimuthal distribution of the intensity of the backscattered cross-polarized light:

$$I_{\perp\text{out}}(\varphi) \sim \frac{\left| \bar{E}_{\text{out}} \right|^2}{2}(1 - \cos 4\varphi). \tag{7.13}$$

As can be seen, this relationship describes the four-lobe structure of the observed polarization patterns, but the appearance of any anisotropy of the boundary will destroy the fourfold symmetry.

It is clear that the traces of the rays emerging at point IIii are typically more complicated than the above-described in-plane sequences. Due to multiple reflections, the outgoing field also depends on nonlocal properties of the scattering "surface-like" medium. These processes are associated with nonplanar reflection sequences, the appearance of relative phase differences that are due to the total internal reflections, etc. However, because of the random nature of multiple-scattering processes, they result in the stochastic depolarization of the outgoing light. In turn, the random depolarization diminishes the contrast of the observed polarization patterns.

In the frames of the considered ray-optics model, if we do not take into account the radial dependence of the cross-polarized backscattered intensity, the depolarizing effects can be described by the introduction of an additional generic rotation γ of the electric vector. Thus, after the ideal cornerlike sequence, the cross-polarized component of the electric field can be written as:

$$\bar{E}_{\perp\text{out}} = \bar{E}_{\text{out}} \sin(2\varphi + \gamma). \tag{7.14}$$

The value of γ is regarded as the fluctuating part of the rotation angle with the probability distribution determined by the statistical properties of the "quasisurface." Under the above mentioned conditions (the isotropy and the statistical homogeneity of the "quasisurface"), the fluctuations of γ can be considered as a zero-mean random process. In this case, evaluation of the ensemble-averaged cross-polarized intensity yields the following relationship:

$$\langle I_{\perp out}\rangle = (\langle|\bar{E}_{out}|^2\rangle/2)(1 + (1 - 2\langle\cos^2\gamma\rangle)\cos 4\varphi). \quad (7.15)$$

This formula also shows the fourfold structure of the azimuthal dependencies of the detected cross-polarized intensity, but the contrast of the observed polarization patterns is reduced and controlled by the value of $\langle\cos^2\gamma\rangle$. By assuming that the rms value of γ is proportional to the average penetration depth of the probe light, the characteristic changes in the experimentally obtained patterns, which result from the increasing absorption of the probed medium, can be easily interpreted.

7.5 Summary

Analysis of polarization decay in weakly ordered media (for instance, biological tissues) can be used as an effective method to characterize the structural properties of scattering systems. In particular, the spectral dependence of the degree of residual polarization of backscattered radiation is strongly influenced by the optical properties of the probed scattering media, such as the values of the reduced scattering coefficient and the absorption coefficient at the wavelength of the probe light. In turn, these parameters are controlled by structural parameters, such as the average size and concentration of scattering particles as well as by the type and concentration of the absorbant components of the probed medium. This allows for the development of relatively simple polarization techniques for functional diagnostics and for the visualization of in vivo biological tissues in the laboratory or clinical setting. Some of the related techniques will be considered in Chap. 11. The high sensitivity of the degree of residual polarization of backscattered light to changes in the absorption or scattering properties of probed media is the physical basis for several novel approaches to medical polarization diagnostics which employ controlled changes in the optical parameters of the scattering system (by adding absorbing agents).

8

Degree of Polarization in Laser Speckles from Turbid Media

8.1 Introduction

The degree of polarization (DOP) of scattered light from biological tissues is an important concept and deserves special attention. As a coherent light source is generally used in the experiments, speckle patterns play significant roles in the polarization measurements. The statistics of laser speckle patterns, including partially polarized speckle patterns, were well described in Goodman's chapter [1]. In fact, partially polarized speckle patterns have been studied extensively in recent years [2–5]. Fercher and Steeger [2,3] determined the theoretical first-order statistics of the Stokes parameters and later verified the theory with experiments. Brosseau et al. [5] studied the statistics of normalized Stokes parameters and discussed potential applications. Freund et al. [6] proposed microstatistics to describe the polarization behavior of a single coherence area in a speckle field. The work was focused on deriving polarization correlation functions for extracting information from the speckle pattern about the direction of the incident polarization. Tarhan et al. [7] further investigated the microstatistics; they measured the intensity at many points in a speckle pattern for a given polarization angle of the incoming laser beam and obtained the probability density distributions for the parameters in the statistics. However, these two studies did not evaluate the DOP at those points in the speckle field, which is a key parameter for the understanding of polarized speckle fields. Elies et al. [8], in a more recent investigation of speckle polarization, observed a speckle field produced by light reflected from a polished aluminum sample with a CCD camera. Their results showed that depolarization among multiple speckle grains increased with sample inclination although each speckle grain remained polarized.

Li et al. [9] reported on an investigation of polarization in a speckle field formed by coherent light transmitted through a surface-scattering medium (a ground-glass plate) or a volume-scattering medium (a wax plate). The degree of polarization, as well as the degree of linear polarization (DOLP) and the degree of circular polarization (DOCP), was measured both within a single coherence area and over multiple coherence areas and modeled theoretically. Although it is widely acknowledged that multiple scattering events in volume-scattering media can depolarize polarized incident light and hence reduce the DOP, this study demonstrated that the measured DOP depends significantly on the conditions of observation. Readers are encouraged to refer to Sect. 6.4 for a related discussion.

8.2 Experiments and Simulation

To extend this work further, Li et al. [9] developed an experimental setup as shown in Fig. 8.1. A diode laser (850 nm) emitted a beam of 1.5 mm in FWHM diameter and of 60 m in coherence length. After passing through an optical isolator and a half-wave retardation plate, the beam was horizontally linearly polarized with a DOP of 0.99 and an intensity fluctuation of ~1%. The isolator and the retardation plate were used to prevent back reflection into the laser and to fine tune the orientation of the polarization, respectively. The beam was incident upon the sample, which produced a speckle field from the transmitted light. An iris was placed close behind the sample to control the average size of the coherence areas in the speckle field. Another iris was used to select a portion of the speckle field for observation. The selected light, after passing through a variable-wave plate, a Glan-Thompson analyzer, and a nonpolarizing beam splitter, was detected by a large-area photoreceiver. The variable wave plate was calibrated to an accuracy of 99% before measurements. A chopper, operating at 900 Hz, modulated the beam intensity, and the output of the photoreceiver was measured with a lock-in amplifier to improve the signal-to-noise ratio. The chopper was set behind the first iris (close to the sample) to ensure that only the light emerging from the sample was modulated and detected. A CCD camera was used to monitor the speckle pattern simultaneously. A 3-mm-thick wax plate was used as a volume-scattering sample, which diffusely scattered the light to be received. The wax sample was sufficiently thick to produce a speckle pattern of a high contrast, approaching the theoretical limit for unpolarized speckles (~70%). For comparison, a ground-glass plate was used as a surface-scattering sample, which deformed the optical wavefront only on the surface of the sample.

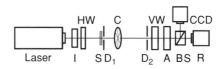

Fig. 8.1. Experimental setup. I, optical isolator; HW, half-wave plate; S, sample; D_1 and D_2, irises; C, chopper; VW, variable-wave plate; A, analyzer; BS, nonpolarizing beam splitter; R, photoreceiver

The average diameter (d) of the coherence areas in the speckle field at the plane of detection, located at the second iris, was estimated by the following expression [10]:

$$d = \frac{2.44\lambda L}{D_1}, \tag{8.1}$$

where L is the distance between the two irises, D_1 is the diameter of the first iris, and λ is the optical wavelength. Equation (8.1) is based on the definition of the diameter of the Airy disk, which represents the minimum speckle size in a speckle pattern [11] and can be used to estimate the average speckle size in a "fully developed" speckle pattern. By definition, a "fully developed" speckle pattern is completely polarized (DOP = 1). Although the speckle patterns in these experiments are not "fully developed" due to the depolarization caused by multiple scattering [12], for simplicity, (8.1) was used as an approximation for the average speckle size. For measurements of a single coherence area (multiple coherence area), D_1 was set to 0.1 mm (2 mm), yielding an average diameter for the coherence areas of 14.8 mm (0.74 mm) at the detection plane with $L = 711$ mm. By varying the area of detection determined by the size of the second iris (D_2), one could select the number of detected coherence areas, which were monitored with the CCD camera.

Figure 8.2a shows the DOP, DOLP, and DOCP measured within a single coherence area as functions of the size of the detection area. For the ground-glass sample, the DOP showed little variation associated with the size of the detection area and remained at ∼0.99, which was approximately the same as that of the laser source. By contrast, for the wax sample, only the DOP of those small areas of detection was close to unity, and the DOP decreased as the area of detection was enlarged. The DOLP and DOCP behaved similarly for the wax sample. For the ground-glass sample, the DOLP and DOCP had nearly constant values: ∼0.99 and ∼0, respectively, which showed that linear-polarization states were maintained in the speckle field. Small fluctuations were seen in the DOCP measured from the ground-glass sample, which were due to low signal-to-noise ratios in the detection of the low-intensity circular-polarized component. Figure 8.2b shows the DOP, DOLP and DOCP measured for multiple coherence areas. For both the ground-glass and the wax samples, the trends in Fig. 8.2a continued. It should be mentioned that the results in Fig. 8.2a, b were not joined together because the measurements were not made under the same conditions as a result of the replacement of the first iris.

From the Stokes vectors obtained with the ground-glass sample, it was found that the horizontally linear polarization state of the laser source was maintained in each measurement. In the measurements of the wax sample, a variation of the relative distribution of speckle intensity was observed with the CCD camera when the analyzer was rotated, indicating that the polarization states in the speckle field were nonuniformly distributed. Based on the effect of scattering on light polarization, it was deduced that the multiple scattering events in the wax sample caused the broadened distribution of polarization in the speckle field.

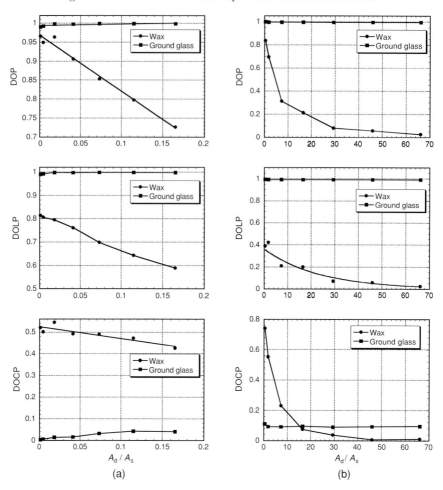

Fig. 8.2. Measured DOP, DOLP, and DOCP as functions of the area of detection. A_d is the area of detection, controlled by the diameter of the second iris (D_2); A_s is the average area of the coherence areas, computed through their estimated average diameter (d). (**a**) Measurements within a single coherence area, where $A_s = 171\,\mathrm{mm}^2$. (**b**) Measurements over multiple coherence areas, where $A_s = 0.43\,\mathrm{mm}^2$

The probability density functions (PDFs) of the Stokes parameters in the speckle field that were generated by the wax sample were investigated. Speckle patterns containing multiple coherence areas were recorded with the CCD camera. The Stokes parameters measured at each CCD pixel were taken for a statistical estimation. Figure 8.3 shows the PDFs of the four Stokes parameters that were measured in the speckle field from the wax sample. The PDF of the first Stokes parameter, I, was similar to that obtained by Goodman [1] for the intensity of the sum of two speckle patterns, which was different from the negative exponential distribution of a fully polarized speckle pattern. The

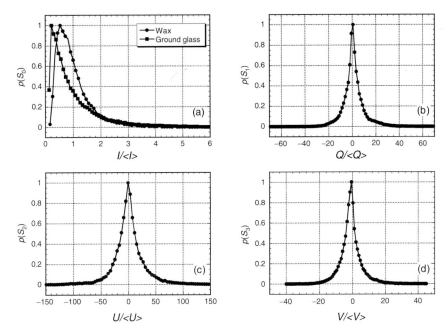

Fig. 8.3. Normalized probability density functions of Stokes parameters, which were measured in the speckle fields generated by the wax sample: (**a**) I, (**b**) Q, (**c**) U, (**d**) V. The probability density function of the first Stokes parameter I in the speckle field generated by the ground-glass sample is also given in (**a**) for comparison. $<>$ represents the average values. See Chap. 3 for the definitions of I, Q, U, V

other three PDFs were symmetrically distributed. For comparison, a PDF of the first Stokes parameter of a speckle pattern from the ground-glass sample is shown in Fig. 8.3a as well. It is seen that the distribution of this PDF is closer to the negative exponential distribution. Note that the DOPs corresponding to the two speckle fields in Fig. 8.3a are $\sim$0.13 and $\sim$0.99, respectively. The variation of the PDF with the DOP agrees with Goodman's theory [1]. According to Fercher et al.'s theory [2] in which the speckle field was described as a superposition of two fully developed uncorrelated linearly polarized speckle fields, the symmetrical distribution of the PDF of the second Stokes parameter indicates that the mean intensities of the two fields are the same.

In addition, to further understand the phenomenon observed, the polarization states in speckle fields from a surface- and a volume-scattering medium, respectively, were theoretically simulated. For the volume-scattering medium, both the polarization state and the phase of the transmitted optical field were assumed to be randomized by multiple scattering events. For the surface-scattering medium, only the phase of the transmitted optical field was assumed to be randomized as a result of the deformation of the phase front. In the simulation, the optical field at the first iris (D_1) was represented by a

Jones vector:

$$\mathbf{E}(\xi, \eta) = \begin{bmatrix} E_x(\xi, \eta) \\ E_y(\xi, \eta) \end{bmatrix} = \begin{bmatrix} E_{x_0}(\xi, \eta)e^{-j\phi_x(\xi, \eta)} \\ E_{y_0}(\xi, \eta)e^{-j\phi_y(\xi, \eta)} \end{bmatrix}, \tag{8.2}$$

where $E_x(\xi, \eta)$ and $E_y(\xi, \eta)$ were two orthogonal components of the field, and (ξ, η) was the coordinate of a point in the plane where the first iris was located. A pupil function was applied to simulate the first iris, which gave the distribution of the optical field in the plane. For the surface-scattering medium, a horizontally linear polarization state with a constant E_{x_0} and a zero E_{y_0} was assumed:

$$E(\xi, \eta) = \begin{bmatrix} E_{x_0}e^{-j\phi_x(\xi, \eta)} \\ 0 \end{bmatrix}, \tag{8.3}$$

and the phase, $\phi_x(\xi, \eta)$, was assumed to be randomized. For the volume-scattering medium, several assumptions were made: $\arctan(E_{y_0}(\xi, \eta)/E_{x_0}(\xi, \eta))$ was randomized between $-\pi$ and π whereas the total optical intensity $(E_{x_0}^2(\xi, \eta) + E_{y_0}^2(\xi, \eta))$ remained constant, and the phases, $\phi_x(\xi, \eta)$ and $\phi_y(\xi, \eta)$, were randomized as well. For both of the media, the phase was evenly randomized between $-\pi$ and π.

The two field components, $E_x(\xi, \eta)$ and $E_y(\xi, \eta)$, were diffracted independently, which generated two independent speckle patterns in the far field. The diffraction processes were simulated by Fourier transforms:

$$E_x(x, y) = F\{E_x(\xi, \eta)\}, \tag{8.4}$$

$$E_y(x, y) = F\{E_y(\xi, \eta)\}, \tag{8.5}$$

where $E_x(x, y)$ and $E_y(x, y)$ denoted the optical fields at point (x, y) on the observation plane, and F{} denoted the Fourier transform. The final speckle pattern was generated by the summation of the two speckle patterns. The Stokes vectors of the speckle pattern were then calculated as follows:

$$\begin{bmatrix} I \\ Q \\ U \\ V \end{bmatrix} = \begin{bmatrix} E_x(x, y)E_x^*(x, y) + E_y(x, y)E_y^*(x, y) \\ E_x(x, y)E_x^*(x, y) - E_y(x, y)E_y^*(x, y) \\ E_x(x, y)E_y^*(x, y) + E_y(x, y)E_x^*(x, y) \\ E_x(x, y)E_y^*(x, y) - E_y(x, y)E_x^*(x, y) \end{bmatrix}. \tag{8.6}$$

The results of the simulation are shown in Fig. 8.4. Fig. 8.4a shows the variations of the DOP, DOLP, and DOCP with the size of the detection area within a single coherence area. Figure 8.4b shows the results over multiple coherence areas. The simulation results qualitatively agree with the experimental observation: a constant DOP of unity for the surface-scattering medium and a decreasing DOP for the volume-scattering medium as the area of detection increases are observed. The DOLP and DOCP decrease with the enlargement of the detection area for the volume scattering medium, whereas they remain constant for the surface-scattering medium. Due to the statistical nature of a speckle field, the experimental results and the simulation results can be compared only qualitatively.

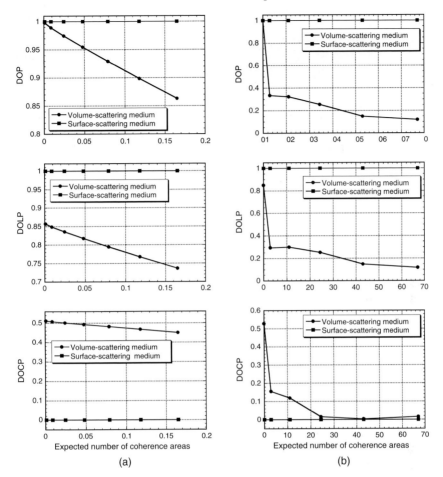

Fig. 8.4. Results of simulation, where the expected number of coherence areas is defined as the area of detection divided by the expected area of the coherence areas. (**a**) The variations of DOP, DOLP, and DOCP within a single coherence area. (**b**) The variations of DOP, DOLP, and DOCP over multiple coherence areas

Figure 8.5 displays the four Stokes-vector components of a segment of the speckle field from the volume-scattering medium corresponding to the maximum area of detection in Fig. 8.4a. It is clearly seen that the profiles are different among the four components. This agrees with the experimental observation from the wax sample and indicates that the Stokes vectors (polarization states) and the DOPs can vary from point to point in the speckle field, even within a single coherence area. This conclusion differs from the previous findings in speckle fields formed by light reflected from surface-scattering media [8].

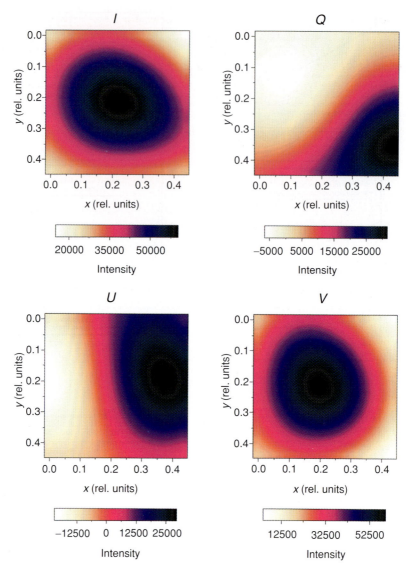

Fig. 8.5. The Stokes-vector components for a volume-scattering medium within a coherence area corresponding to the maximum in Fig. 8.4a

8.3 Discussion and Conclusions

The explanation of the results from the surface-scattering medium is that because the speckle field is formed by the diffraction of an optical field with a single polarization state, consequently, the speckle field maintains the original polarization. For the volume-scattering medium, the independent diffraction processes of E_x and E_y create two orthogonal speckle fields polarized in the x and y directions, respectively. The vector sum of the two orthogonal speckle fields yields the total speckle field. Although the polarization states before diffraction are randomized, each point in the total speckle field has a DOP of unity because the resultant E_x and E_y components have a particular ratio of amplitude and a particular phase relation. Of course, the polarization states at different points in the total speckle field are statistically different from each other because both the ratio of amplitude and the phase between the two orthogonal speckle fields vary from point to point. The Stokes vector for an area containing more than one such "point" is then determined by summing the Stokes vectors of all of the points in the area. As a result, the DOP of the area is less than unity and decreases statistically as the area is enlarged because more points are included in the enlarged area. It is worth noting that because of their statistical nature, polarization states and DOPs can be different even for detection areas of the same size.

It can be concluded that the measured DOP, DOLP, and DOCP in a speckle field that is generated by a volume-scattering medium depends on the size of the detection area; These values decrease with an increasing area of detection, and only the DOP of an area much smaller than a coherence area is close to unity. This conclusion is important for the understanding of polarization phenomena in biomedical optics, where polarized coherent light is applied and a speckle field is generated. When the DOP, DOLP, and DOCP of a speckle field from a scattering medium such as biological tissue are measured, the above properties should be considered, especially if the measurement is made from a small area in the field. The fact that these parameters may vary statistically even for areas of the same size should be taken into account as well. Moreover, if a speckle field was observed in reflection mode from a piece of biological tissue, contributions from both the rough surface and the multiscattered light should be considered.

It is useful to compare the conclusions reported here with those reported on the DOP in a heterodyne detection scheme such as the one used in optical coherence tomography (OCT). It was found that the DOP in OCT maintains a value of unity as long as the scattering sample is stable during data acquisition regardless of how many speckles are detected [13]. OCT is an amplitude-based detection system that uses interference heterodyne. OCT detects the electric field of only the coherent part of the backscattered light. The electric field of the light from various locations on the detector surface is projected onto the analyzing polarization state and then added in amplitude. Equivalently, the electric field vectors of the light from the various locations of the detector are summed, and the vector sum is then projected onto the analyzing polarization state. As a result of this coherent-detection scheme in OCT, a DOP of unity is maintained despite scattering.

Monte Carlo Modeling of Polarization Propagation in Strongly Scattering Media

9.1 Introduction

In optical imaging and diagnostics, the optical properties of the biological tissue samples are unknown; only the incident light and output light can be measured. The goal of optical imaging or diagnostics in biological tissue is to retrieve the optical properties of the sample based on the measurable parameters, which is an inverse problem. The corresponding forward problem involves calculating the measurable output with known input parameters and known optical properties of the sample. A clear understanding of the forward problem is rudimentary for solving the inverse problem because inverse algorithms are often built upon forward solutions. Studying the forward problem facilitates the understanding of light transport in scattering biological tissues.

Monte Carlo simulation has been proven to be an accurate and flexible approach and has been widely used in tissue optics [1–4]. It can be applied to complex tissue structures and compositions. In conventional Monte Carlo simulations, polarization of the incident light is not considered. However, light fields are inherently vector-based, which means polarization is an important feature. In recent years, tissue polarimetry has become an attractive topic in biomedical optics [5]. Besides conventional optical contrast, polarized light furnishes polarization contrasts that are related to specific tissue properties. For example, collagen fibers possess linear birefringence [6] and glucose molecules have circular birefringence (optical activity) [7]. These polarization contrasts cannot be revealed by non-polarized light. In addition to providing polarization contrasts, polarization-based techniques have also been employed to discriminate weakly scattered light from strongly scattered light. It is widely recognized that the original polarization state is lost in diffusely scattered light but is partially preserved in weakly scattered light. For a detailed discussion on degrees of polarization, refer to Chap. 8. By using polarization discrimination, one can perform imaging using weakly scattered photons.

The propagation of polarized light in scattering (turbid) media is a complex process. Parameters, such as the size, shape, and density of the scatterers as well as the polarization state of the incident light, all play important roles [8]. A good understanding of this process is essential for improving the polarization-based techniques. Because the number of scattering events is related to the optical path length or the time of propagation, a time-resolved study is needed to understand the evolution of polarization in turbid media.

In this chapter, a time-resolved Monte Carlo technique is used to simulate the propagation of polarized light in turbid media [9, 10]. Mie theory is used to calculate the single scattering events [11]. A Mueller-matrix approach is applied because it provides a complete description of the polarization properties of the light and the materials [12]. Specifically, the reflection and transmission Mueller matrices of turbid media, and the evolution of the degree of polarization (DOP) in turbid media, are studied. The effects of the size of the scatterers and the polarization state of the source are studied as well. A continuous-wave (CW) Monte Carlo code modeling polarized light interaction with a multiple scattering medium is described in Sect. 4.3 [13].

9.2 Method

Because of the nature of the Monte Carlo simulation used, coherent phenomena, such as laser speckles, are not modeled. Nevertheless, the simulation method can be applied in the non-coherent regime or in cases where the coherent effect is removed, such as ensemble-averaged measurements. A detailed account of a Monte Carlo simulation of non-polarized light transport in scattering media can be found elsewhere [4]. Several groups have used Monte Carlo techniques to simulate the steady-state backscattering Mueller matrix of polarized light from a turbid medium [14, 15]. Whereas an indirect method utilizing the symmetry of the backscattering Mueller matrix was used in [14], the direct tracing method [15] is used here. The turbid medium is assumed to have a slab structure, on which a laboratory coordinate system is defined (Fig. 9.1). A pencil beam is incident upon the origin of the coordinate system at time zero along the Z axis.

The basic idea is to track the Stokes vector of each photon packet. The derivation of the single scattering Mueller matrix from Mie theory is shown in the Appendix. The coordinate transformation equations used in the simulation are also shown in the Appendix. The flow chart of the program is shown in Fig. 9.2.

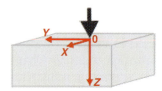

Fig. 9.1. Laboratory coordinate system for the Monte Carlo simulation

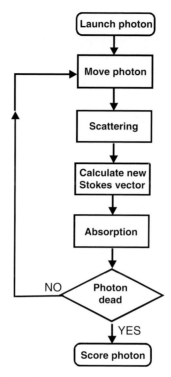

Fig. 9.2. Flow chart of Monte Carlo simulation of propagation of polarized light in turbid media

In the simulation, the Stokes vector and the local coordinates of each incident photon packet are traced statistically. At each scattering event, the incoming Stokes vector **S** of the photon packet is first transformed into the scattering plane through a rotation operator **R** and then through a scattering matrix **M**:

$$\mathbf{S'} = \mathbf{M}(\theta)\mathbf{R}(\phi)\mathbf{S}, \tag{9.1}$$

where **S** is the Stokes vector before scattering; $\mathbf{R}(\phi)$ is the rotation matrix (see (A.11)); $\mathbf{S'}$ is the Stokes vector of the scattered photon; θ is the polar angle; ϕ is the azimuth angle; and **M** is the single-scattering Mueller matrix, given by Mie theory as follows [11] (see also (3.30)):

$$\mathbf{M}(\theta) = \begin{bmatrix} M_{11} & M_{12} & 0 & 0 \\ M_{12} & M_{11} & 0 & 0 \\ 0 & 0 & M_{33} & M_{34} \\ 0 & 0 & -M_{34} & M_{33} \end{bmatrix}. \tag{9.2}$$

The element M_{11} satisfies the following normalization requirement:

$$2\pi \int_0^\pi M_{11}(\theta) \sin(\theta)\mathrm{d}\theta = 1. \tag{9.3}$$

The I' component of $\mathbf{S}' = \{I', Q', U', V'\}$ represents the scattered intensity. Therefore, I'/I represents the joint probability density function (PDF) of the polar angle θ and the azimuth angle ϕ, which is a function of the incident Stokes vector $\mathbf{S} = \{I, Q, U, V\}$:

$$\rho(\theta, \phi) = M_{11}(\theta) + M_{12}(\theta)[Q \cos(2\phi) + U \sin(2\phi)]/I. \tag{9.4}$$

Here, the polar angle θ is first sampled according to $M_{11}(\theta)$, and the azimuth angle ϕ is sampled with the following function:

$$\rho_\theta(\phi) = \frac{\rho(\theta, \phi)}{2\pi M_{11}(\theta)}. \tag{9.5}$$

It is worth noting that a biased sampling technique was used in [15]. The Stokes vectors of all the output-photon packets are transformed to the laboratory coordinate system and then accumulated to obtain the final Stokes vector. The Mueller matrix of the scattering media can be calculated algebraically from the Stokes vectors of four different incident polarization states [16]. The DOP is calculated by

$$\mathrm{DOP} = \sqrt{Q^2 + U^2 + V^2}/I. \tag{9.6}$$

To accelerate the computation, the single-scattering Mueller matrix and the PDFs of the scattering angles are calculated and stored in arrays before the photon packets are traced. The path length of the photon packets is recorded to provide pathlength- or time-resolved information. For purposes of illustration, the scatterers are assumed to be spherical; the thickness of the scattering slab is taken to be 2 cm; the temporal resolution is 1.33 ps, corresponding to 0.4 mm in real space; the wavelength of light is 543 nm; the absorption coefficient is $0.01\,\mathrm{cm}^{-1}$; the index of refraction of the turbid medium is unity, matching that of the ambient. The dimensions of the pseudocolor images in the following section are 4, 4, and 2 cm along the X, Y, and Z axes, respectively.

9.3 Results

Figure 9.3 shows the reflection and the transmission Mueller matrices of a turbid medium with a scattering coefficient of $4\,\mathrm{cm}^{-1}$ and a scatterer radius of $0.102\,\mu\mathrm{m}$. The calculated Mueller-matrix elements are normalized to the M_{11} element to compensate for the radial decay of intensity. Each of the images is displayed with its own color map to enhance the image contrast. The size of each image is $4 \times 4\,\mathrm{cm}^2$.

The patterns of the reflection Mueller matrix are identical to those reported previously [14, 15]. The symmetries in the patterns can be explained by the symmetries in the single-scattering Mueller matrix and the turbid

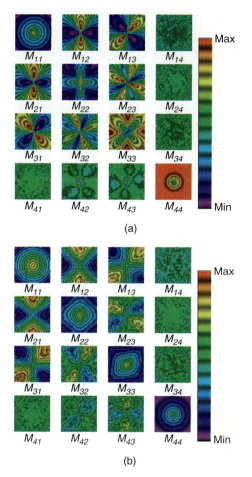

Fig. 9.3. (a) Reflection Mueller matrices and (b) transmission Mueller matrices of a slab of turbid medium

medium [14]. The transmission Mueller matrix has different patterns from the reflection Mueller matrix. One of the noticeable differences appears in elements M_{31} and M_{13}, which are antisymmetric in the reflection Mueller matrix but symmetric in the transmission Mueller matrix. This difference is caused by the mirror effect in the reflection process of the scattered light.

Fig. 9.4 shows the time-resolved DOP propagation in the turbid medium with right-circularly (R) and horizontal-linearly (H) polarized incident light. The scattering coefficient is $1.5\,\mathrm{cm}^{-1}$, and the radius of the scatterers is 0.051 μm. The scattering anisotropic factor $\langle\cos(\theta)\rangle$ is 0.11. The transport mean free path is calculated to be 0.74 cm. In the simulation, the Stokes vectors of the forward propagating photons are accumulated to calculate the DOP. As shown in the movies, the DOPs at the expanding edges of the distributed light remain

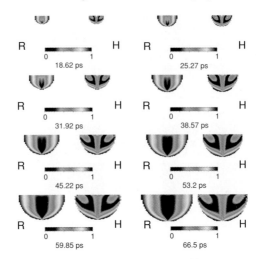

Fig. 9.4. Time sequences of the DOP propagation in the slab $(0.051\,\mu m$ particle). The X axis is along the horizontal direction, and the Z axis is along the vertical direction. R: right-circularly polarized incident light. H: horizontal-linearly polarized incident light

near unity because these photons experience few scattering events. As the light propagates in the medium, the DOP in some regions decreases significantly. The DOP patterns are dependent on the single-scattering Mueller matrix and the density of the scattering particles.

The DOP images in Fig. 9.4 show different patterns for the R- and the H-polarized incident light. As expected, such a difference appears in the DOP images of transmitted light as well. As shown in Fig. 9.5, the DOP images of the transmitted light are rotationally symmetric for circularly polarized incident light, whereas such symmetry does not exist for linearly polarized incident light. This difference is related to the dependence of the scattering probability on the incident Stokes vector (9.4) and (9.5). The Stokes vectors of the R- and the H-polarized light are $\{1, 0, 0, 1\}$ and $\{1, 1, 0, 0\}$, respectively. According to (9.5), the R-polarized light has a uniform PDF for the azimuth angle, whereas the H-polarized light has a non-uniform one. Hence, the single-scattering pattern depends on the polarization state of the incident light. As changes in the DOP are related strongly to single scattering events, it is understandable that the DOP images have different features for different incident Stokes vectors.

To demonstrate the dependence of the evolution of the DOP on the number of scattering events, we record the time-resolved images of the average number of scattering events of the transmitted light (Fig. 9.6). The simulation parameters are the same as those for Figs. 9.4 and 9.5. As it can be seen, the transmitted photons experience more and more scattering events as time elapses. If we compare Figs. 9.5 and 9.6, it is clear that the patterns

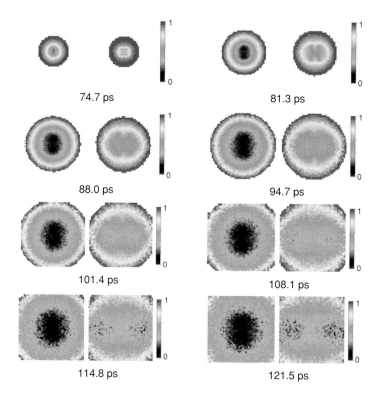

Fig. 9.5. Time sequences of the DOP of the transmitted light with R-polarized (left side images) and H-polarized (right side images) incident light (0.051 μm particle). The X axis is along the horizontal direction, and the Y axis is along the vertical direction

of the DOP are related directly to the patterns of the scattering counts—the DOP decreases as the number of scattering events increases. Nevertheless, the number of scattering events does not solely determine the change in the DOP. Two dark regions are clearly visible in Fig. 9.5b, but they are inseparable in Fig. 9.6b. The change in the DOP must also depend on the nature of each scattering event as determined by the single-scattering Mueller matrix.

To study the effect of the size of the scatterers, we simulate the evolution of the DOP in a scattering medium with a different radius (1.02 μm). The scattering coefficient of the medium is 14 cm^{-1}, and the anisotropic factor is 0.91. The transport mean free path is 0.76 cm, which is similar to the value for Figs. 9.4–9.6. The time-resolved propagation of the DOP in the medium is shown in Fig. 9.7. The DOP movie of the transmitted light is shown in Fig. 9.8.

Note the different patterns in Figs. 9.7 and 9.8 and Figs. 9.4 and 9.5. The key difference is that the DOP of R-polarized incident light is preserved much

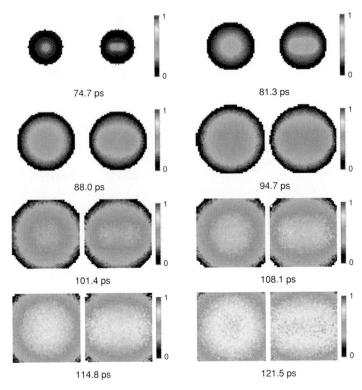

Fig. 9.6. Time sequences of the weighted-averaged numbers of scattering events for R-polarized (left side images) and H-polarized (right side images) incident light. The numbers of scattering events are normalized to a maximum value of 7 for the plots. The X axis is along the horizontal direction, and the Y axis is along the vertical direction

better than the DOP of H-polarized incident light for the large scatterers, as shown in Figs. 9.7 and 9.8. In contrast, the DOP of H-polarized incident light is preserved better than the DOP of R-polarized incident light for the small scatterers, as shown in Figs. 9.4 and 9.5. This observation is consistent with previous experimental and theoretical findings.

Another significant difference is that the DOP patterns for the large scatterers become rotationally symmetric even when the incident light is H-polarized. This phenomenon can be easily understood if we examine the PDFs of the scattering angle for different particle sizes. The scattering angle θ is determined by M_{11}, and its PDF $\rho(\theta)$ is $2\pi M_{11} \sin(\theta)$. The PDF of the azimuth angle ϕ is function of both ϕ and the incident Stokes vector, as defined in (9.5). The contribution of the ϕ-dependent term is proportional to $|M_{12}/M_{11}|$. The curves of $\rho(\theta)$ and $|M_{12}/M_{11}|$ are shown in Fig. 9.9. When the size of the scatterer is small, $\rho(\theta)$ is approximately homogeneous and the photon is likely to be scattered into $60 - 120°$ (Fig. 9.9a). At these angles, the

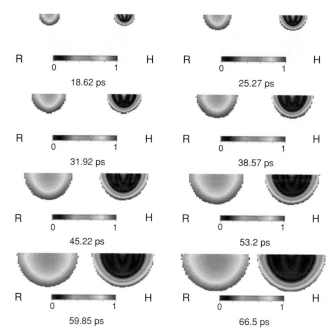

Fig. 9.7. Time sequences of the DOP propagation in the slab (1.02 μm particle). The X axis is along the horizontal direction, and the Z axis is along the vertical direction. R: right-circularly polarized incident light. H: horizontal-linearly polarized incident light

$|M_{12}/M_{11}|$ ratio has large values and (9.5) depends strongly on ϕ. When the size of the scatterer is large, most of the photons are scattered into smaller angles (Fig. 9.9b). The $|M_{12}/M_{11}|$ ratio is small at small scattering angles, which means that the homogeneous-distribution term is dominant in the probability distribution function of the ϕ angle. As a consequence, the scattering process becomes rotationally symmetric for the larger particle sizes.

As observed in Figs. 9.5 and 9.8, the DOP at the center of the 2D time-resolved images is smaller than that in the outer area. However, the DOP decreases radially in the time-integrated images (Fig. 9.10), which is in agreement with previous experimental results [8]. This is because the photons exiting at the early times have a limited span in space and hence have a dominant weight in the central area. These early exiting photons preserve the DOP better because they experience fewer scattering events than the photons exiting at later times. Consequently, a combination of time-gating and polarization discrimination [17] is better at rejecting multi-scattered photons than either technique alone. Figure 9.10 also reveals that the DOP decreases as the scattering coefficient increases. The radial distribution of the DOP becomes flat at a large scattering coefficient because of the increasing number of multi-scattered photons.

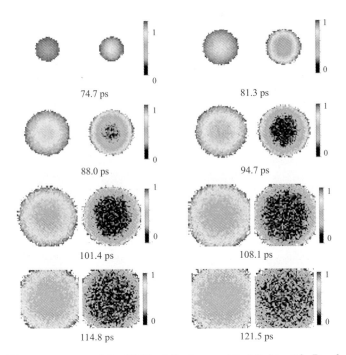

Fig. 9.8. Time sequences of the DOP of the transmitted light with R-polarized (left side images) and H-polarized (right side images) incident light (1.02 µm particle). The X axis is along the horizontal direction, and the Y axis is along the vertical direction

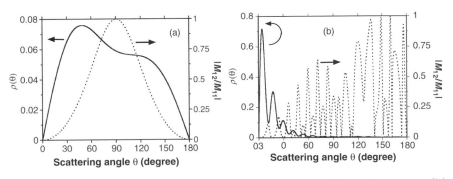

Fig. 9.9. PDF $\rho(\theta)$ and $|M_{12}/M_{11}|$ at a particle radius of (**a**) 0.051 µm and (**b**) 1.02 µm

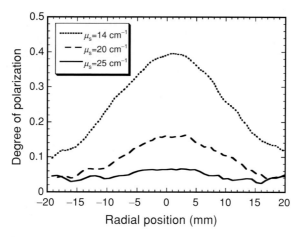

Fig. 9.10. Radial distribution of the DOP of the transmitted light for different scattering coefficients. The particle radius was $1.02\,\mu\mathrm{m}$. The incident light was H polarized

9.4 Summary

In this chapter, a Monte Carlo technique was employed to simulate the time-resolved propagation of polarized light in scattering media. Results are consistent with prior experimental findings. Hence, time-resolved simulation is a useful tool for understanding the essential physical processes of polarization propagation in turbid media. Because of the nature of the Monte Carlo simulation, coherent phenomena, such as laser speckles, are not modeled. Nevertheless, the simulation method can be applied in the non-coherent regime or in cases where the coherent effect is removed, such as ensemble-averaged measurements.

10

Polarization-Sensitive Optical Coherence Tomography

10.1 Introduction

As a result of its noninvasive characteristic, its high spatial resolution and its easy optical fiber implementation, optical coherence tomography (OCT) is emerging as an important optical imaging modality. Various technical approaches have been developed to increase its spatial resolution [1, 2], imaging rate [3,4], and image quality [5,6]. To completely retrieve the information carried by backscattered light fields, both amplitude and polarization information need to be recorded. Conventional OCT systems record the amplitude but not the polarization information from scattered light. In contrast, polarization-sensitive OCT can capture the polarization states of backscattered light and, as a result, can reveal the polarization properties, such as birefringence, of a sample, which cannot be recovered by conventional OCT [7–16]. Birefringence is related to various biological components such as collagen, muscle fibers, myelin, retina, keratin, and glucose. Consequently, polarization can provide novel contrast mechanisms for imaging, diagnosis, and sensing. In Mueller calculus, the polarization state of light can be completely characterized by a Stokes vector, and the polarization transforming properties of an optical sample can be completely characterized by a Mueller matrix. The combination of Mueller calculus and OCT offers a unique way to acquire the Mueller matrix of a scattering sample with OCT resolution [10, 11]. Yao and Wang [10] were the first to report on two-dimensional depth-resolved Mueller-matrix images of biological tissues measured with OCT based on 16 combinations of source and detection polarization states. The relatively time-consuming nature of the measurement process, however, limited the application of the technique to stable samples such as bones. Jiao et al. [11] further demonstrated that the degree of polarization (DOP) of backscattered light measured by OCT is unity throughout the detection range and that a DOP of unity indicates that the measured Mueller matrix is nondepolarizing. This conclusion allows the use of a Jones matrix, instead of a Mueller matrix, in OCT.

To minimize motion artifacts in soft-tissue imaging, a system that can determine the Jones matrix with a single depth scan (A-scan) has been developed [17, 18]. In other words, this system can acquire a Jones matrix as fast as its conventional OCT counterpart can acquire a regular image. The measured Jones matrix can be further transformed into an equivalent Mueller matrix to take advantage of the latter's superiority in revealing polarization-independent information.

10.2 Experimental System: Serial Implementation

As described in Chap. 2, a 4×4 Mueller matrix has 16 independent elements in the most general cases; therefore, at least 16 independent measurements must be acquired to determine a full Mueller matrix. As will be discussed in the subsequent sections, the number of necessary measurements can be reduced in polarization-sensitive OCT to construct a Jones or Mueller matrix.

In an OCT system, the interference signal generated by the light beams from the reference arm and the sample arm is

$$
\begin{aligned}
I_{\text{OCT}} &= 2\text{Re}\left\{< \mathbf{E}_{\text{s}}(l_{\text{s}}) \cdot \mathbf{E}_{\text{r,A}}^{*}(l_r) >\right\} \\
&= 2[I_{\text{s,A}}(l_{\text{s}})I_{\text{r,A}}]^{1/2} |V(\Delta l)| \cos(k_0 \Delta l),
\end{aligned}
\tag{10.1}
$$

where $\mathbf{E}_{\text{s}}$ denotes the sample electric field, $\mathbf{E}_{\text{r,A}}$ denotes the reference electric field with a polarization state A, l_{s} and l_r are the optical path lengths of the sample arm and the reference arm, respectively, $I_{\text{s,A}}$ denotes the light intensity from the sample arm projected onto polarization state A, $I_{\text{r,A}}$ denotes the intensity of light in polarization state A from the reference arm, V is the temporal coherence function of the field, Δl represents the path-length difference between the sample and reference arms, and k_0 is the magnitude of the average wave vector.

The path-length difference can be approximately converted to the optical depth, which is the physical depth multiplied by the index of refraction, in the sample in the ballistic or quasiballistic regime. For light reflected from a given optical depth in the sample, the following quantity is used to substitute for the time-resolved intensity in conventional polarimetry that would otherwise be directly measured by a noninterference analyzer of polarization state A:

$$
I_{\text{s,A}} \propto I_{\text{OCT}}^2 / I_{\text{r,A}}
\tag{10.2}
$$

Figure 10.1 shows a schematic of the OCT system that quantifies the Mueller matrix of biological tissue with a series of polarization measurements. A superluminescent diode (SLD) with a center wavelength of 850 nm and a full-width-at-half-maximum (FWHM) bandwidth of 26 nm is used as the light source. The light intensity after the polarizer P is 400 μW. After passing through polarizer P, the half-wave plate HW, and the quarter-wave plate

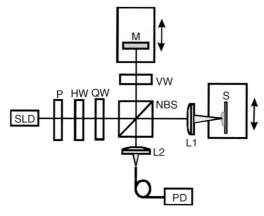

Fig. 10.1. Schematic of the serial Mueller OCT system: SLD, superluminescent diode; P, polarizer; HW, zero-order half-wave plate; QW, zero-order quarter-wave plate; NBS, non-polarization beam splitter; VW, variable wave plate; M, mirror; L1, L2, lens; PD, photodiode

QW, the light is split by a nonpolarization beam splitter (NBS). The sample beam is focused into the sample by an objective lens (L1). The reference beam passes through a variable wave plate (VW) and is reflected back. The reflected beams from the reference and sample arms are coupled into a single-mode fiber and detected by a silicon photodiode. The depth resolution of this system is $10\,\mu$m. The step size of the lateral scan is also $10\,\mu$m. The focal spot size of the objective lens (L1) is $6.9\,\mu$m in air and larger in tissue. The lateral resolution is expected to be around $10\,\mu$m.

Four different incident polarization states, H, V, $+45°$, and R are achieved by rotating the half-wave plate (HW) and the quarter-wave plate (QW) in the source arm. For each of these four incident polarization states, the variable wave plate (VW) in the reference arm is adjusted to sequentially achieve the H, V, P, and R polarization states. The light intensities of both the source arm and the reference arm are measured for each of the 16 combinations of polarization states in the source and reference arms. The source intensity is measured for calibration purposes. The reference intensities are used to convert the OCT signals for calculations of Stokes vectors and Mueller matrices. A total of 16 polarization-sensitive OCT images are acquired and processed to obtain the 16 Mueller matrix images $[M_{ij}]$. Alternatively, if the Stokes vector of the backscattered light is sought for a given incident polarization state, only four measurements need be acquired by varying the reference polarization state.

The OCT system was carefully calibrated and validated. The four incident polarization states, as well as the four reference polarization states associated with each incident polarization state, were examined in terms of polarization purity. The polarization purity is defined as I_{min}/I_{max}, where I_{max} is the signal intensity of the designed polarization state and I_{min} is the intensity of the orthogonal polarization state. As the beam splitter is not an ideal polarization-independent optical element, the Mueller matrix of the beam splitter was measured for calibration.

10.3 Jones Calculus and Mueller Calculus

A Jones matrix ($\mathbf{J}$) transforms an input Jones vector ($\mathbf{E}_{IN}$) into an output Jones vector ($\mathbf{E}_{OUT}$) while a Mueller matrix ($\mathbf{M}$) transforms an input Stokes vector ($\mathbf{S}_{IN}$) into an output Stokes vector ($\mathbf{S}_{OUT}$):

$$\mathbf{E}_{OUT} = \begin{bmatrix} E_{OH} \\ E_{OV} \end{bmatrix} = \mathbf{J}\mathbf{E}_{IN} = \begin{bmatrix} J_{11} & J_{12} \\ J_{21} & J_{22} \end{bmatrix} \begin{bmatrix} E_{iH} \\ E_{iV} \end{bmatrix}, \tag{10.3}$$

$$\mathbf{S}_{OUT} = \begin{bmatrix} I \\ Q \\ U \\ V \end{bmatrix} = \mathbf{M}\mathbf{S}_{IN} = \begin{bmatrix} M_{11} & M_{12} & M_{13} & M_{14} \\ M_{21} & M_{22} & M_{23} & M_{24} \\ M_{31} & M_{32} & M_{33} & M_{34} \\ M_{41} & M_{42} & M_{43} & M_{44} \end{bmatrix} \begin{bmatrix} I_i \\ Q_i \\ U_i \\ V_i \end{bmatrix}, \tag{10.4}$$

where E_{OH} and E_{OV} are the horizontal and vertical components of the electric vector of the output light field, E_{iH} and E_{iV} are the horizontal and vertical components of the electric vector of the input light field, I, Q, U, and V are the elements of the Stokes vector of the output light, and I_i, Q_i, U_i, and V_i are the elements of the Stokes vector of the input light, respectively. I and I_i are the intensity of the output and input light, respectively. In an OCT system, I_i represents the intensity of the incident light in the sample arm, and I represents the detected intensity of the backscattered light. In (10.4), we can clearly see that M_{11} represents the intensity transformation property of the sample and contains no polarization information.

The Jones matrices of a homogenous partial polarizer ($\mathbf{J}_P$) and a homogenous elliptical retarder ($\mathbf{J}_R$) can be expressed as

$$\mathbf{J}_P = \begin{bmatrix} P_1 \cos^2 \alpha + P_2 \sin^2 \alpha & (P_1 - P_2) \sin \alpha \cos \alpha \, e^{-i\Delta} \\ (P_1 - P_2) \sin \alpha \cos \alpha \, e^{i\Delta} & P_1 \sin^2 \alpha + P_2 \cos^2 \alpha \end{bmatrix}, \tag{10.5}$$

$$\mathbf{J}_R = \begin{bmatrix} e^{i\varphi/2} \cos^2 \theta + e^{-i\varphi/2} \sin^2 \theta & (e^{i\varphi/2} - e^{-i\varphi/2}) \sin \theta \cos \theta \, e^{-i\delta} \\ (e^{i\varphi/2} - e^{-i\varphi/2}) \sin \theta \cos \theta \, e^{-i\delta} & e^{i\varphi/2} \sin^2 \theta + e^{-i\varphi/2} \cos^2 \theta \end{bmatrix}, \tag{10.6}$$

where P_1, P_2 are the principal coefficients of the amplitude transmission for the two orthogonal polarization eigenstates, α is the orientation of $\mathbf{J}_P$, φ and θ are the retardation and orientation of $\mathbf{J}_R$, respectively, and Δ and δ are the phase differences for the vertical and horizontal components of the eigenstates of $\mathbf{J}_P$ and $\mathbf{J}_R$, respectively.

A retarder is called elliptical when its eigenvectors are those of elliptical polarization states. A linear retarder is a special case where the eigenpolarizations are linear; and a Faraday rotator is another special case where the eigenpolarizations are circular. When two or more linear retarders are cascaded, the overall retarder is generally elliptical.

A polarizing element is called homogeneous when the two eigenvectors of its Jones matrix are orthogonal [19, 20]. Linear polarizers and linear and circular retarders are typical homogeneous polarizing optical elements. A typical example of inhomogeneous polarizing elements is a circular polarizer, whose Jones matrix is $\frac{1}{2}\begin{bmatrix} 1 & 1 \\ i & i \end{bmatrix}$, which is constructed by using a linear polarizer set at $45°$ followed by a $\lambda/4$ plate with its fast axis set at horizontal. The eigenvectors of such a circular polarizer are $\frac{1}{\sqrt{2}}\begin{bmatrix} 1 \\ -1 \end{bmatrix}$ for a $-45°$ linear polarization state and $\frac{1}{\sqrt{2}}\begin{bmatrix} 1 \\ i \end{bmatrix}$ for a right circular polarization state; these are not orthogonal.

For an intensity-based noninterference detection system, a turbid medium is generally depolarizing unless the detector is small; in other words, when a completely polarized light beam (DOP = 1) is scattered by the medium, the output light becomes partially polarized (DOP < 1) unless the area of the detector is much less than the average size of speckles. This point is covered in more detail in Chap. 8. However, OCT is an amplitude-based detection system using interference heterodyne, which detects the part of the backscattered electric field that is coherent with the reference beam, regardless of whether the overall backscattered light is partially polarized or not. The OCT signal, I_{OCT}, received by a detector of a finite area can be considered as the sum of the contributions from the backscattered optical fields, $\mathbf{E}_{si}$, reaching the various points of the detector:

$$
\begin{aligned}
I_{OCT} &= \mathbf{E}_r \cdot \mathbf{E}_{s1} + \mathbf{E}_r \cdot \mathbf{E}_{s2} + \mathbf{E}_r \cdot \mathbf{E}_{s3} + \dots \\
&= \mathbf{E}_r \cdot (\mathbf{E}_{s1} + \mathbf{E}_{s2} + \mathbf{E}_{s3} + \dots) \\
&= \mathbf{E}_r \cdot \mathbf{E}_s,
\end{aligned}
\tag{10.7}
$$

where $\mathbf{E}_r$ represents the reference optical field, $\mathbf{E}_s$ is an equivalent total backscattered optical field, and the dot product represents the interference signal (apart from a constant factor). As shown in (10.7), each backscattered optical field from the sample contributes to the OCT signal by projecting onto the reference optical field, $\mathbf{E}_r$. Equivalently, the backscattered optical fields reaching the various points of the detector can be summed in vector, and the vector sum, $\mathbf{E}_s$, is then projected onto the reference optical field to yield the OCT signal. One can imagine this is equivalent to shrinking the full field over the area of detection to a single point before interfering with the reference beam. If all of the $\mathbf{E}_{si}$ share the same polarization state, $\mathbf{E}_s$ will have the same polarization state; otherwise, $\mathbf{E}_s$ will have a net apparent polarization state. In either case, the measured $\mathbf{E}_s$ will have a unique polarization state.

As a result, the DOP measured by OCT will be unity. In an intensity-based noninterference detection system, by contrast, the backscattered optical fields reaching the various points of the detector would add in intensity. In this case, if all of the $\mathbf{E}_{si}$ do not share the same polarization state, the DOP will be less than unity.

Unlike a Mueller matrix, which is suitable for all kinds of optical systems, a Jones matrix can only be applied to a nondepolarizing optical system. A Jones matrix can completely characterize the polarization properties of a nondepolarizing optical system. In other words, for a nondepolarizing optical system, a Jones matrix is equivalent to a Mueller matrix. A Jones matrix has four complex elements, in which one phase is arbitrary and, consequently, seven real parameters are independent. Equivalently, there are seven independent parameters in a nondepolarizing Mueller matrix.

When the two matrices are equivalent, the Jones and Mueller matrices have different advantages. A Jones matrix has fewer elements and the physical meanings of the matrix elements are clearer. On the other hand, a Mueller matrix uses only real numbers; and the intensity transformation property of a sample is explicitly expressed in its M_{11} element, which provides an image of the sample without the influence of its polarization property. M_{11} contains no polarization artifact such as is usually encountered in a conventional OCT image when the sample contains birefringence. Therefore, a Mueller matrix clearly separates the structural information from the polarization information of a sample.

The Jones matrix of a nondepolarizing optical system can be transformed into an equivalent nondepolarizing Mueller matrix by the following relationship [21]:

$$
\begin{aligned}
\mathbf{M} &= \mathbf{U}(\mathbf{J} \otimes \mathbf{J}^*)\mathbf{U}^{-1} \\
&= \mathbf{U} \begin{bmatrix} J_{11}\mathbf{J}^* & J_{12}\mathbf{J}^* \\ J_{21}\mathbf{J}^* & J_{22}\mathbf{J}^* \end{bmatrix} \mathbf{U}^{-1} \\
&= \mathbf{U} \begin{bmatrix} J_{11}J_{11}^* & J_{11}J_{12}^* & J_{12}J_{11}^* & J_{12}J_{12}^* \\ J_{11}J_{21}^* & J_{11}J_{22}^* & J_{12}J_{21}^* & J_{12}J_{22}^* \\ J_{21}J_{11}^* & J_{21}J_{12}^* & J_{22}J_{11}^* & J_{22}J_{12}^* \\ J_{21}J_{21}^* & J_{21}J_{22}^* & J_{22}J_{21}^* & J_{22}J_{22}^* \end{bmatrix} \mathbf{U}^{-1},
\end{aligned}
\tag{10.8}
$$

and a Jones vector of a light field can be transformed into a Stokes vector by

$$
\mathbf{S} = \sqrt{2}\mathbf{U}(\mathbf{E} \otimes \mathbf{E}^*) = \sqrt{2}\mathbf{U} \begin{bmatrix} E_H\mathbf{E}^* \\ E_V\mathbf{E}^* \end{bmatrix} = \sqrt{2}\mathbf{U} \begin{bmatrix} E_H E_H^* \\ E_H E_V^* \\ E_V E_H^* \\ E_V E_V^* \end{bmatrix},
\tag{10.9}
$$

where $\otimes$ represents the Kronecker tensor product and $\mathbf{U}$ is the 4×4 Jones–Mueller transformation matrix:

$$\mathbf{U} = \frac{1}{\sqrt{2}} \begin{bmatrix} 1 & 0 & 0 & 1 \\ 1 & 0 & 0 & -1 \\ 0 & 1 & 1 & 0 \\ 0 & i & -i & 0 \end{bmatrix}. \tag{10.10}$$

At least two independent incident polarization states, which are not necessarily orthogonal, are needed to fully determine a Jones matrix.

10.4 Experimental System: Parallel Implementation

Figure 10.2 shows a schematic of the OCT system that quantifies a Jones matrix of biological tissue with multiple polarization measurements in parallel. Two super luminescent diodes (SLD) are employed as low-coherence light sources and are amplitude modulated at 3 and 3.5 kHz by modulating the injection current. The two light sources are in horizontal and vertical polarization states, respectively, and each delivers about 200 μw of power to the sample. The central wavelength, FWHM bandwidth, and the output power of the light sources are 850 nm, 26 nm, and 3 mw, respectively. The Jones vectors of the two sources are $[1, 0]^{\mathrm{T}}$ and $[0, 1]^{\mathrm{T}}$, respectively, where the superscript T transposes the row vectors into column vectors. The two source

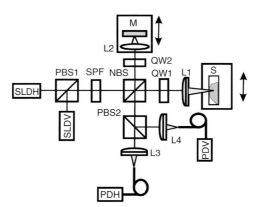

Fig. 10.2. Schematic of the parallel Jones OCT system: SLDH and SLDV, superluminescent diodes, horizontally polarized (H) and vertically polarized (V), respectively; PBS1 and PBS2, polarizing beam splitters; SPF, spatial filter assembly; NBS, nonpolarizing beam splitter; QW1 and QW2, zero-order quarter-wave plates; M, mirror; L1, L2 L3 and L4, lenses; PDH and PDV, photodiodes for H and V polarization components

beams are merged by a polarizing beam splitter (PBS1), filtered by a spatial filter assembly and then split into the reference arm and the sample arm by a nonpolarizing beam splitter (NBS). The sample beam passes through a quarter-wave plate ($\lambda/4$ plate), the fast axis of which is oriented at $45°$ and is focused into the sample by an objective lens (L1: $f = 15\,\mathrm{mm}$ and $\mathrm{NA} = 0.25$). The Jones vectors of the sample beam at the sample surface for the two sources are $[1, i]^T$ and $[1, -i]^T$, which are right-circularly and left-circularly polarized, respectively. The reference arm consists of a $\lambda/4$ plate, the fast axis of which is oriented at $22.5°$, a lens (L2), and a mirror. After retro-reflection by the reference mirror and double passing through the $\lambda/4$ plate, the horizontal polarization (H) of the incident light is converted into $45°$ polarization, $[1, 1]^T$, while the vertical polarization (V) of the incident light is converted into $-45°$ polarization, $[1, -1]^T$, and then the reference beam combines with the backscattered sample beam through the NBS. The combined light is split into two orthogonal polarization components, i.e., the horizontal and vertical components of the Jones vector, by a polarization beam splitter PBS2. The two components are coupled into two single-mode fibers with objective lenses. The two polarization components are detected by photodiodes PDH and PDV, respectively. A data-acquisition board (DAQ board), sampling at $50\,\mathrm{kHz}$ per channel, digitizes the two signals. The scan speed of the reference arm is $0.5\,\mathrm{mm\,s}^{-1}$ generating a Doppler frequency of about $1.2\,\mathrm{kHz}$. The carrier frequencies, 1.8, 2.3, 4.2, and $4.7\,\mathrm{kHz}$, are the beat and harmonic frequencies between this Doppler frequency and the modulation frequencies of the light sources.

The two function generators (DS345, Stanford Research Systems), which are used for the modulation of the two light sources, respectively, are synchronized and share the same time base. Burst mode was used to ensure that the initial phases of the two modulation signals are fixed for each A-scan. The time delay between the scanning of the two channels of the DAQ board is $10\,\mu\mathrm{s}$. The phase difference between the two channels caused by this time delay for each beat and harmonic frequency was compensated for during signal processing.

For OCT signals based on single-backscattered photons, the incident Jones vector $\mathbf{E}_i$ in the sample arm is transformed to the detected Jones vector $\mathbf{E}_o$ by

$$\begin{aligned}
\mathbf{E}_o &= \mathbf{J}_{\mathrm{NBS}}\mathbf{J}_{\mathrm{QB}}\mathbf{J}_{\mathrm{SB}}\mathbf{J}_{\mathrm{M}}\mathbf{J}_{\mathrm{SI}}\mathbf{J}_{\mathrm{QI}}\mathbf{E}_i \\
&= \mathbf{J}_{\mathrm{NBS}}\mathbf{J}_{\mathrm{QB}}\mathbf{J}\mathbf{J}_{\mathrm{QI}}\mathbf{E}_i = \mathbf{J}_T\mathbf{E}_i,
\end{aligned} \tag{10.11}$$

where $\mathbf{J}_{\mathrm{QI}}$ and $\mathbf{J}_{\mathrm{QB}}$ are the Jones matrices of the $\lambda/4$ plate for the incident and the backscattered light, respectively, $\mathbf{J}_{\mathrm{SI}}$ and $\mathbf{J}_{\mathrm{SB}}$ are the Jones matrices of the sample for the incident and backscattered light, respectively, $\mathbf{J}_{\mathrm{M}}$ is the Jones matrix of the single backscatterer – the same as the one for a mirror, $\mathbf{J}_{\mathrm{NBS}}$ is the Jones matrix of the reflecting surface of the nonpolarizing beam splitter, $\mathbf{J}$ is the combined round-trip Jones matrix of the scattering medium, and $\mathbf{J}_T$ is the overall round-trip Jones matrix.

In (10.11), the output Jones vector $\mathbf{E}_o$ is constructed for each light source from the measured horizontal and vertical components of the OCT signal. Upon acquiring the output Jones vectors and knowing the input Jones vectors, the overall round-trip Jones matrix $\mathbf{J}_T$ can be calculated. The Jones matrix $\mathbf{J}$ of the sample can be extracted from $\mathbf{J}_T$ by eliminating the effect of the Jones matrices of the quarter-wave plate, the mirror and the beam splitter.

In a commonly used convention, $\mathbf{J}_M$ transforms the polarization state of the forward light expressed in the forward coordinate system into the polarization state expressed in the backward coordinate system. Similarly, $\mathbf{J}_{NBS}$ transforms the polarization state of the backward light into the polarization state expressed in the detection coordinate system. However, in this work we express the polarization states of both the forward and backward light in the forward coordinate system. In this convention, $\mathbf{J}_M$ and $\mathbf{J}_{NBS}$ are identity matrices:

$$\mathbf{J}_M = \mathbf{J}_{NBS} = \begin{bmatrix} 1 & 0 \\ 0 & 1 \end{bmatrix}. \tag{10.12}$$

In each A-scan, the optical paths for the forward and backward light are the same, and, therefore, the Jones' reversibility theorem can be applied [22]. The Jones reversibility theorem indicates that the Jones matrices, $\mathbf{J}_{BWD}$ and $\mathbf{J}_{FWD}$ of an ordinary optical element for the backward and forward light propagations, have the following relationship if the same coordinate system is used for the Jones vectors: $\mathbf{J}_{BWD} = \mathbf{J}_{FWD}^T$. Therefore, we have the following relationships:

$$\mathbf{J}_{SB} = \mathbf{J}_{SI}^T, \mathbf{J}_{QB} = \mathbf{J}_{QI}^T = \frac{1}{\sqrt{2}} \begin{bmatrix} 1 & -i \\ -i & 1 \end{bmatrix}, \tag{10.13}$$

$$\mathbf{J} = \mathbf{J}_{SB}\mathbf{J}_M\mathbf{J}_{SI} = \mathbf{J}_{SI}^T\mathbf{J}_{SI} = \mathbf{J}^T, \tag{10.14}$$

$$\mathbf{J}_T = \mathbf{J}_{NBS}\mathbf{J}_{QB}\mathbf{J}\mathbf{J}_{QI} = \mathbf{J}_{QI}^T\mathbf{J}\mathbf{J}_{QI} = \mathbf{J}_T^T. \tag{10.15}$$

In other words, matrices $\mathbf{J}$ and $\mathbf{J}_T$ are transpose symmetric. Because of this symmetry, the number of independent parameters in the Jones matrix is reduced from seven to five.

As reported by Yao and Wang, who used Monte Carlo simulation [23], the light backscattered from the sample can be divided into two parts: Class *I* and Class *II*. Class *I* light provides a useful signal, which is scattered by the target layer in a sample and the path-length difference of which from the reference light is within the coherence length of the light source. Class *II* light is the part scattered from the rest of the medium, whose path-length difference from the reference light is also within the coherence length of the light source. Class *II* light contributes to the background noise of the OCT signal. The weight of Class *II* light in the detected OCT signal increases with depth and will exceed that of the Class *I* signal beyond some critical depth. An increase in the weight of the Class *II* light deteriorates the resolution and signal-to-noise ratio and thus limits the effective imaging depth. The Class *I* signal also contains multiply scattered photons, but owing to the requirement of matching the optical path-lengths, these multiple scattering events must be small-angle scattering.

For the multiply scattered photons, (10.11) still holds if the probabilities for photons to travel along the same round-trip path, but in opposite directions, are equal, which is a valid assumption when the source and detector have reciprocal characteristics. As these photons are coherent, the round-trip Jones matrix of the sample $\mathbf{J}$ is the sum of the Jones matrices of all the possible round-trip paths; and for each possible path – for example, the kth path – the round-trip Jones matrix is the sum of the Jones matrices for the two opposite directions $[\mathbf{J}_i(k)$ and $\mathbf{J}_r(k)]$. Consequently, we have

$$\mathbf{J} = \sum_k [\mathbf{J}_i(k) + \mathbf{J}_r(k)] = \sum_k \left\{ \mathbf{J}_i(k) + [\mathbf{J}_i(k)]^T \right\} = \mathbf{J}^T. \qquad (10.16)$$

In other words, $\mathbf{J}$, as well as $\mathbf{J}_T$, still possesses transpose symmetry even if multiple scattering occurs, as long as the source and the detector meet the condition.

After calculation, (10.11) can be expressed as

$$
\begin{bmatrix} E_{oH} \\ E_{oV} \end{bmatrix} =
\begin{bmatrix} \dfrac{i}{2}(J_{11} - 2iJ_{12} - J_{22}) & \dfrac{1}{2}(J_{11} + J_{22}) \\ \dfrac{1}{2}(J_{11} + J_{22}) & \dfrac{i}{2}(-J_{11} - 2iJ_{12} + J_{22}) \end{bmatrix}
\times
\begin{bmatrix} E_{iH} \\ E_{iV} \end{bmatrix}
$$
$$
=
\begin{bmatrix} J_{T11} & J_{T12} \\ J_{T12} & J_{T22} \end{bmatrix}
\times
\begin{bmatrix} E_{iH} \\ E_{iV} \end{bmatrix},
$$
$$(10.17)$$

where J_{ij} and J_{Tij} $(i, j = 1, 2)$ are the elements of $\mathbf{J}$ and $\mathbf{J}_T$, respectively. For two light sources with independent polarization states, (10.17) can be rearranged as

$$
\begin{bmatrix} E_{oH1} & E_{oH2} \\ E_{oV1} & E_{oV2} \end{bmatrix} =
\begin{bmatrix} J_{T11} & J_{T12} \\ J_{T12} & J_{T22} \end{bmatrix}
\times
\begin{bmatrix} E_{iH1} & E_{iH2}e^{i\beta} \\ E_{iV1} & E_{iV2}e^{i\beta} \end{bmatrix}, \qquad (10.18)
$$

where E_{oH1} and E_{oH2}, E_{oV1} and E_{oV2} are the elements of the output Jones vectors for source 1 and source 2, respectively, and β is a phase difference related to the characteristics of the two light sources. If the two light sources are identical, β is zero. In practice, the spectral characteristics of the two light sources are close but not identical. $\mathbf{J}_T$ can be calculated from (10.18) as

$$
\begin{bmatrix} J_{T11} & J_{T12} \\ J_{T12} & J_{T22} \end{bmatrix} =
\begin{bmatrix} E_{oH1} & E_{oH2} \\ E_{oV1} & E_{oV2} \end{bmatrix}
\times
\begin{bmatrix} E_{iH1} & E_{iH2}e^{i\beta} \\ E_{iV1} & E_{iV2}e^{i\beta} \end{bmatrix}^{-1}
$$
$$
= \dfrac{1}{D}
\begin{bmatrix} E_{oH1} & E_{oH2} \\ E_{oV1} & E_{oV2} \end{bmatrix}
\times
\begin{bmatrix} E_{iV2}e^{i\beta} & -E_{iH2}e^{i\beta} \\ -E_{iV1} & E_{iH1} \end{bmatrix},
$$
$$(10.19)$$

as long as the determinant

$$D = \begin{vmatrix} E_{iH1} & E_{iH2}e^{i\beta} \\ E_{iV1} & E_{iV2}e^{i\beta} \end{vmatrix} = e^{i\beta} \begin{vmatrix} E_{iH1} & E_{iH2} \\ E_{iV1} & E_{iV2} \end{vmatrix} \neq 0, \tag{10.20}$$

i.e., the two light sources are not in the same polarization state. The random phase difference β can be eliminated with the transpose symmetry of $\mathbf{J}_T$:

$$e^{i\beta} \left(E_{oH1} E_{iH2} + E_{oV1} E_{iV2} \right) = \left(E_{oV2} E_{iV1} + E_{oH2} E_{iH1} \right). \tag{10.21}$$

Equation (10.21) can be solved when $(E_{oH1} E_{iH2} + E_{oV1} E_{iV2}) \neq 0$. Once $\mathbf{J}_T$ is found, $\mathbf{J}$ can then be determined from $\mathbf{J}_T$. Six real parameters of $\mathbf{J}$ can be calculated, in which one phase is arbitrary and can be subtracted from each element, and eventually five independent parameters are retained.

When $(E_{oH1} E_{iH2} + E_{oV1} E_{iV2}) = 0$, it is impossible to eliminate the random phase by using the transpose symmetry. This situation happens if the sample arm does not alter the polarization states of the two incident beams in addition to producing a mirror reflection. For example, this situation occurs if (1) a horizontal or vertical incident beam is used, (2) a $\lambda/4$ plate is not inserted in the sample arm, and (3) the fast axis of a birefringent sample is horizontal or vertical. The use of the $\lambda/4$ plate at a 45° orientation in the sample arm can ameliorate the situation. However, there are still some drawbacks with this configuration. For example, when the round-trip Jones matrix $\mathbf{J}$ is equivalent to one of a half-wave plate with its fast axis oriented at 45° and thus $\mathbf{J}_T$ is equivalent to an identity matrix, we have $(E_{oH1} E_{iH2} + E_{oV1} E_{iV2}) = 0$. To overcome this drawback, we can employ two nonorthogonal incident polarization states; for example, one source will be in a horizontal polarization state and the other source in a 45° polarization state.

The interference signals are band-pass filtered with central frequencies of 4.2 kHz and 4.7 kHz and a bandwidth of 10 Hz – the harmonic frequencies of the interference signals of source H and source V, respectively – to extract the interference components of each light source. The interference components are used to form the imaginary parts of the elements of the output Jones vectors, $E_{x,y}(t)$, where x and y represent the detected polarization state (H or V) and the source polarization state (H or V), respectively; the corresponding real parts are obtained through inverse Hilbert transformation [24, 25]:

$$\mathrm{Re}\{E_{x,y}(t)\} = \frac{1}{\pi} P \int_{-\infty}^{\infty} \frac{\mathrm{Im}\{E_{x,y}(t)\}}{\tau - t} d\tau, \tag{10.22}$$

where P stands for the Cauchy principal value of the integral. Unlike other transforms, the Hilbert transformation does not change the domain. A convenient method of computing the Hilbert transform is by means of the Fourier transformation. If $u(t)$ and $v(t)$ are a Hilbert pair of functions, i.e.,

$$u(t) \overset{H}{\Longleftrightarrow} v(t), \tag{10.23}$$

and $U(w)$ and $V(w)$ are the Fourier transforms of $u(t)$ and $v(t)$, the following algorithm can be used to calculate the Hilbert transform [25]:

$$u(t) \overset{F}{\Rightarrow} U(w) \Rightarrow V(w) = -\mathrm{i} \cdot sgn(w)U(w) \overset{F^{-1}}{\Rightarrow} v(t), \qquad (10.24)$$

$$v(t) \overset{F}{\Rightarrow} V(w) \Rightarrow U(w) = \mathrm{i} \cdot sgn(w)U(w) \overset{F^{-1}}{\Rightarrow} u(t), \qquad (10.25)$$

where F and F^{-1} denote the Fourier and inverse Fourier transformations, respectively; $sgn(w)$ is the signum function defined as

$$sgn(w) = \begin{cases} +1 & w > 0 \\ 0 & w = 0 \\ -1 & w < 0. \end{cases} \qquad (10.26)$$

The real and imaginary parts of each interference component are combined to form the complex components of the output Jones vectors. Upon determining the output Jones vector, when the input Jones vectors are known, the elements of the Jones matrix $\mathbf{J}$ of the sample can then be calculated from (10.17).

10.5 Experimental Results

The system is often tested by measuring the matrix of a standard sample – a $\lambda/4$ wave-plate at various orientations in combination with a mirror. Figure 10.3a shows the amplitude of the vertical components of the measured Jones vector versus the orientation of the wave-plate, where the amplitude of each Jones vector was normalized to unity. Figure 10.3b shows the phase differences between the vertical components and the horizontal components of the Jones vectors. The calculated results were averaged over 1,000 points centered at the peak of the interference signals, where 1,000 points correspond to $10\,\mu\mathrm{m}$ – the resolution of the system. The results show that the measured data agree very well with the theoretical values.

The parallel OCT system was applied to image soft tissue – a piece of porcine tendon. The tendon was mounted in a cuvette filled with saline solution. The sample was transversely scanned with a step size of $5\,\mu\mathrm{m}$, and multiple A-scan images were taken. The digitized interference signals were first band-pass filtered with software and Hilbert transformed to extract the analytical signals of each polarization component. For each A scan, the pixels were formed by averaging the calculated elements of the Jones matrix over segments of 1,000 points. Two-dimensional (2D) images were formed from these A-scan images and then median filtered. The final 2D images are shown in Fig. 10.4.

Clear band structures can be seen in some of the images, especially in M_{24}, M_{33}, M_{34}, M_{42}, M_{43}, and M_{44}. The period of the band structure is $\sim 0.13\,\mathrm{mm}$. There is no such band structure present in the M_{11} image, which is the image

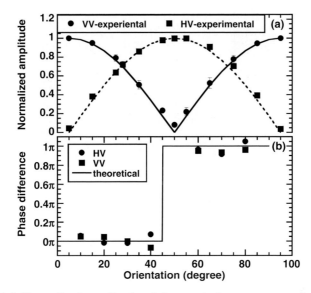

Fig. 10.3. (a) Normalized amplitude of the vertical components of the measured Jones vectors of a quarter-wave plate versus the orientation of the fast axis. HV is for the horizontally polarized incident light and VV is for the vertically polarized incident light. The lines represent the expected theoretical values. (b) Phase differences between the vertical and the horizontal components of the Jones vectors of the same quarter-wave plate. The standard deviations are smaller than the symbols

based on the intensity of the back-scattered light. In other words, the M_{11} image is free of the effect of polarization. We believe that the band structure is generated by the birefringence of the collagen fibers in the porcine tendon. The band structure distributes quite uniformly in the measured region; therefore, the birefringence is also uniform in the measured area.

Although all the polarization properties of a sample are contained in the Mueller matrix implicitly, explicit polarization parameters can be extracted from the matrix. The nondepolarizing Mueller matrix $\mathbf{M}$ can be decomposed by polar decomposition [19, 26]:

$$\mathbf{M} = \mathbf{M_P}\mathbf{M_R}, \qquad (10.27)$$

where $\mathbf{M_P}$ and $\mathbf{M_R}$ are the Mueller matrices of a diattenuator and an elliptical retarder, respectively. In biological tissues, it is reasonable to believe that the orientations of the diattenuator and retarder are the same. In this case, $\mathbf{M}$ is homogenous in the polarization sense [26] and the order of $\mathbf{M_P}$ and $\mathbf{M_R}$ in (10.27) is reversible. Only linear birefringence is considered in $\mathbf{M}$ and circular birefringence is not taken into account. The magnitude of birefringence and the diattenuation are related to the density and the properties of collagen fibers in the sample, whereas the orientation of the birefringence indicates the orientation of the collagen fibers.

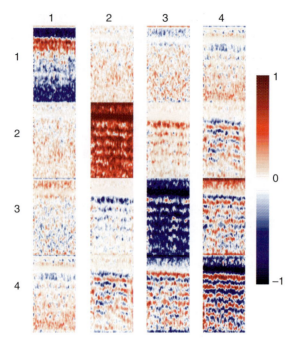

Fig. 10.4. 2D Mueller-matrix images of a piece of porcine tendon. Each image except M_{11} is pixel-wise normalized with the M_{11} element and shares the same color table. The size of each image is $0.5\,\text{mm} \times 1\,\text{mm}$

We extracted polarization information from a piece of porcine tendon set at various orientations. The rotation axis of the sample is collinear with the optical axis (direction of incidence). The measurements were made at five different orientations with an interval of $10°$. The M_{42} and M_{43} elements in (10.27) for a Mueller matrix that contains linear birefringence and linear or circular diattenuation can be expressed as

$$M_{42} = P(P_1, P_2)\sin(2\theta)\sin(\delta),$$
$$M_{43} = -P(P_1, P_2)\cos(2\theta)\sin(\delta), \tag{10.28}$$

where P is a function of the principal coefficients of the amplitude transmission, P_1 and P_2, for the two orthogonal polarization eigenstates of $\mathbf{M_P}$, θ and δ are the orientation of the fast axis and the phase retardation of the retarder, respectively.

To increase the signal-to-noise ratio, every 20 adjacent A scans of M_{42} and M_{43} in the calculated 2D images were laterally averaged and fitted for a physical depth of $0.4\,\text{mm}$ (optical depth divided by the refractive index, which is assumed to be 1.4) from the surface. The calculated birefringence from the fitted data is $(4.2 \pm 0.3) \times 10^{-3}$, which is comparable to the previously reported value of $(3.7 \pm 0.4) \times 10^{-3}$ for bovine tendon [7]. The calculated orientations of the fast axis are $(0 \pm 4)°$, $(9 \pm 2.9)°$, $(20.9 \pm 1.9)°$, $(30 \pm 2.8)°$ and $(38 \pm 4.3)°$ after subtracting an offset angle. The small angular offset is due to the discrepancy between the actual and the observed fiber orientations. The results are very good considering that the tendon was slightly deformed when it was mounted in the cuvette and the rotation axis for the sample may not have been exactly collinear with the optical axis. The diattenuation, defined as

$$D = \left(P_1^2 - P_2^2\right)\Big/\left(P_1^2 + P_2^2\right) = \sqrt{M_{12}^2 + M_{13}^2 + M_{14}^2}\Big/M_{11},$$

was averaged over all the orientations and linearly fitted over a depth of $0.3\,\text{mm}$. The fitted D versus the round-trip physical path length increases with a slope of $0.26\,\text{mm}^{-1}$ and reaches 0.075 ± 0.024 at the depth of $0.3\,\text{mm}$ after subtracting an offset at the surface.

10.6 Other Implementations

Optical fiber based polarization-sensitive OCT has been implemented [27]. Fiber-based systems possess advantages in alignment and handling. However, the fibers distort the polarization states of the optical beams, which require calibration. As the polarization distortion varies with time in imaging, dynamic calibration is required [28].

The serial implementation presented in Sect. 10.2 has also been extended to spectral interferometer based OCT (Fig. 10.5) [29]. The mode-locked Ti:sapphire laser produces a 150-fs light pulse with a central wavelength of 775 nm and a bandwidth of ± 13 nm. The back-reflected probe and reference beams are combined and then measured by a spectrometer consisting of a grating and a Fourier-transform lens. The spectral interferometric fringes are projected onto a CCD camera. It can be shown that the inverse Fourier transform of the spectral fringes yield the axial structure of the sample. Therefore,

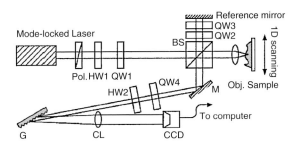

Fig. 10.5. Schematic of a polarization-sensitive spectral interferometric OCT system. BS, beam splitter; M, mirror; G, grating; CL, cylindrical lens; Obj., objective; Pol., polarizer; HW1, HW2, half-wave plates; QW1–QW4, quarter-wave plates; 1D, one-dimensional. (Permission of reprint of this figure was obtained from Optical Society of America.)

the advantage of the spectral interferometer based OCT over the conventional axial scanning OCT is that it acquires one-dimensional images with a single shot measurement. The principle of constructing a Mueller matrix using this system is identical to that presented in Sect. 10.3.

10.7 Summary

In this chapter, we have presented various methods for the implementation of Mueller or Jones matrix measurements based on OCT. The Jones matrix can be used because of the coherent detection scheme in OCT. However, the Mueller matrix is often used for the final presentation because the upper-left element M_{11} of the matrix provides polarization independent information related to backscattering only, i.e., it separates the contributions from polarization and backscattering. If a Jones or Mueller matrix is measured, the quantification of polarization is complete; otherwise, it is partial. This technology is expected to find immediate application in burn imaging and glaucoma detection.

11

Biomedical Diagnostics and Imaging

11.1 Introduction

This chapter is dedicated to consideration of the various polarization-sensitive techniques that can be applied to the imaging and functional diagnosis of biological tissue. The polarization discrimination of a probe light that has been scattered by, or transmitted through, a probed tissue is the common basis for all of the considered methods. This principle is realized in a simple fashion in the polarization imaging of structural inhomogeneities that are hidden in the probed tissue volume. Typically, the probed tissue site is irradiated by linearly polarized light, and the tissue structure is imaged with scattered linearly polarized light that has an orthogonal polarization direction. Despite the obvious simplicity of this approach, it appears to be a powerful tool for functional diagnostics and for the imaging of diseased fragments of skin tissue and, in some cases, subcutaneous tissues. This feature, in combination with relatively inexpensive commercially available instrumentation and nonsophisticated image processing algorithms (e.g., normalization and subtraction of images), offers an opportunity for the successful implementation of these methods in clinical practice. The addition of spectral selection of scattered radiation provides a novel quality to the polarization-sensitive methods and significantly improves their diagnostic potential.

Polarization-based technologies can also be applied to other types of functional diagnostics, such as concentration analysis of various tissue components (e.g., glucose sensing). Some of these methods are also discussed in this chapter.

11.2 Imaging through Scattering Media and Tissues with Use of Polarized Light

In recent years, the principle of polarization discrimination of multiply scattered light has been fruitfully applied by many research groups to morphological analysis and to visualization of subsurface layers in strongly scattering tissues [1–8]. One of the most popular approaches to polarization imaging in heterogeneous tissues is based on using linearly polarized light to irradiate the object (the chosen area of the tissue surface) and to reject the scattered light with the same polarization state (co-polarized radiation) by the imaging system. Typically, such polarization discrimination is achieved simply by placing a polarizer between the imaging lens and the object. The optical axis of the polarizer is oriented perpendicularly to the polarization plane of the incident light. Thus, only the cross-polarized component of the scattered light contributes to the formation of the object image. Despite its simplicity, this technique has been demonstrated to be an adequately effective tool for functional diagnostics and for the imaging of subcutaneous tissue layers. Moreover, the separate imaging of an object with co-polarized and cross-polarized light permits separation of the structural features of the shallow tissue layers (such as skin wrinkles, the papillary net, etc.), and the deep layers (such as the capillaries in derma). The elegant simplicity of this approach has stimulated its widespread application in both laboratory and clinical medical diagnostics.

A typical scheme of instrumentation for polarization imaging using the above-discussed approach is presented in Fig. 11.1.

In the imaging system developed by Demos et al. [5], a dye laser with Nd:YAG laser pumping is used as the illumination source. The probe beam diameter is 10 cm, and the average intensity is approximately equal to $5\,\mathrm{mW\,cm^{-2}}$. A cooled CCD camera with a 50 mm focal length lens is used to

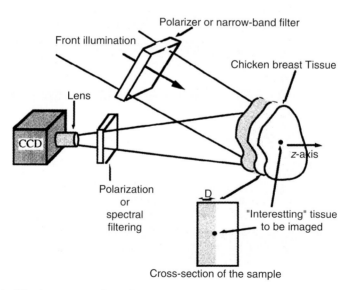

Fig. 11.1. The instrument for selective polarization or spectral imaging of subsurface tissue layers [5]

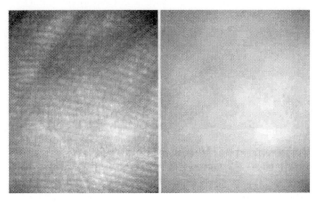

Fig. 11.2. The co-polarized and cross-polarized images of the human palm [5]

detect retroreflected light and to capture the image. A first polarizer, placed after the beam expander, is used to ensure illumination with linearly polarized light. A second polarizer is positioned in front of the CCD camera with its polarization orientation perpendicular or parallel to that of the illumination.

The efficiency of selective polarization imaging is illustrated in Fig. 11.2, where the human palm images obtained from the parallel orientation of the second polarizer (left panel, surface image, papillary pattern is clearly seen) and from the perpendicular orientation (subsurface image) are presented. A 580-nm polarized laser light illumination is used in this case.

Another example is the cross-polarized images of a tissue phantom, such as chicken breast tissue containing a fat lesion 3.5 mm in size beneath the surface. Panels (a) and (b) show images obtained with 580 nm and 630 nm illumination, respectively. Both images are then normalized, and the 580 nm image subtracted from the 630 nm image. The result is shown in panel (c). For comparison, the 3.5 mm outer layer of the chicken breast tissue is removed and the image of the fat tissue lesion is obtained separately (panel (d)) (Fig. 11.3).

A similar camera system, but one that uses an incoherent white light source such as a xenon lamp, is described in [9], where results are presented of a pilot clinical study of various skin pathologies using polarized light. The image processing algorithm that is used is based on evaluation of the degree of polarization $(I_{\mathrm{par}} - I_{\mathrm{per}}) / (I_{\mathrm{par}} + I_{\mathrm{per}})$, which is then considered the imaging parameter. The polarization images of pigmented skin sites (freckles, tattoos, pigmented nevi) and unpigmented skin sites (nonpigmented intradermal nevi, neurofibromas, actinic keratosis, malignant basal cell carcinomas, squamous cell carcinomas, vascular abnormalities (venous lakes), and burn scars) are analyzed to find the differences caused by various skin pathologies. Also, the point-spread function of the backscattered polarized light is analyzed for images of a shadow cast from a razor blade onto a forearm skin site. This function describes the behavior of the degree of polarization at the imaging parameter near the shadow edge. It was discovered that near the shadow edge, the degree

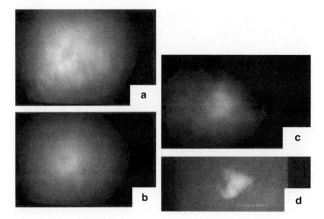

Fig. 11.3. Polarization images of a fat tissue lesion hidden in chicken breast tissue [5]

of polarization approximately doubles in value because no I_{per} photons are superficially scattered into the shadow-edge pixels by the shadow region while I_{par} photons are directly backscattered from the superficial layer of these pixels. This result suggests that the point-spread function in skin for cross-talk between pixels of the polarization image has a half-width-half-max of about $390\,\mu\mathrm{m}$.

Comparative analysis of polarization images of normal and diseased human skin has shown the ability of the above approach to emphasize image contrast based on light scattering in the superficial layers of the skin. The polarization images can visualize disruption of the normal texture of the papillary and upper reticular layers caused by skin pathology. Polarization imaging has proven itself an adequately effective tool for identifying skin cancer margins and for guiding surgical excision of skin cancer. Various modalities of polarization imaging are also considered in [10]. In particular, polarization-difference imaging techniques are demonstrated to improve the detectability of target features that are embedded in scattering media. The improved detectability occurs with both passive imaging in moderately scattering media (<5 optical depths) and with active imaging in more highly scattering media. These improvements are relative to what is possible with equivalent polarization-blind, polarization-sum imaging under the same conditions. In this study, the point-spread functions for passive polarization-sum and polarization imaging in single-scattering media are studied analytically, and Monte Carlo simulations are used to study the point-spread functions in single- and moderately multiple-scattering media. The obtained results indicate that the polarization-difference point-spread function can be significantly narrower than the corresponding polarization-sum point-spread function, implying that better images of target features with high-spatial-frequency information can be obtained by using differential polarimetry in scattering media. Although the analysis is performed using passive imaging at moderate optical depths, the results have potential implications for mitigating the effects of multiple scattering in experiments performed in more highly scattering media with active imaging methods.

11.3 Transillumination Polarization Techniques

The polarization discrimination of light that is passed through multiply scattering media also provides high efficiency in the detection and imaging of inhomogeneities embedded in the probed medium. In particular, the transillumination polarization diaphanography of a heterogeneous scattering object is considered in [11]. This technique makes it possible to locate and to image absorbing objects hidden in a strongly scattering medium. The method uses modulation of the polarization azimuth of a linearly polarized laser beam and lock-in detection of the light transmitted through the object. The instrumentation is schematically depicted in Fig. 11.4.

The scattering sample is probed by an Ar-ion laser beam. The orientation of the polarization plane of the probe beam is modulated by a Pockels cell as follows: during the first half-period of the modulating signal, it is not changed and during the second half-period, it is rotated by 90°. Transmitted (depolarized) and forward-scattered (polarized) components of the probe light are collimated by two diaphragms and divided in two channels by a polarizing beam-splitter. Movable black rods are used as phantom absorbing objects to estimate the capability of this imaging technique.

The signal processing algorithm is based on discrimination of the scattered component of the detected light (S) that is considered totally depolarized in comparison with the polarized transmitted component (U).

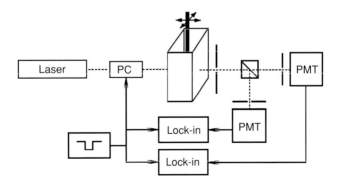

Fig. 11.4. Schematic of the setup for transillumination polarization diaphanography of scattering objects with absorbing inclusions. PC, pockels cell; PMT, photomultiplier tube [11].

Output signals of lock-in amplifiers are:

$$L_A = U_1 + \frac{1}{2}S_1 - \frac{1}{2}S_2,$$
$$L_B = U_1 - \frac{1}{2}S_1 + \frac{1}{2}S_2. \tag{11.1}$$

Thus, summation of the output signals gives only the transmitted (nonscattered) component:

$$L_A + L_B = U_1 + U_2, \tag{11.2}$$

which can be used to form the shadow image of the probed absorbing inhomogeneity, if the scanning procedure is provided.

The dependencies of the PMT output signal (before the lock-in processing) on the absorber position are shown in Fig. 11.5. This case corresponds to conventional diaphanography.

However, the lock-in processing of the detected signals in the two polarization channels provides much better quality in the shadow image than in the image obtained for a nonscattering background medium (Fig. 11.6). It was found that polarization-modulation diaphanography allows one to generate shadow images of a hidden object in a highly dense medium which is characterized by up to 29 average scattering processes [11].

Comparisons of polarization imaging with applications of partially depolarized transmitted light and conventional transillumination visualization are carried out in [12]. In this case, the absorbing inhomogeneity, such as the blackened plate within a scattering slab, is probed by a linearly polarized laser beam (Fig. 11.7) and the shadow images are reconstructed from the profiles of the intensity and the degree of polarization P of the transmitted light (Fig. 11.8). Note that the dependencies of the degree of linear polarization on the edge position exhibit an increase in P in the vicinity of the edge. The explanation of this peculiarity is similar to that proposed by Jacques et al. [9]

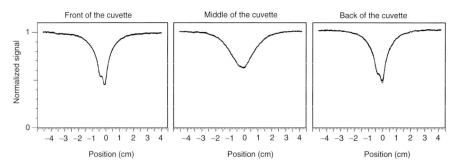

Fig. 11.5. Cross-sections of the shadow images of the absorbing rod embedded in the strongly scattered medium (10% Intralipid solution) [11]. The images were obtained for different positions of the rod (see notations above each panel) without lock-in processing

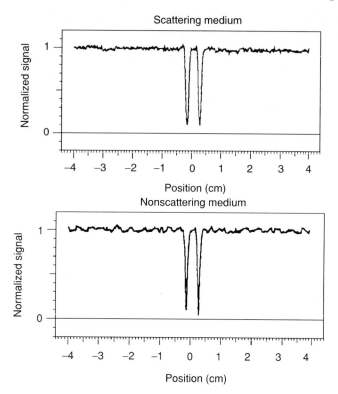

Fig. 11.6. Shadowgrams of a pair of absorbing rods with a 1 mm diameter. The scattering medium is Intralipid – 10%; the nonscattering medium is water. The cuvette thickness is 30 mm [11]

for the polarization-sensitive detection of backscattered light (see Sect. 11.2). Analysis of the quality of the shadow images of the object obtained with the use of the conventional transillumination imaging technique and polarization visualization allows us to conclude that the latter approach provides a better quality of shadowgrams in cases involving moderately scattering media.

11.4 Potentialities and Restrictions of Polarization Imaging with Backscattered Light

In this subsection, the effects of the optical properties of the scattering media and of the scattering geometry on the quality of the polarization images of a hidden inhomogeneity are discussed [13]. Polarization imaging is produced by employing the polarization characteristics (the normalized intensities of the co- and cross-polarized components and the degree of polarization) of the backscattered radiation with the initial linear polarization as the visualization

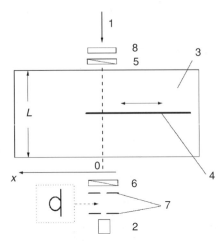

Fig. 11.7. Schematic sketch of the experimental setup for transillumination polarization imaging. (1) linearly polarized beam of He–Ne laser; (2) detector; (3) flat glass cuvette filled by diluted milk; (4) absorbing half-plane; (5, 6) polarizers; (7) collimating diaphragms or light-collecting optical fiber; (8) chopper

parameters. The model represents the scattering medium as a layer containing a plate-like absorbing heterogeneity. This plate is parallel to the boundaries of the medium. The depth of the inhomogeneity position within the scattering layer, at which the polarization imaging retains its efficiency, is estimated.

Our evaluation of the quality of the polarization images is based primarily on the above-described presentation of multiply scattered depolarized light (see Chap. 7) as a superposition of partial contributions characterized by different values of the optical paths s in the scattering medium. The statistical properties of the ensemble of partial contributions are described by the probability density function of the optical paths $\rho(s)$, whereas the statistical moments of the scattered light are represented by the integral transforms of $\rho(s)$ with properly chosen kernels. Following from the analysis presented in Chap. 7, the degree of polarization of multiply scattered radiation with initial linear polarization can be approximately represented in the form of the Laplace transform of $\rho(s)$:

$$P_L^r = \frac{I_\parallel - I_\perp}{I_\parallel + I_\perp} \approx \frac{3}{2} \int_0^\infty \exp\left(-\frac{s}{\xi_L}\right) \rho(s)\, \mathrm{d}s, \qquad (11.3)$$

where $I_\parallel$ and $I_\perp$ are, respectively, the intensities of the co- and cross-polarized components of the scattered light. The parameter ξ_L is the depolarization length for linearly polarized light.

By considering the polarization visualization of the absorbing macroheterogeneity, with the degree of polarization of backscattered light as the visualization parameter, we can define the contrast of the polarization

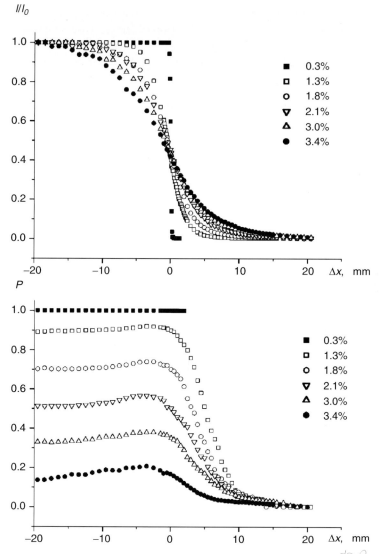

Fig. 11.8. Experimental dependencies of the normalized intensity (*left panel*) and degree of linear polarization (*right panel*) of transmitted light on the half-plane edge position at different concentrations of the background scattering medium (diluted-milk)

image as:

$$V = \frac{P_L^{r,in} - P_L^{r,back}}{P_L^{r,in} + P_L^{r,back}}, \qquad (11.4)$$

where $P_{\mathrm{L}}^{\mathrm{r,in}}$ is the degree of residual linear polarization of the backscattered light detected in the region of the localization of the heterogeneity and $P_{\mathrm{L}}^{\mathrm{r,back}}$ is the analogous quantity determined far from the region of localization. By considering the above-mentioned model of an inhomogeneous scattering medium (Fig. 11.9), we can estimate the contrast V of the reconstructed polarization images as a function of the scattering layer thickness H, the depth of inhomogeneity position $\tilde{H}$, the transport mean free path l^* of the scattering medium, the scattering anisotropy g, and the depolarization length ξ_{L}. The probability density function of the optical paths $\rho(s)$, depending on the detection conditions (in the region of inhomogeneity localization or far from it), can be obtained with a Monte Carlo (MC) simulation. It should be noted that modern analytic methods based on various approximations of the radiative transfer equation (RTE) do not allow one to obtain an adequate description of the path length density $\rho(s)$ with a short-path limit (i.e., for small values of s), whereas the portion of short-path photons in the detected scattered light is significant under the backscattering condition. The MC code used to simulate path-length distributions for the model discussed is basically similar to that described in [14].

In the course of the simulation, the values of $\rho(s)$ were obtained, under the condition of illumination of a scattering layer with an absorbing rear boundary, by using a broad collimated beam and detecting the backscattered radiation. The Laplace transform values are calculated from the obtained values of $\rho(s)$ for the specified depolarization length ξ_{L}, and the contrast of the polarization images is then evaluated for different parameters of the model.

The theoretical dependencies of V on the normalized depth of inhomogeneity position H/l^*, for different values of the dimensionless thickness of the scattering medium $\tilde{H}/l^*$, are plotted in Fig. 11.10. A scattering medium with an expressed scattering anisotropy ($g = 0.85$) is assumed. The depolarization length is set equal to $1.5l^*$. The values of the polarization image contrast are also presented for different values of the transport mean free path under

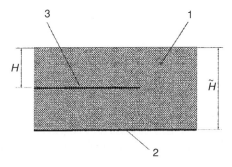

Fig. 11.9. Model used for analyzing the quality of the polarization images of an object embedded in a scattering medium. (1) scattering medium layer; (2) absorbing rear boundary of the layer; (3) absorbing half-plane

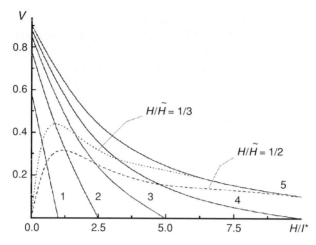

Fig. 11.10. Theoretical dependencies of the contrast in the test object image obtained with the degree of polarization as a visualization parameter on the normalized depth of the absorber position. $H/l^* = (1)1, (2)2.5, (3)5, (4)10$, and (5) 20. D*ashed* and *dotted* curves represent the dependencies of the contrast on H/l^* for the fixed values of $H/\tilde{H}$

condition $H/\tilde{H} = \mathrm{const}$. This corresponds to variations in the concentration of scattering particles with a fixed scattering geometry. The analysis of the obtained dependencies allows us to conclude that the maximum value of contrast in the polarization images, obtained with the use of P_L^r as the visualization parameter, is reached at the depth of an inhomogeneity position on the order of $(0.3–1.5)l^*$. In particular, for $H/\tilde{H} = 1/3$, the maximum value of the contrast (≈ 0.45) corresponds to $H/l^* \approx 0.9$. An increase in the depth of the heterogeneity position results in a shift of the contrast maximum to the region of smaller values of the parameter $\tilde{H}/l^*$, that corresponds to a decrease in the concentration of the scattering particles.

To compare the efficiency of the various polarization imaging modalities, the following experimental setup is used (Fig. 11.11). The total normalized intensity (the intensity of the co-polarized and cross-polarized components) or the degree of residual linear polarization of the backscattered light are considered as the visualization parameters. The scattering medium consists of a water–milk emulsion at low volume concentrations of milk in a rectangular glass cell ($180 \times 260 \times 260 \,\mathrm{mm^3}$). The side walls, as well as the rear wall, of the cell are blackened. The absorbing object – a rectangular plate with blackened rough surfaces – is positioned in the central part of the cell at different distances H (from 10 to 40 mm) from the transparent front wall.

Light from a nonmonochromatic source (a halogen lamp) was used as the probe radiation. The probe light was linearly polarized perpendicular to the plane of incidence. To eliminate the negative effect of specular reflection from the front wall of the cell, the illuminating beam was directed at an angle of 30°

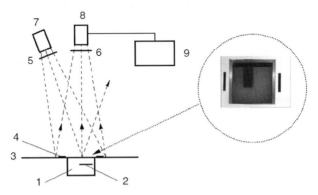

Fig. 11.11. Schematic diagram of the polarization imaging experiment. (1) A cell with a scattering medium, (2) absorbing plate, (3) white marker, (4) black markers, (5) polarizer, (6) analyzer, (7) source of nonmonochromatic light (a halogen lamp), (8) CCD camera, and (9) PC

relative to the normal to the wall. The capture of the backscattered radiation is performed with a color CCD camera (Panasonic NV-RX70EN). A manually rotated polarizer is placed between the camera and the object to provide the selection of co-polarized and cross-polarized components of backscattered light. To preclude the influence of the automatic built-in system that controls image brightness, normalization of the dynamic range of the image being recorded is realized at a fixed intensity of the source by using black and white markers in the visual field (Fig. 11.11). The capture of images is done using a Miro DC20 frame grabber (product of MiroVideo, Germany).

The color 8-bit images of the object are captured with 647×485 resolution for each of the three chromatic coordinates (R, G, and B) with the use of co-polarized and cross-polarized backscattered light. The brightness distributions for each of the R, G, B image components along an arbitrarily chosen line of the image (Fig. 11.12) are applied to reconstruct the images of the absorbing heterogeneity with different visualization parameters (the backscattered light intensity, the intensities of the co- and cross-polarized components, and the degree of polarization of the backscattered radiation). The values of the contrast and the edge sharpness are determined for each obtained image depending on the experimental conditions. In this case, the contrast is defined as:

$$V = \frac{T_{\max} - T_{\min}}{T_{\max} + T_{\min}}, \tag{11.5}$$

where $T_{\max}$ is the average value of the visualization parameter over a group consisting of 10 pixels outside the heterogeneity zone, and $T_{\min}$ is the corresponding value determined in the heterogeneity zone.

The image sharpness is determined as the quantity $(\Delta x)^{-1}$, which is the inverse of the edge width for the inhomogeneity image. The edge width is

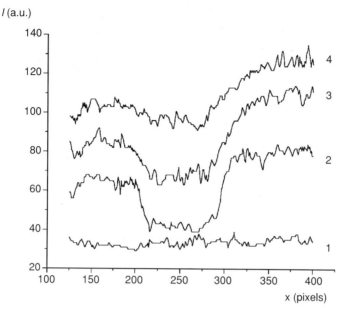

Fig. 11.12. Distributions of backscattered radiation intensity along an arbitrarily chosen line of the image for different volume concentrations of milk emulsion: (1) 0%, (2) 0.66%, (3) 1.96%;, and (4) 5.51% (R-component of the color image)

estimated by the number of pixels, for which the value of the visualization parameter changed from 0.1 to 0.9 of its maximum value.

Figure 11.13 demonstrates the distributions of the backscattered intensity (R-component), along an arbitrarily chosen line of the object image, for the different milk concentrations. In the absence of a scattering medium and, hence, the backscattered radiation, the image contrast is equal to zero (Fig. 11.13a, curve 1). Increase in the milk concentration results in a sharp increase in the image contrast, which was estimated for the different characteristics of backscattered radiation that are used as the visualization parameters, up to the maximum values with the subsequent monotonic decrease (Fig. 11.13a). A similar behavior is also observed for the G- and B-components of the object image.

An obvious explanation of the contrast decrease in regions of high scatter concentration is related to the transition from the low-step scattering regime, in the case of $H \cong l^*$, to the strong scattering regime when $H \gg l^*$. This transition occurs in the region of scattering volume between the absorbing heterogeneity and the front wall of the cell.

Comparison of the image contrast values obtained with the different visualization parameters allows us to conclude that the maximal value of V (≈ 0.48) is obtained for the degree of residual polarization $P_{\mathrm{L}}^{\mathrm{r}}$. In this case, the maximum of the contrast is shifted to the region of high scatter concentration

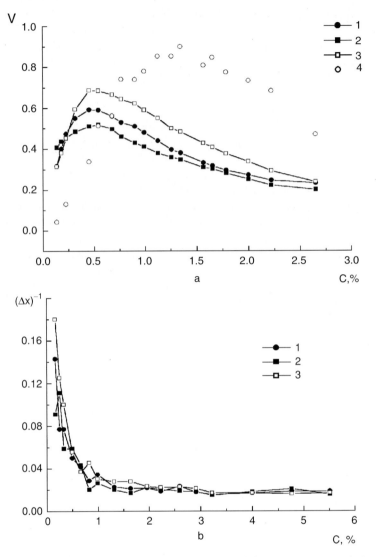

Fig. 11.13. Dependencies of (**a**) the contrast and (**b**) the sharpness of the image edge on the volume concentration of milk emulsion when the normalized intensity of the (1) unpolarized light, (2) co-polarized components and (3) crosspolarized components, and (4) the degree of polarization of backscattered radiation are used as the visualization parameter

in comparison with the other visualization parameters (I_{tot}, I_{II}, and $I_{\perp}$, Fig. 11.13a). As mentioned above, the MC simulation gives the maximal contrast in images reconstructed with $P_{\mathrm{L}}^{\mathrm{r}}$ for $H \sim (0.3\text{--}1.5)l^*$ (in the dependence on the scattering medium parameters). In the case discussed here, the maximal contrast value is obtained experimentally for H at a milk concentration of c equal to 1.5%. By using the experimentally established dependence of the scattering coefficient for water–milk emulsion on the milk concentration: $\mu_s \approx 0.082\ mm^{-1} \times c\ (\%)$ [12], and taking into account the relation $l^* = [\mu_s\,(1-g)]^{-1}$, we find that the maximal contrast is obtained at $H \approx 0.4l^*$, which agrees satisfactorily with the results of the MC simulation. The maximal contrast of the polarization image (≈ 0.48) is also in fair agreement with the MC results (contrast on the order of 0.55–0.65 for small values of the ratio $H/\tilde{H}$). Increasing depth in the inhomogeneity position H causes a shift in the contrast maximum to a region of lower concentrations, which was also revealed in the MC simulation (decreasing scatter concentration corresponded to a decrease in the parameter $\tilde{H}/l^*$).

Comparison of the experimental data and the MC results allows us to conclude that maximal contrast in the polarization image is obtained at the depth of an inhomogeneity position on the order of $(0.25–0.6)\xi_L$ (depending on the degree of residual polarization in the backscattered background component detected outside the region of the inhomogeneity localization). In particular, this conclusion agrees with data on polarization visualization of skin, which points to the efficiency of polarization visualization for epidermis and upper layers of papillary derma (100–150 μm) [16]

The dependencies presented in Fig. 11.13a are obtained with the absorbing object placed at a distance $L = 20\,mm$ from the entrance window of the cell. Similar behavior is observed with a decreasing wavelength of the probe radiation, when instead of a R-component of the object image, its G- and B-components are processed.

Figure 11.13b illustrates the effect of scatter concentration on the sharpness of the polarization image. The rapid decrease of $(\Delta x)^{-1}$ in the region of low-step scattering with increasing scatter concentration and its saturation with strong scattering in the probed medium should be noted. In addition, the value of the image sharpness is practically independent of the visualization parameter used for image reconstruction.

Thus, polarization visualization can provide certain advantages in the case of intermediate scattering regimes, when the modal value of the photon path length in the scattering medium is comparable with the depolarization length. Therefore, polarization images should be the most sensitive to morphological changes in tissue structure, especially in subcutaneous tissue layers. The optimal correlation between the depth of light penetration into the tissue volume and the depolarization length can be obtained by the selection of the appropriate wavelength for probe radiation.

11.5 Polarized Reflectance Spectroscopy of Tissues

One of the promising approaches to early cancer diagnosis is based on analysis of a single scattered component of light perturbed by tissue structure. The wavelength dependence on the intensity of the radiation, elastically scattered by the tissue structure, appears sensitive to changes in tissue morphology that are typical of pre-cancerous lesions. In particular, it has been established that specific features of malignant cells, such as increased nuclear size, increased nuclear/cytoplasmic ratio, pleomorphism, etc. [16], are markedly manifested in the elastic light scattering spectra of probed tissue [17]. A specific fine periodic structure in the wavelength of backscattered light has been observed for mucosal tissue [18]. This oscillatory component of light scattering spectra is attributable to a single scattering from surface epithelial cell nuclei and can be interpreted within the framework of Mie theory. Analysis of the amplitude and frequency of the fine structure allows one to estimate the density and size distributions of these nuclei. It should be noted, however, that the extraction of a single scattered component from the masking multiple scattering background is a problem. Also, absorption of stroma related to the hemoglobin distorts the single scattering spectrum of the epithelial cells. Both of these factors should be carefully taken into account when interpreting the measured spectral dependencies of backscattered light.

The negative effects of a diffuse background and of hemoglobin absorption can be significantly reduced by the application of a polarization discrimination technique in the form of illumination of the probed tissue with linearly polarized light followed by separate detection of the elastic scattered light at parallel and perpendicular polarization states (i.e., the co-polarized and cross-polarized components of the backscattered light) [19,20]. This approach, called polarized elastic light scattering spectroscopy, or polarized reflectance spectroscopy (PRS), will potentially provide a quantitative estimate not only of the size distributions of cell nuclei but also of the relative refractive index of the nucleus. These potentialities, which have been demonstrated in a series of experimental works with tissue phantoms and in vivo epithelial tissues [17–20], allow one to classify the PRS technique as a new step in the development of noninvasive optical devices for real-time diagnostics of tissue morphology and, consequently, for improved early detection of pre-cancers in vivo. An important step in the further development of the PRS method will be the design of portable and flexible instrumentation applicable to in situ tissue diagnostics. In particular, fiber optic probes are expected to "bridge the gap between benchtop studies and clinical applications of polarized reflectance spectroscopy" [21].

11.6 Glucose Sensing

Measurement of glucose concentration within the human body is of widespread interest in the health care of diabetics [22–41]. Automatic noninvasive monitoring of glucose concentration is also important for controlling growing cell cultures in tissue engineering, primarily for the production of implantable in vitro tissues and organs [29, 30]. A wide range of optical technologies have been designed in attempts to develop more robust noninvasive methods for glucose sensing. The methods include infrared [28, 29, 33, 36, 37], Raman [29, 34], and fluorescent [29, 34, 35] spectroscopies, as well as polarimetric [22–24, 27, 29, 30, 37–39], heterodyning [26, 31, 32], and optical coherence tomography (OCT) [40, 41] techniques.

The polarimetric quantification of glucose is based on the phenomenon of optical rotatory dispersion (ORD), whereby a chiral molecule in an aqueous solution will rotate the plane of linearly polarized light passing through the solution ch11:bib [22–24, 27, 29, 30, 34, 37, 38]. The angle of rotation depends linearly on the concentration of the chiral species, the pathlength through the sample, and a constant for the molecule that is called the specific rotation. The net rotation in degrees is expressed as [29, 37]:

$$\phi = \alpha_\lambda LC, \tag{11.6}$$

where α_λ is the specific rotation for the species in $\deg \mathrm{dm}^{-1} \mathrm{g}^{-1} \mathrm{l}$ at wavelength λ, L is the pathlength in dm, and C is the concentration in $\mathrm{g\,l}^{-1}$.

The specific rotation for any wavelength can be determined based on two-wavelength measurements in the spectral range free of absorption bands of a given chiral molecule using the expression [38]:

$$\alpha_\lambda = \frac{k_0}{\lambda^2 - \lambda_0^2}, \tag{11.7}$$

where the constants k_0 and λ_0 are computed by determining the specific rotation at two different wavelengths. The specific rotation of a particular chiral molecule depends also on the pH and temperature of the medium. At a fixed pH and temperature, this equation allows for separate evaluation of the contribution of the particular analyte (glucose) on the background of the other analytes if multispectral measurements and the corresponding regression model are provided [38].

Glucose in the body is dextrorotatory (rotates light in the right-handed direction) and has a specific rotation from +75.0 to +27.5 $\deg \mathrm{dm}^{-1} \mathrm{g}^{-1} \mathrm{l}$ with a wavelength change from 500 to 800 nm [37]. At the sodium D-line of 589 nm, it is equal to +52.6 $\deg \mathrm{dm}^{-1} \mathrm{g}^{-1} \mathrm{l}$, and with the He:Ne laser at 633 nm often used in polarimeters, $\alpha_\lambda = +45.6$ $\deg \mathrm{dm}^{-1} \mathrm{g}^{-1} \mathrm{l}$. For example, for the measurement on a wavelength of 633 nm at physiological concentrations (a normal blood glucose level of $1 \mathrm{g\,l}^{-1}$ and a pathlength of about 1 cm, the optical rotation due to glucose is 4.56 mdeg [29].

In turbid samples, both the rotation of the linearly polarized light and the level of polarization (linear and circular) preservation vary with the glucose concentration [42, 43]. Unfortunately, the optical rotation in typical physiological measurements, which equals approximately 10^{-3} deg, is about 40 times smaller than the current detection limit for turbid tissue. The chirality-induced increase in the preservation of polarization is also not visible in the background of the refractive index matching effect that is caused by glucose.

Therefore, using a polarization sensitive optical technique makes it difficult to measure in vivo glucose concentration in blood through the skin because of the strong light scattering which causes light depolarization. A tissue thickness of 4 mm is sufficient to prompt about 95% depolarization [29]. For this reason, the anterior chamber of the eye has been suggested as a sight well suited for polarimetric measurements [22–24, 29, 32, 34, 37–39], since scattering in the eye is generally very low compared to that in other tissues, and a high correlation exists between the glucose in the blood and in the aqueous humor. The age-dependent steady-state glucose concentration in the aqueous humor is about 70% of that of blood. A time-delay between blood and aqueous humor glucose concentrations caused by glucose diffusivity within the blood vessel walls and eye tissues is in the range of 20–30 min [23, 32, 44]. These values fit well the rate of glucose diffusion in connective tissues of corresponding thickness and temperature [45]. Figure 11.14 depicts an optical sensing scheme with an involved anterior chamber (the fluid-filled space directly below the cornea – the aqueous humor).

The high accuracy of anterior eye chamber measurements is also due to the low concentration of optically active aqueous proteins ($0.13\,\mathrm{g\,l^{-1}}$ [44]) within the aqueous humor, which is a result of diffusion filtering in the surrounding tissues. Figure 11.15 [38] illustrates that at a given pH and temperature, the contribution to polarization rotation made by the major aqueous chiral protein – albumin and molecules of ascorbic acid (evaluated at their average physiological levels) – is very low on the background of rotation induced by glucose. Moreover, these two components are contrarotatory and thus will

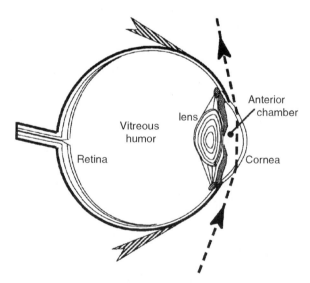

Fig. 11.14. Optical sensing of glucose in the anterior chamber of the eye. Light passing through the anterior chamber interacts with the aqueous humor. A commonly proposed beam path is shown [37]

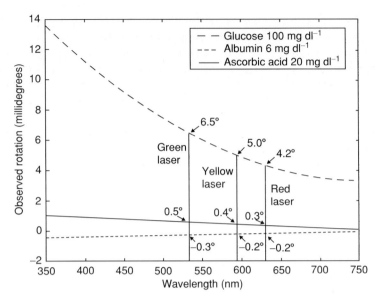

Fig. 11.15. Observed optical rotations for physiological concentrations of aqueous humor analytes, glucose, albumin, and ascorbic acid for a 1 cm pathlength [38]

partially cancel each other out. As also follows from Fig. 11.15 and (11.7), a multiwavelength measuring system will take into account the compensation of any confounding effects from other chiral analytes.

A number of techniques for acquiring measurements with the required high degree of accuracy exist and generally fall into two categories: those which utilize crossed polarizers to measure rotation via amplitude changes, and those which measure the relative phase shift of modulated polarized light passing through the sample [37]. Figure 11.16 illustrates each of these approaches schematically. It presents the optical system structures for the amplitude and phase techniques as well as the resulting polarization and intensity signals that contain information about the optical rotation.

First, an open-loop, amplitude based polarimeter is suggested for optical glucose measurements in the eye [22, 23]. In order to increase the signal-to-noise ratio of the polarization measurements, a phase polarimeter is proposed and tested in in vitro studies [24]. A further increase in sensitivity, that can provide measurements at the level of a few millidegrees, is demonstrated in in vitro studies for a close-loop feedback controlled system, where the stability is improved with digital feedback control usage and tested on glucose measurements in aqueous cell culture media [29, 30, 37].

One of the prototypes of the digital close-loop controlled polarimeter for in vitro and in vivo measurements is presented in Fig. 11.17 [38]. A diode laser emitting 3.5 mW of power at 635 nm is used as the light source. The laser beam is linearly polarized at a high degree of 100,000:1 by the initial Glan-Thompson

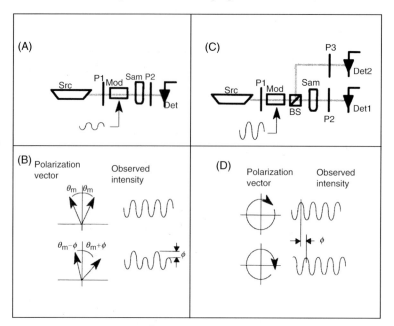

Fig. 11.16. Amplitude-and phase-based polarimetric measurements [37]. (**A**) In the amplitude approach, light from a monochromatic source (Src) is passed through a linear polarizer (P1), a polarization modulator (Mod), a sample (Sam), and a second linear polarizer perpendicular to the first (P2) before being recorded by a detector (Det). (**B**) The resulting polarization vector and observed intensity are symmetric when no optically active sample is present and asymmetric if the sample is optically active with net rotation ϕ. (**C**) In the phase approach, the polarization modulated sample and reference beams are split by a beam splitter (BS), passed through crossed linear polarizers (P2, P3), and recorded by separate detectors (Det1, Det2). (**D**) A rotation of polarization in the sample causes a phase shift between the intensity signals recorded by the two detectors

polarizer (P). Modulation of the polarization vector is then provided by the Faraday rotator (FR) driven by a sinusoidal function generator at a frequency of 1.09 kHz and a modulation depth of approximately $\pm 1°$. This modulated optical signal propagates through a test tube [eyecoupling device (ED)] that is designed with plane parallel windows and filled with saline for index matching. The ED surrounds the anesthetized rabbit's eye to allow propagation of light directly through the anterior chamber of the eye. A Faraday compensator (FC) is used to provide feedback compensation within the system; it nullifies any rotation due to the optically active sample. The Glan-Thompson polarizer, which is cross-polarized to the incident polarization, then serves as an analyzer (A) which transforms the modulation of the polarization vector into intensity modulation that is detected by a photodiode (D). A digital lock-in amplifier,

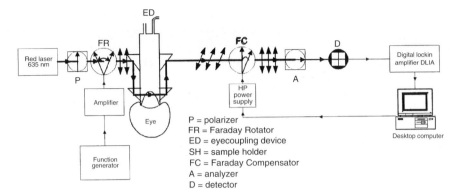

Fig. 11.17. Block diagram of the digital closed-loop controlled polarimeter, in which the sample holder is used for in vitro samples and the eye coupling device is used for in vivo studies [38]

computer and FC power supply are used to drive the FC and, therefore, to nullify the system.

The operation of the system is described by the following equation [38]:

$$I \propto E^2 = \left(\phi^2 + \frac{\theta_m^2}{2} \right) + 2\phi\theta_m \sin(\omega_m t) - \frac{\theta_m^2}{2} \cos(2\omega_m t), \qquad (11.8)$$

where θ_m is the depth of the Faraday modulation, ω_m is the modulation frequency, and ϕ is the rotation due to the optically active sample subtracted from any feedback rotation due to the compensation of the FC. It follows from (11.8) that without an optically active sample and with the DC term removed, the signal detected consists only of the double frequency ($2\omega_m$) term. When the optically active sample is presented, the signal detected becomes an asymmetric sinusoid, which contains both the fundamental (ω_m) and the double frequency components.

As mentioned above, potential problems with the polarimetric sensing of glucose in the eye include the presence of other optically active confounders in the aqueous humor (see Fig. 11.15). In addition, linear and circular birefringence of the cornea as well as eyeball motion artifacts may significantly affect glucose measurements [37]. Problems related to corneal and other analytes rotation may potentially be solved by using multispectral polarimeters. Dual-wavelength polarimetric systems, based on He–Ne lasers with wavelengths of 594 and 633 nm and on diode lasers with wavelengths of 670 and 830 nm, were recently designed [37]. To overcome corneal linear birefringence inclusion and to extract glucose specific rotation, appropriate optical elements [37], closed-loop multispectral probing [38], or full Jones or Mueller matrix measurements [34] could be used.

The major advantage of the optical heterodyne polarimeter described in [31] and [32] is its low sensitivity to light scattering due to antenna properties of optical heterodyning [46]. Such a polarimeter is rather simple and provides experimental sensitivity on the order of 10^{-4} deg for the optical rotation angle. A two-frequency Zeeman laser that provides two orthogonal linearly polarized states (P and S), with a temporal frequency difference of 2.6 MHz, and a Glan-Thompson analyzer at a fixed azimuth angle θ_s are the main optical elements of the system. Two optical frequencies are mixed on a photodetector and amplified; then, the amplitude of the heterodyne signal is measured by a digital voltmeter.

If laser light is incident on an optical active media, such as the aqueous humor that rotates the P- and S-polarized states by angle ϕ simultaneously, then the output intensity, after the analyzer, is expressed as [32]

$$I_s = a_1 a_2 \sin 2(\theta_s + \phi) \cos(\Delta \omega t), \qquad (11.9)$$

where a_1 and a_2 are the amplitudes with respect to the laser eigenmodes, and $\Delta \omega = \omega_1 - \omega_2$ is the beat frequency between the two modes.

To calibrate zero concentration, the azimuth angle of the analyzer is generally set at 1°. Consequently, for the small rotating angles of interest, (11.9) can be reduced to

$$I_s \cong 2a_1 a_2 (\theta_s + \phi) \cos(\Delta \omega t). \qquad (11.10)$$

With zero glucose concentration testing, $I_s \equiv I_0$, i.e., $I_0 = 2a_1 a_2 \theta_s \cos(\Delta \omega t)$; thus the difference intensity for nonzero concentration is proportional to the rotation angle ϕ,

$$|\Delta I| = |I_s - I_0| = 2a_1 a_2 \phi. \qquad (11.11)$$

By using an optical heterodyning technique, the sensitivity of the detectable increment of the aqueous glucose concentration in the rabbit eye of $4\,\mathrm{mg\,dl}^{-1}$ within the 100–170-$\mathrm{mg\,dl}^{-1}$ range at a signal-to-noise ratio of 7 was measured.

Some other less-sophisticated methods for polarization sensitive detection of glucose in the eye anterior chamber are possible. A theoretical analysis of a method based on Brewster reflection off the ocular lens was recently conducted [39]. It shows that circular incident polarization performed better than linear polarization with this method. This is attributed to the high sensitivity to lens reflection error of the linear entrance polarization and to the need for an accurate adjustment of the linear polarization orientation.

A new approach to glucose sensing based on the detection of polarized fluorescence radiation from stretch-oriented reference film and a fluorophore, which changes intensity in response to glucose, was also recently suggested [35]. A glucose-sensitive fluorescent signal is provided by a glucose/galactose binding protein (which decreases fluorescence intensity upon glucose binding). Micromolar glucose concentration sensitivity results from the combination of the glucose-sensitive protein and the polarization-sensitive reference film. Using UV light diodes, one can construct simple and economical devices for glucose sensing.

11.7 Cytometry and Bacteria Sensing

Cytometry, which is based on the polarization properties of elastically scattered light, provides valuable structural information about biological particles, which is helpful for more precise cell or microorganism differentiation [47–61]. Polarization measurements can be collected for flow cytometry (when scattered light from an individual cell or microorganism in a flow is detected) and for suspensions of biological particles (when scattering fields from a collection of cells or microorganisms are analyzed). Several techniques for polarized light scattering are described in [47–61]. In general, full amplitude (Jones) or intensity (Mueller) scattering matrix measurements (see Chap. 1) can be used. Simplicity and time limitations dictate that only certain characteristic matrix elements or their combinations can be evaluated.

Experimental studies on the application of polarization-scattering techniques to the physiological monitoring of cells, bacteria, and other microorganisms are described widely in the literature [47–61]. Mueller matrix measurements have been used to examine the formation of liposome complexes with plague capsular antigens [57] and various particle suspensions, e.g., those of spermatozoid spiral heads [49,50]. It is found that element M_{34} is the most specific for the identification of various biological microorganisms, because of its sensitivity to small morphological alterations in the scatterers.

A single photoelastic modulator can be used for such measurements [47,48]. Stable distinctions are revealed in the values of the normalized element M_{34}/M_{11} for spores of two mutant varieties of bacteria, which can be distinguished by variations in their specific structure but are invisible by means of traditional scattering techniques [47,48]. It has also been proven that the M_{34} measurement is suitable for determining the diameter of rod-shaped bacteria (*Escherichia coli* cells) that are difficult to measure using other techniques [53]. The angular dependencies of the normalized element M_{34}/M_{11} of different bacteria turn out to be oscillating functions whose maxima positions are very sensitive to the varying sizes of the bacteria [52–54]. An analysis of the sensitivity of different matrix elements to variation in the scatterer shape and size for suspensions of biological particles has shown that in the backward scattering direction the values of elements M_{33} and M_{44} may serve as indicators of particle nonsphericity [51].

One promising polarization technique is the so-called phase differential scattering (PDS) method, which is relatively simple and free of experimental errors [49, 50]. As already mentioned, the two-frequency $(\omega_1 + \omega_2)$ Zeeman laser, with two collinear orthogonally linear polarized laser beams, is suitable for polarization measurements [31,32]. Linear or circular birefringence can be easily and accurately measured using such a laser [49,50]. When light from the Zeeman laser is scattered by a sample, the phase and amplitude of the beat frequency signal $(\Delta\omega = \omega_1 - \omega_2)$, produced by laser beams that are mixed on the photodetector, contain information about the structural properties of the scatterer. The only optical component required for such PDS measurements is an analyzing polarizer placed in front of the photodetector.

Measurements for the various scattering angles that provide valuable structural information can also be easily acquired. The transmission axis of the analyzer is oriented at an angle, η, with respect to the horizontal scattering plane. Usually, $\eta = 0°$, $45°$, or $90°$. All other angles give redundant information. The precise orientation of the transmission axis of the analyzer is critical only for $\eta = 0°$ or $90°$.

At a given scattering angle, the particle may interact with the two incident polarizations differently. The specificity of the interactions depends on the size, refractive index, morphology, internal structure, and optical activity of the scatterer. Three different modes of measurements are possible. (1) Measurement of the amplitude of the beat frequency signal ($\Delta\omega$) at a setting of the transmission axis of the analyzer $\eta = 45°$, which characterizes the difference in efficiency of scattering for horizontal and vertical polarizations of the incident light. (2) Measurement of the phase of the beat frequency signal when the analyzer is oriented at $\eta = 45°$, which gives information on the retarding of one of the scattered polarizations relative to the other. (3) Measurements of the amplitude and phase of the beat frequency signal when the analyzer is at $\eta = 0°$ and $90°$, which characterizes the conversion of one orthogonal polarization into the other at light scattering.

The quantitative theory for PDS is based on the scattering amplitude matrix (Jones matrix) formalism [49,50]. For scattering of the two frequencies of Zeeman laser radiation by a particle with fixed orientation, the relationship between the scattered and the incident electric fields is given by the following equation [49] (see (3.4)):

$$
\begin{bmatrix} E_{\|s} \\ E_{\perp s} \end{bmatrix} = \frac{e^{ik(r-z)}}{-ikr} \begin{bmatrix} S_2 & S_3 \\ S_4 & S_1 \end{bmatrix} \begin{bmatrix} E^0_{\|i} e^{-i\omega_1 t} \\ E^0_{\perp i} e^{-i\omega_2 t} \end{bmatrix}, \tag{11.12}
$$

where the incident beam is propagating in the z direction; $E^0_{\|i} e^{-i\omega_1 t}$ is the incident electric field parallel to the scattering plane; $E^0_{\perp i} e^{-i\omega_2 t}$ is the incident electric field perpendicular to the scattering plane; $E_{\|s}$ and $E_{\perp s}$ are the scattered electric fields parallel and perpendicular to the scattering plane; S_{1-4} are Jones matrix elements; r is the distance from the scatterer to the detector; k is the mean wave number; and ω_1 and ω_2 are the laser frequencies, $\Delta\omega = \omega_1 - \omega_2 \ll \omega_1, \omega_2$.

After the analyzer, the detected scattered intensity is described by

$$
I = \mathrm{DC} + \Gamma \cos(\Delta\omega t + \gamma), \tag{11.13}
$$

where DC is a time-independent constant, Γ is the amplitude of the beat frequency signal, and γ is the phase of the beat frequency or PDS phase.

For a single scatterer and three different orientations of the analyzer, at $\eta = 0°$, $45°$, and $90°$, the PDS phase can be calculated using (11.12) with $S_{1-4} = A_{1-4}e^{i\psi_{1-4}}$ [49]

$$\gamma(0°) = \psi_2 - \psi_3, \tag{11.14}$$

$$\gamma(45°) = \tan^{-1}\frac{A_2\sin\psi_2 + A_4\sin\psi_4}{A_2\cos\psi_2 + A_4\cos\psi_4} - \tan^{-1}\frac{A_1\sin\psi_1 + A_3\sin\psi_3}{A_1\cos\psi_1 + A_3\cos\psi_3}, \tag{11.15}$$

$$\gamma(90°) = \psi_4 - \psi_1. \tag{11.16}$$

These equations for the phase and corresponding equations for the PDS amplitude Γ can be expressed in the terms of the Mueller matrix (see (3.5), (3.10), and (3.11)) [49]:

$$\gamma(0°) = \tan^{-1}\frac{M_{14} + M_{24}}{M_{13} + M_{23}}, \tag{11.17}$$

$$\gamma(45°) = \tan^{-1}\frac{M_{14} + M_{34}}{M_{13} + M_{33}}, \tag{11.18}$$

$$\gamma(90°) = \tan^{-1}\frac{M_{14} - M_{24}}{M_{13} - M_{23}}, \tag{11.19}$$

$$\Gamma(0°) \propto \left[(M_{13} + M_{23})^2 + (M_{14} + M_{24})^2\right]^{1/2}, \tag{11.20}$$

$$\Gamma(45°) \propto \left[(M_{13} + M_{33})^2 + (M_{14} + M_{34})^2\right]^{1/2}, \tag{11.21}$$

$$\Gamma(90°) \propto \left[(M_{13} - M_{23})^2 + (M_{14} - M_{24})^2\right]^{1/2}. \tag{11.22}$$

Equations (11.14)–(11.16) are valid for a single particle with a fixed orientation or for a collection of noninteracting spherical particles with an isotropic dielectric constant. Relying on the fact that the Mueller matrix for a collection of particles equals the sum of the matrices for the individual particles, and using (11.17)–(11.22), the PDS amplitude and phase can be calculated for a collection of non-spherical particles.

The PDS amplitude at analyzer orientations $\eta = 0°$ or $90°$ is sensitive to the sample chirality, because of the zero values of the orientationally averaged elements M_{13}, M_{23}, M_{14}, and M_{24} for the nonchiral particles [62]. To get additional PDS information, instead of a linear analyzer, a circular polarization sensitive analyzer can be used. To irradiate an object by two orthogonal circular polarizations, a quarter-wave plate can also be used just after the Zeeman laser. This also gives additional information about the object's structure [49,50].

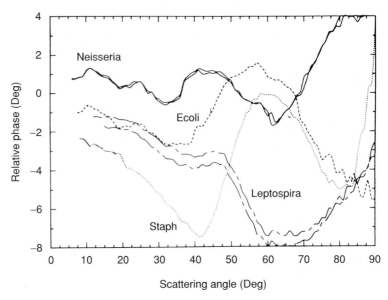

Fig. 11.18. Phase differential scattering (PDS) from pure aqueous suspensions of different viable bacteria [49,50]. Phase of PDS signal, $\gamma\,(45°)$, is shown plotted as a function of scattering angle, θ. The bacteria are *Neisseria lactamica (solid curves)*, *Leptospira biflexa (chain-dash curves)*, *E. coli B (dashed curve)*, and *Staphylococcus aureus (dotted curve)*, all at a concentration of approximately 5×10^7 bacterial cells per ml. The two *Neisseria* samples were aliquots from the same culture. They were run 5 h apart. Before the measurements, one of the samples was stored at room temperature. The two *Leptospira* samples were obtained from separate cultures grown 3 weeks apart

Figures 11.18 and 11.19 illustrate angular dependences of the PDS phase for bacterial aqueous suspensions. Measurements were taken at the transmission axis of the linear analyzer at 45° to the scattering plane. Such data are approximately equal to $\arctan(M_{34}/M_{33})$ for orientationally averaged scatterers (see (11.18)). Therefore, they are somewhat related to M_{34} measurements [47,48]. However, PDS measurements are simpler and less sensitive to experimental artifacts, in particular, to scattering cell birefringence.

PDS angular dependences are very sensitive to bacterial structure (see Figs. 11.18 and 11.19). It is easy to differentiate such viable bacteria as *Neisseria lactamica, Leptospira biflexa, E. coli B*, and *Staphylococcus aureus*. In contrast, the PDS phase is not very sensitive to reasonable changes in the storage time (a few hours) of the same bacterial culture or to separate cultures of the same bacteria type grown a few weeks apart. Nevertheless, the sensitivity is rather high for differentiating some structural changes that appear, for instance, due to different growing conditions (encapsulated and unencapsulated *B. subtilis* bacteria). Pure aqueous suspensions of bacteria have been

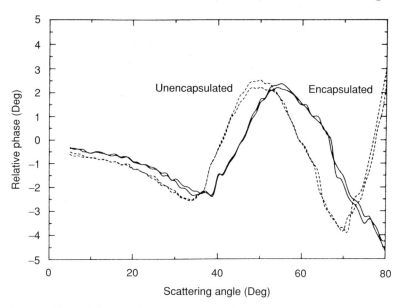

Fig. 11.19. Phase differential scattering (PDS) from pure aqueous suspensions of viable *B. subtilis* bacteria at a concentration of 5×10^7 bacterial cells per ml [49,50]. The phase of the PDS signal, $\gamma(45°)$, is shown plotted as a function of scattering angle, θ. The *solid curve* represents the PDS phase data for the encapsulated bacteria, and the *dashed curve* is a replicate run (the same sample stored at room temperature and run 20 min later). The *chain-dash* curve represents an identical sample of bacteria grown under conditions that preclude encapsulation. The *dotted curve* is a replicate run (the same sample stored at room temperature and run 20 min later)

distinguished by their PDS signature at concentrations as low as 5×10^5 bacterial cells per ml. At concentrations above $5 \times 10^7 \, \mathrm{ml}^{-1}$, the multiple scattering, which hides the fine structure in the PDS data, is expected to be important [63]. The backscattering mode of measurement is also possible.

Laser flow cytometry is a sensitive tool, which can serve to identify and separate various populations of cells, particularly white blood cells. Therefore, it is of great interest in medical diagnosis [49,50,55,56,59]. A flow cytometer (see Fig. 11.20) consists of a cuvette, in which a cell suspension is forced to flow and to be centered due to the application of a hydrofocusing technique. Cells in flow interact one by one with a perpendicular-to-cell-flow focused laser beam and produce various signals – fluorescence, scattering in forward, perpendicular, or backward directions. Fluorescence measurements usually require cell staining. Scattering techniques are free of any effect on the cell, which makes them more attractive for use. Polarization scattering technologies, which are sensitive to the details of cell morphology (nonsphericity, inhomogeneity, existence of internal granules, membrane roughness, encapsulation, etc.), work well in combination with other scattering techniques. The basis for combined

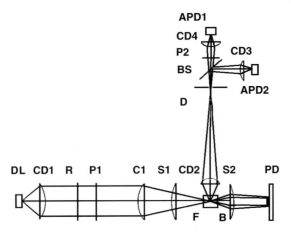

Fig. 11.20. Optical scheme of a portable solid state flow cytometer [64]. DL is the diode laser (780 nm); CD1 – CD4 are the lenses from the CD player; R is the retarder – a half wavelength plate; P1 and P2 are the linear polarizers; C1 and C2 are the cylindrical lenses; F is the time-of-flight cuvette; B is the direct light beam shutter; S1 and S2 are spherical lenses; PD is the photodiode; D is the diaphragm; BS is the beam splitter; APD1 and APD2 are the avalanche photodiodes

scattering flow cytometry is concurrent measurements of scattered light intensity for different wavelengths for different scattering and detecting angles and for different states of polarization in incident and scattered light. For example, the major components of the human leukocyte population (lymphocytes, monocytes, and granulocytes) are easily differentiated in combined measurements of the scattering intensity in the forward (within limits of $0.5°$–$2.0°$) and perpendicular ($90° \pm 30°$) directions (see Fig. 11.21a) or depolarized scattering under angle of $90°$ (see Fig. 11.21b). In this latter case, the differentiation of granulocytes into two fractions – neutrophils and eosinophils – is possible.

The optical scheme of a portable and inexpensive time-of-flight cytometer, which is based on a single mode diode laser (780 nm, 20 mW) and two avalanche photodetectors for the detection of weak scattering polarized and unpolarized components with angles smaller than $90°$ is presented in Fig. 11.20 [64].

11.8 Polarization Microscopy and Tissue Clearing

Polarized light microscopy has been used in biomedicine for more than a century to study optically anisotropic biological structures that may be difficult, or even impossible, to observe using a conventional light microscope. Obviously, polarization microscopy is a routine technique. A number of commercial microscopes are available on the market, and numerous investigations of biological objects have been made using polarization microscopy. Nevertheless,

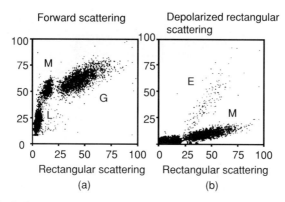

Fig. 11.21. The light-scattering histograms of human white blood cells, as measured in a flow cytometer and presented in coordinates "Forward scattering – rectangular scattering (90°)" **(a)** and "Depolarized rectangular scattering (90°) – rectangular scattering (90°)" **(b)** [64]. Areas marked as L, M, G, N, and E show, respectively, the distributions of lymphocytes, monocytes, granulocytes, neutrophils, and eosinophils. Each point of the distributions corresponds to scattering by a separate cell flowing through the cytometer

modern polarization microscopy has the potential to enable us to acquire new and more detailed information about biological cells and tissue structures. Over the years, the equipment for polarized light microscopy has been perfected so that now it is possible to detect optical path differences of even less than 0.1 nm [65–70]. Such sensitivity as well as the capability to examine scattering samples are due to recent achievements in video, interferential, and multispectral polarization microscopy. Full Mueller matrix measurements and other combined techniques, such as polarization/confocal and polarization/OCT microscopy, promise new capabilities for polarization microscopy that involve receiving more precise structural information about the objects plus the ability to provide in vivo measurements.

In this section, we shall discuss only a few of the recent studies and novel techniques that have the potential to examine the anisotropic properties of scattering samples. One of the examples is the multispectral imaging micropolarimeter (MIM) technique, which can detect the birefringence of the peripapillary retinal nerve fiber layer (RNFL) in glaucoma diagnosis [68,69]. The optical scheme of MIM is presented in Fig. 11.22. Light from a tungsten–halogen lamp, followed by an interference filter (band of 10 nm), provides monochromatic illumination to an integrating sphere (IS). Lens L_1 ($F = 56$ mm) collimates the beam incident onto a polarizer (P). Use of an integrating sphere assures that the output intensity of the polarizer varies less than 0.2% as it rotates 360°. Lens L_2 ($F = 40.5$ mm, NA $= 0.13$) focuses the image of the exit aperture of the integrating sphere onto a specimen (SP) in a chamber (CB) with a flat entrance and exit windows. Lens L_3 ($F = 60$ mm, NA $= 0.07$)

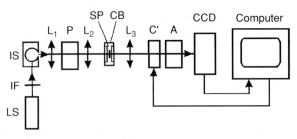

Fig. 11.22. Optical scheme of the multispectral imaging micropolarimeter used in transmission mode [69]. LS is the light source; IF is the interference filter; IS is the integrating sphere; P is the linear polarizer; SP is the specimen; CB is the chamber; L_1, L_2, and L_3 are the lenses; C′ is the linear retarder; A is the linear analyzer; and CCD is the charge-coupled device

focuses the specimen image onto a cooled CCD camera that provides a pixel size of about $4\,\mu m$ on a specimen in an aqueous medium (magnification ≈ 5.8). Although the lenses are achromatic, the wide spectral range (440–830 nm) requires only small changes in the detection optics position (moving the lens L_3 and CCD together within a 0.5 mm range) to adjust the focus for each wavelength. A liquid crystal linear retarder (C′), followed by a linear analyzer (A), is used to measure the output Stokes vector of the specimen. Both polarizer and analyzer are Glan-Taylor polarization prisms. The azimuth and retardance of the retarder are set for a few discrete values, and the azimuth of the analyzer is always fixed at 45°. Each setting of the retarder (respectively, azimuth and retardance – (1) 0°, 90°; (2) 0°, 200°; (3) 22.5°, 207°; (4) –22.5°, 207°) – is characterized by a 1×4 measurement vector. The four retarder/analyzer settings together are characterized by a 4×4 matrix **D** with each row corresponding to one measured vector.

A Stokes vector $\bar{S}$ can be calculated as

$$\bar{S} = \mathbf{D}^{-1}\bar{R}, \tag{11.23}$$

where $\mathbf{D}^{-1}$ is the inverse of the measurement matrix and $\bar{R}$ is a 4×1 response vector corresponding to the four retarder/analyzer settings [71].

To evaluate the linear retardance of a specimen, the Mueller matrix should be found from the measurements of the incident $\bar{S}_{inc}$ and the output $\bar{S}$ Stokes vectors (see Chap. 3):

$$\bar{S} = K\mathrm{M}(\rho, \delta)\bar{S}_{inc}, \tag{11.24}$$

where the factor K accounts for the losses of intensity in transmission and ρ and δ are, respectively, the azimuth and retardance of the specimen. This expression includes four equations for the three unknowns, K, ρ, and δ. In most cases it is useful to overdetermine the system of equations in (11.24) by using more than one $\bar{S}_{inc}$.

The retardance and azimuth of a living and fixed rat's RNFLs are measured over a wide spectral range [69]. It is found that the RNFL behaves as a linear retarder and that the retardance is approximately constant at a wavelength range from 440 to 830 nm. The average birefringence measured for a few unfixed rat RNFLs, with an average thickness of $13.9 \pm 0.4\,\mu m$, is $0.23 \pm 0.01(nm\,\mu m^{-1}) \equiv 2.3 \times 10^{-4}$. The influence of the polarization properties of the retina on the RNFL's anisotropic properties is stated. Images presented in Fig. 11.23 illustrate the importance of correcting for the polarization properties of the retina and for the distributions of retardance and azimuth within the sample.

Another technique, which is related to quantitative polarized light microscopy, is based on a video microscopy technique, which is applied to measure variations in the orientations of the collagenous fibers arranged in lamellae within eye corneal tissue [65]. The lamellar structure of the cornea and sclera is very visible in Figs. 2.5 and 2.6. Within a lamella, the fibrils are parallel, but the fibrils of adjacent lamellae do not, in general, run in the same direction. They may have a relative orientation at any angle between $0°$ and $180°$.

As was discussed in Sect. 2.7, in the corneal stroma, the fibrils have a diameter of 25–39 nm while the mean diameter of scleral fibrils is equal to 100 nm. Therefore, the individual fibrils cannot be resolved with light microscopy, but due to the intrinsic birefringence of collagen fibrils and its dependence on the angle of their orientation from lamella to lamella, they can be recognized. Along its fiber axis, collagen is highly birefringent, so those lamellae that are cut parallel to the fiber axes ($\theta = 0°$, see Fig. 11.24a) appear brighter under polarized light when the polarizer and analyzer are crossed and the length of the tissue section is oriented at $45°$ to the polarizer/analyzer axis (see Fig. 11.24b). Collagen is not birefringent perpendicular to its fiber axis ($\theta = 90°$, see Fig. 11.24a); so those lamellae that are cut perpendicular to their fibril direction appear completely dark in this section (see Fig. 11.24b). The variation of intensity along the transect X–Y across the cornea section (see

Fig. 11.23. Estimated retardances (*arrows' length*) and slow axes (*arrows' direction*) of the bundle and gap areas of rat RNFLs [69]. The images are at 440 nm. Sizes of images: (a) $222 \times 199\,\mu m$; (b) $187 \times 177\,\mu m$. Nerve fiber bundles appear as brighter bands. Each *arrow* starts in the center of the area measured. The calibration bar is 1 nm of retardance. (**a**) The *white arrows* represent measurements that are not corrected for retinal polarization ability and the *black arrows* are the corrected ones. (**b**) The *black arrows* are corrected bundle retardances; the *white small arrows* in the gaps show the variation of residual retardances, also after correction

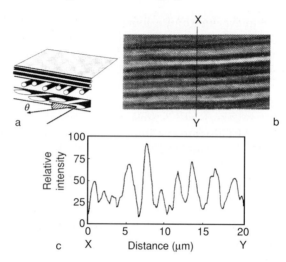

Fig. 11.24. Polarization microscopy of a collagenous tissue structure [65]. (**a**) Schematic diagram of a tissue section containing three lamellae (the number of fibrils is greatly reduced and the relative fibril diameter greatly exaggerated; an actual electronic micrograph is shown in Fig. 2.6); angle θ is the angle of inclination of the fibrils relative to the plane of sectioning. (**b**) Digital photomicrograph of a section through part of a rabbit cornea viewed under polarized light ($\times500$). (**c**) The variation in intensity along the transect X–Y across the cornea section of a photomicrograph (**b**)

Fig. 11.24b) is caused by the different angular orientation of the particular lamella (totally about 15 lamellae are seen) and presented in Fig. 11.24c.

Because of the regular arrangement of the lamellae, the angle θ is all that is necessary to define the three-dimensional orientation of the fibrils in sections of normal cornea. Nevertheless, to find this angle distribution for a specific tissue section, the lamellar birefringence of form that contributes about 67% to the total birefringence should be accounted for [65]. For sections of disrupted pathological cornea and for sections of sclera and limbus (the region where the cornea and sclera fuse), the situation is more complicated because the lamellae have a much less ordered "wavy" arrangement (see Fig. 2.6b for sclera).

Many tissues possess very complex patterns of alignment of the structure-forming elements. Some promising polarization-microscopic methods for generating alignment maps for such tissues have been developed (see, for example, [72–74]). The method presented in [72, 73], as well as the method used in [65], is a useful tool in cases where the tissue structure along the direction of probing light propagation can be considered to be a uniform one.

In [72,73], a microscopic polarimetry method for generating fiber alignment maps, which can be used for characterization of the structure of fibrous tissues, tissue phantoms, and other fibrillar materials, is considered. This method is based on probing the sample with elliptically polarized light from a rotated quarter-wave plate and an effective circular analyzer. Nonlinear regression techniques are implemented for estimating the optical parameters of the optic train and the sample. The processing of the sequence of images obtained with different mutual orientations of the rotated quarter-wave plate and the analyzer, which is based on the fast harmonic analysis, permits the recovery of an alignment direction map and a retardation map. These maps describe a spatial distribution of a sample's local linear birefringence and, therefore, can be used for morphological analysis of tissues with an expressed structural anisotropy that have linear birefringence as their dominant optical property. The potential of this method for accurately generating alignment maps for samples that act as linear retarders to within a few degrees of retardation, has been demonstrated in experiments with an in vitro sample of a porcine heart valve leaflet.

The method proposed in [74] can also be used in more general situations when the orientation of the structure-forming elements (for instance, the collagen fibers) varies along the probing direction. Such variation should be taken into account when thick ($> 50\,\mu$m) tissue layers (e.g., dermis) are analyzed. In the method discussed [74], a standard polarization microscope arranged with a video camera can be used. The measurements are usually carried out with wide-spectral-range color filters and without a quarter-wave plate. The following expression describes the dependence of the detected signal at any detection point on the angles of orientation of the polarizer (ϑ) and the analyzer (ϑ') of the microscope:

$$i_C \approx B_0 + B_1 \cos\eta + B_2 \cos\varsigma + B_3 \sin\eta + B_4 \sin\varsigma,$$
$$\eta = 2\,(\vartheta - \vartheta')\,,\quad \varsigma = 2\,(\vartheta + \vartheta')\,, \tag{11.25}$$

where $B_i\,(i = 0, 1, 2, 3, 4)$ are the coefficients which depend on the local optical properties of the sample in the probed region and the spectral properties of incident light. As was shown in [74], the measured values of B_i are capable of providing important information about the sample structure. They can also be used for characterization of specific features of light propagation in the sample; in particular, they allow us to recognize the so-called adiabatic regime of light propagation in the studied medium. This means that in the adiabatic regime, the orientation of the local optical axis changes smoothly in the probed region of the tissue. Note that for fibrous tissues, the direction of the local optical axis typically coincides with the local preferred direction of fiber orientation. If the adiabatic regime is realized, then the angles v and ϕ, which are calculated from the obtained values of B_i as $v = (1/4)\mathrm{arctg}(B_2/B_4)$ and $\phi = (1/2)\mathrm{arctg}(B_1/B_3)$, provide information about the structure of the sample. The angle ϕ is equal to the angle between the azimuthal projections of the local optical axes of the medium at the upper and lower boundaries of the sample, and the angle v defines the orientation of the bisector of the angle between these projections. Analysis of the experimentally obtained "B_i-maps," as well as the spatial distributions of v and ϕ, can be proposed as an effective tool for tissue structure characterization [74]. In particular, Fig. 11.25 displays the B_0-map (this coefficient characterizes the local transmittance of the sample for nonpolarized light), the v-map, and the ϕ-map for the in vitro sample of human epidermis (stratum corneum), which was obtained from a cuticle of nail. The epidermis layer is placed in a drop of glycerol and covered with cover-glass. The maps are recovered from seven digital images of the sample obtained at different orientations of the polarizer and an analyzer of the microscope. The spectral selection is carried out using wide-range green and red glass filters. During the image processing,

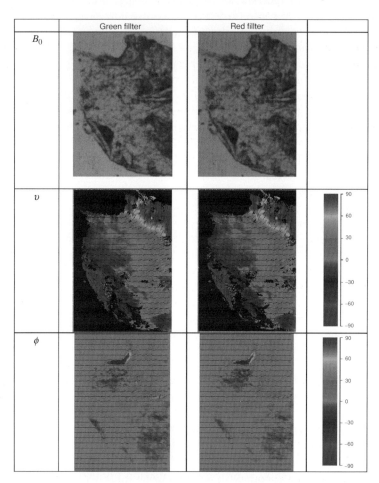

Fig. 11.25. The experimentally obtained B_0-, v-, and ϕ-maps for an in vitro sample of human epidermis. For the ϕ-maps, the horizontal orientation of the indicating lines corresponds to $\phi = 0$. The identity of the v-, and ϕ-maps obtained with different color filters, indicates that the adiabatic regime of polarized light propagation in the sample takes place [74]

corrections are made to exclude the influence of polarization imperfections on the optical elements, the temperature drift and fluctuations of the dark current on the video camera, and the fluctuations of the illumination source.

The study of collagen structure and function is important for understanding a wide range of pathophysiological conditions, including aging. One of the prospective laser techniques, which can provide in vivo microscopic monitoring of collagen structure, is polarized second harmonic microscopy [75, 76]. The backscattered second harmonic signal generated by a 100 fs Ti:Sapphire laser, with a mean wavelength of 800 nm, a maximum energy of 10 nJ, and a pulse repetition rate of 82 MHz, is measured by a polarized second harmonic scanning confocal microscope [76]. The microscope objective has a transverse resolution of about 1.5 µm and an axial resolution of about 10 µm. It should be noted that inside the scattering media, both numbers increase. The maximum intensity in the sample is about $4 \times 10^{11} \mathrm{W\,cm}^{-2}$. To avoid sample damage, a continuous scanning technique is used.

A systematic analysis of type I collagen in a rat-tail tendon fascicle was conducted using the microscope described above [76]. Type I collagen from the fascicles provides one of the strongest second harmonic generation (SHG) signals of all of the various tissues analyzed by the authors in [76]. They hypothesize that such high SHG efficiency is due to the collagen's highly ordered architecture. The polarization properties of collagen are also defined by its ordering. It was shown experimentally that the second harmonic signal intensity varies by about a factor of 2 across a single cross-section of the rat-tail tendon fascicle [76]. The signal intensity depends both on the collagen organization and the backscattering efficiency. To characterize collagen structure, both intensity and polarization dependent SHG signals should be detected. Actually, axial and transverse scans for different linear polarization angles of the input beam show that SHG in the rat-tail tendon depends strongly on the polarization of the input laser beam. In contrast to SHG signal intensity, the functional form of the polarization dependence does not change significantly over a single cross-section of the sample, and it is not affected by backscattering efficiency.

The measured data are in good agreement with an analytical model developed for a SHG signal at linear polarized excitation. They are used to determine the fibril orientation and the ratio between the only two nonzero, independent elements in the second-order nonlinear susceptibility tensor, $\gamma \approx -(0.7\text{–}0.8)$ [76]. The small range of values observed for γ in a tendon fascicle suggests that there is structural homogeneity. This parameter might, therefore, be useful in characterizing different collagen structures noninvasively [76].

The main problem encountered in the in situ microscopy of tissues is multiple scattering, which randomizes the direction, coherence, and polarization state of incident light. A number of optical gating methods have been proposed to filtrate ballistic and least-scattering photons, which carry information about the object structure [46]. One of these is the polarization gating method and its modifications, which are described elsewhere in this book. The fundamental limitation of all optical gating methods, including the polarization one, is the fact that only a small number of ballistic and least-scattering photons take part in the formation of an object image. Therefore, polarization-gating techniques in combination with image reconstruction methods can be useful for improving the image resolution in the case of a highly scattering object [77, 78]. Both reflection-mode and transmission-mode polarization gating scanning microscopes have been analyzed [77, 78].

An optical immersion technique, based on matching the refractive index of the tissue scatterers and the surrounding ground (interstitial) medium, allows one to essentially control the scattering properties of a tissue [46]. Usually the refractive index of the ground medium is controlled. This is accomplished by impregnating the tissue with a biocompatible chemical agent, like glucose, glycerol, propylene glycol, or X-ray contrasting medicals. Due to the fact that the refractive index of the applied agent is higher than that of the tissue ground substance, which is close to the index of water, the refractive index of the ground increases and scattering decreases. Most of the applied agents are osmotically active; therefore, they can produce a temporal and local dehydration of the tissue which also leads to an increase in the refractive index of the interstitial space. Reduction of the scattering at optical immersion makes it possible to detect the polarization anisotropy of tissues more easily and to separate the effects of light scattering and intrinsic birefringence on the tissue polarization properties. It is also possible to study birefringence of form with optical immersion, but when the immersion is strong, the average refractive index of the tissue structure is close to the index of the ground media, and the birefringence of form may be too small to see.

The dynamics of tissue optical clearing and the manifestation of tissue anisotropy at the reduction of scattering are characteristic features, which correlate with clinical data [46, 79–81]. Figures 11.26 and 11.27 show experimental results on the temporal transmittance of linear polarized light through tissue sections measured by a white-light video-digital polarization microscope on the application of an immersion agent (X-ray contrasting agent – trazograph-60) [80]. The immersion solution was heated up to $(36–40)°C$ and simply dropped on the tissue sample surface. Sections of the various connective and vascular tissues of an area of $1 \times 1\,cm^2$ and a thickness of 0.1–1.5 mm were studied. The temporal image contrast $C(t)$ and its rate $V(t)$ are used for a quantitative description of the diffusion process of the agent in the tissue

$$C(t) = T(t)/T_{\mathrm{max}}, \qquad (11.26)$$
$$V(t) = \mathrm{d}C(t)/\mathrm{d}t, \qquad (11.27)$$

$T(t)$ is the current sample brightness; T_{max} is the maximal sample brightness.

The experimental image contrast rate $V(t)$ is well described by the following empirical equation which is valid for various studied tissue samples:

$$V(t) = A + B\exp(-G \times t), \qquad (11.28)$$

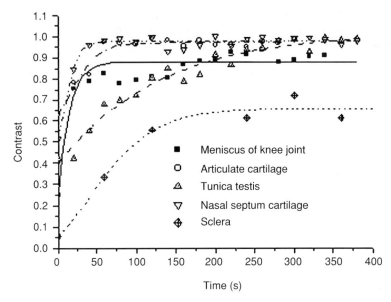

Fig. 11.26. Experimental temporal dependencies for a normalized transmittance $C(t) = T(t)/T_{\max}$ of linear polarized light through tissue sections measured by a white-light video-digital polarization microscope at the application of an immersion agent (trazograph-60) [80]

where A, B, and G are the empirical parameters associated with the tissue structure and the chemical agent diffusion ability. Different types of tissues have quite different rates of optical clearing which is influenced by their structural peculiarities. For example, for human sclera, $A = 0.70, B = 2.59$, and $G = 6.4$; and for meniscus of the knee joint, $A = 0, B = 9.83$, and $G = 17.2$.

Both Figs. 11.26 and 11.27 show rates of tissue optical clearing for different tissues when the tissues change from an initial turbid (multiple scattering) at $t = 0$ to a less depolarized and more transparent state (less scattering), $C(t) \to 1$. Evidently, this difference depends on the tissue structure, which defines the initial (natural) turbidity, and the efficiency of the chemical agent's interaction with the tissue. For example, vein and aorta samples have approximately the same initial turbidity and degree of linear depolarization, but they interact quite differently with an immersion agent (see Fig. 11.27). The agent is less able to penetrate the more dense aorta than the vein; therefore, its action on the aorta takes several hours while 10 min is enough time to complete the clearing of the vein.

At the reduction of scattering, tissue birefringence can be measured more precisely. In particular, the birefringence of form and material can be separated. For example, in a translucent human scleral sample that is impregnated with a highly concentrated glucose solution (about 70%), the measured optical anisotropy $\Delta n = (n_e - n_o)$ is equal to $\approx 10^{-3}$ [81]. This is 1.5–4.5 folds

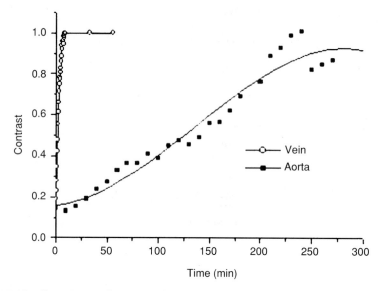

Fig. 11.27. Experimental temporal dependences for a normalized transmittance $C(t) = T(t)/T_{\max}$ of linear polarized light through vascular tissue sections (aorta and vein-*Vena cava inferior*) measured by a white-light video-digital polarization microscope at the application of immersion agent (trazograph-60) [80]

less than the other birefringent tissues described in Sect. 2.5, which can be explained primarily by the reduction in the inclusion of birefringence of form in the optical immersion.

Additional measurements of the collimated transmittance allow for estimation of the refractive index of the ground substance of the translucent tissue n_2 using expressions derived from radiative transfer and Mie theory [46]

$$I \approx I_0 \exp(-\mu_s d); \mu_s \propto (n_1/n_2 - 1)^2, \qquad (11.29)$$

where I_0 is the intensity of the incident light, μ_s is the scattering coefficient, d is the sample thickness, and n_1 is the refractive index of the collagen fibers. Index n_2 was evaluated as 1.39. Using this value and the value of the refractive index of hydrated collagen, $n_1 = 1.47$ estimated in Sect. 2.7, the collagen volume fraction, f_1, can be calculated from (1.9), as

$$f_1(1 - f_1) \cong \frac{\Delta n}{(n_1 - n_2)^2} n_2, \qquad (11.30)$$

$f_1 \cong 0.32$, which correlates well with the estimation made in Sect. 2.7.

Figure 11.28 illustrates the reversibility of the optical immersion effect. A polarization-speckle microscope working in transmittance mode was used to carry out these measurements [79]. The sample was irradiated by a linear polarized focused laser beam which was scanned along the trace of 1.5 mm

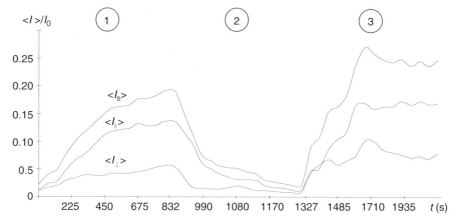

Fig. 11.28. Time-dependent mean speckle intensity $< I_s >$, averaged over the scanning trace (1.5 mm), and its polarization components $< I_{||} >$ and $< I_{\perp} >$, measured in the paraxial region for a human sclera sample ($d = 0.4$ mm) at $\lambda = 633$ nm using a polarization-speckle microscope [79]. (1), (2) and (3) – the subsequent measurements for a sample kept first in trazograph-60 solution (1), then in physiological solution (0.9% NaCl) (2), and finally again in trazograph-60 solution (3)

to average the speckle modulation in the far zone, where the analyzer and photodector were placed. Two orthogonal linear polarized components of the transmitted light were detected.

It can be seen that initially the sample had poor transmittance with the equal intensity components $<I_{||}> = <I_{\perp}>$ and that multiple scattering takes place. When the immersion agent acts in the 14th min, $<I_{||}>$ prevails substantially over $<I_{\perp}>$, and the tissue becomes less scattering. The subsequent action of the physiological solution, which washes out the immersion agent, returns the tissue to its normal (initial) state, and it becomes turbid again in the 22nd min with no measured difference between the intensities of the orthogonally polarized components. The secondary application of the immersion agent again makes the tissue less scattering with a maximum reached at the 28th min. Practically all healthy connective and vascular tissues show the strong or weak optical anisotropy typical of either uniaxial or biaxial crystals [80, 81]. Pathological tissues show isotropic optical properties [82].

Polarization microscopy is also helpful for investigating individual cells, in particular, for evaluating the amount of glycated hemoglobin in erythrocytes that could be an early diagnostic marker of hyperglycemia in diabetic patients [70]. Hemoglobin glycation causes changes in the cell's refraction index. By using polarizing-interference microscopy, it is possible to measure the light refractive index in an individual erythrocyte [70]. The refractive index of hemoglobin or a red blood cell, containing about 95% hemoglobin, varies approximately linearly with a change in glucose concentration – it saturates only under strong hyperglycemic conditions [41]. A Nomarsky polarizing-interference microscope, MPI-5 (Poland), is used for measurements of light phase retardation [70]. Using a Wollaston prism mounted on an object, the erythrocyte images for ordinary and extraordinary light beams are completely separated. In the thickest erythrocyte region, the first interference maxima are visually adjusted to the eye sensitive purple color for ordinary and extraordinary images by shifting the second Wollaston prism placed in the rear focus of the object. For each erythrocyte measured, the Wollaston prism displacement renders a second value. From the whole interference bandwidth h and the measured Wollaston prism displacement $2d$, the phase retardation Φ and the refractive index are calculated for each erythrocyte [70]:

$$n = n_v + \frac{\Phi}{t} = n_v + \frac{d\lambda}{ht}, \tag{11.31}$$

where $n_v = 1.5133 \pm 0.0001$ is the refractive index of the embedding media, t is the thickness of the erythrocyte, $\lambda = 550 \, \text{nm}$. Separate measurements of the erythrocyte thickness using two embedded media with different refractive indices gives $t = 0.89 \, \mu\text{m}$. Using this value, the refractive index is calculated with a standard deviation of ± 0.0005.

A robust z-polarized confocal microscope employing only one or two binary phase plates with a polarizer has been suggested by Huse et al. [82]. The major advantage of the microscope having a significant longitudinal field component is that it is then possible to image the z-polarized features in randomly oriented agglomerations of molecules of biomedical interest.

11.9 Digital Photoelastic Analysis

Photoelasticity is an established experimental technique that has been applied to study the biomechanics of hard tissues like bone and tooth [83–86]. The photoelastic measuring technique is based on the stress-induced optical birefringence effect, which for plane stress analysis is described by the following stress-optic law [87]:

$$\sigma_1 - \sigma_2 = \frac{\theta}{2\pi} \frac{f_\sigma}{h} = \frac{N f_\sigma}{h}, \tag{11.32}$$

where $(\sigma_1 - \sigma_2)$ is the difference in the in-plane principle stress, θ is the resultant optical phase generated due to stress-induced birefringence in the sample, f_σ is the material fringe value, and h is the thickness of the specimen. Since the values of f_σ and h are constants for the mechanical stresses, recording the optical phase (θ) or fringe order ($N = \theta/2\pi$) at every point of interest on the fringe pattern allows for analysis of the stress distribution [84–86].

As an example, we shall consider the results of photomechanical studies of post endodontically rehabilitated teeth, using a conventional circular polariscope and an image processing system, which are the basis for the digital phase shift photoelastic technique described in [84–86]. A special loading device has been manufactured that applies loads along the long axis (0°) and 60°

lingual to the long axis of the tooth. Using the polariscope, four phase-stepped images are obtained for the sample at each load by rotating the analyzer at 0°, 45°, 90°, and 135° angles with respect to the polarizer. The fringe patterns obtained are acquired using a high-resolution (753 × 244 pixels) CCD camera; these are then stored and processed by a computer. The four images are evaluated using a traditional phase stepping algorithm to obtain a wrapped phase map [84, 85]. Phase unwrapping is done on selected lines to make the fringe modulation continuous and to get information on the nature of the stress distribution.

Figure 11.29 shows a phase-wrapped image of the rehabilitated tooth model, loaded at 125 N at an angle of 60° in the direction of the long axis of the tooth. It was found that there is a significant (up to three-fold) increase in the magnitude of the stress within the rehabilitated tooth model in comparison with the model of the intact tooth. Increased bending stress is identified in the cervical region and in the middle region of the root. This results in higher compressive stress in the cervical region (facial side) and higher tensile stress in the mid region (lingual side).

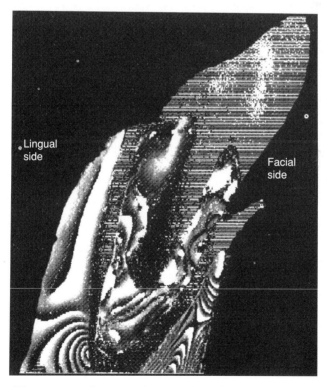

Fig. 11.29. Phase-wrapped image obtained from four-phase shifted images in a rehabilitated tooth model, loaded at 125 N, 60° lingual to the long axis of the tooth [86]

The designed digital phase-shift photoelastic technique is of importance for the investigation of hard tissue elasticity distributions; here, for instance, it highlighted the behavior of a post-core rehabilitated tooth to functional forces.

11.10 Fluorescence Polarization

Fluorescence polarization measurements are used to estimate various parameters of the fluorophore environment [88]; therefore, they have a potential role to play in biomedical diagnosis, in particular, in discriminating between normal and malignant tissues [89–91]. At polarized light excitation, the emission from a fluorophore in a none-scattering media becomes depolarized because of the random orientation of the fluorophore molecules and the angular displacement between the absorption and emission dipoles of the molecules [88]. These intrinsic molecular processes that result in additional angular displacement of the emission dipoles are sensitive to the local environment of the fluorophore. As was already shown in preceding chapters, light depolarization in tissues is determined by multiple scattering; therefore, both excitation and emission radiations should be depolarized in scattering media [88–92]. Polarization state transformation in scattering media depends on the optical parameters of the medium: the absorption coefficient, μ_a; the scattering coefficient, $\mu_s(4.7)$; and the scattering anisotropy factor, $g(3.28)$. Due to the different structural and functional properties of normal and malignant tissues, the contribution of multiple scattering to depolarization may be different for these tissues. The reduced (transport) scattering coefficient, $\mu_s'(4.17)$, or the transport mean free path (MFP), $l_t(4.18)$, in particular, determines the characteristic depolarization depth for different tissues (see Figs. 5.5–5.8).

Thus, fluorescence polarization measurements may be sensitive to tissue structural or functional changes, which are caused, for instance, by tissue malignancy at the molecular level (the sensitivity of excited molecules to the environmental molecules) or at the macrostructural level (the sensitivity of propagating radiation to tissue scattering properties).

Mohanty et al. [91] consider a fluorophore located at a distance z from the surface of a turbid medium. The homogeneous distribution of the fluorophores and the validity of the diffusion approximation for light transport in a scattering medium are assumed. The average number of scattering events experienced by the excitation light before it reached the fluorophore, and by the emitted light before it exited the medium, are described, respectively, as

$$N_1(z) = z \times \mu_s^{\text{ex}} \tag{11.33}$$

$$N_2(z) = z \times \mu_s^{\text{em}}. \tag{11.34}$$

The fluorescence polarization ability is characterized by polarization anisotropy (r) which is a dimensionless quantity independent of the total fluorescence intensity of the object [88]

$$r = \frac{I_{||} - I_{\perp}}{I_{||} + 2I_{\perp}}. \tag{11.35}$$

It is defined as the ratio of the polarized component to the total intensity and is connected with the light polarization value P:

$$r = \frac{2P}{3 - P}. \tag{11.36}$$

The polarization, measured as

$$P = \frac{I_{||} - I_{\perp}}{I_{||} + I_{\perp}}, \tag{11.37}$$

is an appropriate parameter for describing a light source when a light ray is directed along a particular axis. The polarization of this light is defined as the fraction of light that is linearly polarized. In contrast, the radiation emitted by a fluorophore is symmetrically distributed around this axis, and the total intensity is not given by $I_{||} + I_{\perp}$, but rather by $I_{||} + 2I_{\perp}$ (see Sect. 10.4 of [88]).

Assuming that each scattering event reduces the fluorescence polarization anisotropy r by a factor of $A (A = 0–1)$, the anisotropy of fluorescence that is due to a fluorophore embedded at a depth z can be written as

$$r(z) = r_0 \times A^{[N_1(z)+N_2(z)]}, \tag{11.38}$$

where r_0 is the value of the fluorescence anisotropy without any scattering.

For a homogeneous distribution of fluorophores in a tissue of thickness d, the observed value of the fluorescence anisotropy is defined by each ith tissue layer:

$$r_{\text{obs}} = \sum_i (I_i^{\text{f}} r_i) / \sum_i I_i^{\text{f}}, \tag{11.39}$$

where I_i^{f} is the contribution to the observed fluorescence intensity from the ith layer of thickness dz at a depth, z and r_i is the value of the fluorescence anisotropy for this layer.

For the broad-beam illumination of a flat tissue surface, the propagation of excitation (ex) light beyond a few MFPs (see (4.19)) is well described by one-dimensional diffusion theory (see (4.13)). In this approximation, and taking into account $\mu_{\text{e}}^{\text{ex}} \gg \mu_{\text{d}}^{\text{ex}}$, which is valid for many tissues (see (4.7) and (4.14)), the excitation intensity reaching depth z is expressed as

$$I(z) \cong C_{\text{ex}} \exp(-\mu_{\text{d}}^{\text{ex}} z), \tag{11.40}$$

where C_{ex} is proportional to the excitation intensity and is the function of the tissue optical parameters at the wavelength of the excitation light [27].

The fluorescence from the fluorophores, embedded at depth z from the tissue surface, reaching the same surface will, therefore, be

$$I^f(z) \approx [C_{ex} \exp(-\mu_d^{ex} z)] \times \varphi[C_{em} \exp(-\mu_d^{em} z)], \qquad (11.41)$$

where C_{em} and μ_d^{em} for the emission wavelength are defined similarly as C_{ex} and μ_d^{ex} for the excitation wavelength and φ is the fluorescence yield.

By substituting the values I_i^f from (11.41) and r_i from (11.38) in (11.39), the observed fluorescence anisotropy is expressed as

$$r_{obs} = r_0 \frac{\int_0^d \exp(-\mu_d^{tot} z) \times A^{[N_1(z)+N_2(z)]} dz}{\int_0^d \exp(-\mu_d^{tot} z) dz}, \qquad (11.42)$$

where

$$\mu_d^{tot} = \mu_d^{ex} + \mu_d^{em}, \qquad (11.43)$$

$$\mu_s^{tot} = \mu_s^{ex} + \mu_s^{em}. \qquad (11.44)$$

Integration of (11.42) gives

$$r_{obs} = r_0 \frac{\mu_d^{tot}}{\mu_d^{tot} - \ln(A) \times (\mu_s^{tot})} \times \frac{1 - \exp(-\mu_d^{tot} d) \times (A)^{\mu_s^{tot} d}}{1 - \exp(-\mu_d^{tot} d)}. \qquad (11.45)$$

Fluorescence anisotropy measurements are usually provided by commercially available spectrometers, the sensitivity of which is different for two orthogonal polarization states. Therefore, all measured fluorescence spectra should be corrected for the system response [88]:

$$r = \frac{I_{||} - GI_{\perp}}{I_{||} + 2GI_{\perp}}, \qquad (11.46)$$

where G is the ratio of the sensitivity of the instrument to the vertically and the horizontally polarized light.

Typical G-corrected polarized fluorescence spectra at 340-nm excitation from malignant and normal breast tissue with thickness $\approx 2\,mm$ are shown in Fig. 11.30 [91]. Collagen, elastin, coenzymes (NADH/NADPH), and flavins contribute to these spectra and the spectra received at a longer wavelength of 460 nm [93–95]. The contribution of NADH dominates with excitation at 340 nm and different forms of flavins dominate with excitation at 460 nm. In Fig. 11.30, a blue shift in the polarized fluorescence spectra maximum can be clearly seen in the malignant, as compared to the normal, tissue. A similar shift of 5–10 nm was observed also for 460 nm excited fluorescence. This shift is associated with the accumulation of positively charged ions in the intracellular environment of the malignant cell [89]. Some differences, in particular, a spectral shift of the maximum, between the parallel and cross-polarized fluorescence spectra observed for rather thick tissue layers ($\approx 2\,mm$), may be

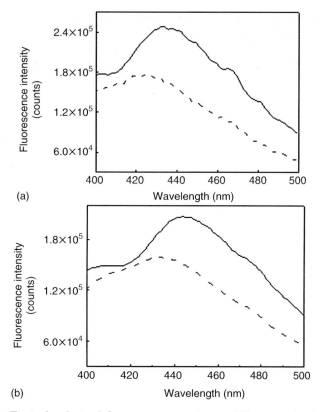

Fig. 11.30. Typical polarized fluorescence spectra at 340-nm excitation of human breast tissue samples of 2-mm thickness [91]. *Solid curves*, spectra with excitation and emission polarizers oriented vertically ($I_{||}$); *dashed curves*, spectra with crossed excitation and emission polarizers ($I_{\perp}$). (**a**) Malignant tissue; (**b**) normal tissue

associated with wavelength-dependent scattering and the absorption properties of the tissue (see Fig. 11.30).

The mean fluorescence anisotropy values for normal and malignant human breast samples of tissue varying from $10\,\mu$m to $2\,$mm in thickness, determined with $440\,$nm emission and 340-nm excitation, are presented in Fig. 11.31 [91]. The theoretical fit to experimental data using (11.45) and the parameter of single scattering anisotropy reduction $A = 0.7$ gives the following data for the anisotropy and optical parameters: $r_0 = 0.34$; $\mu_s^{tot} = 59\,$mm^{-1}; $\mu_d^{tot} = 5.35\,$mm^{-1} for malignant tissue and $r_0 \approx 0.25$; $\mu_s^{tot} = 47\,$mm^{-1}; $\mu_d^{tot} = 3.45\,$mm^{-1} for normal tissue. The anisotropy values are higher for malignant tissues as compared to normal for very thin tissue sections, $d \leq 30\,\mu$m. By contrast, in thicker sections, the malignant tissue shows smaller fluorescence anisotropy than the normal tissues.

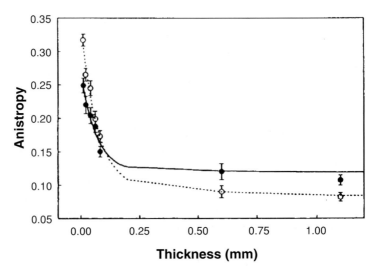

Fig. 11.31. The polarized fluorescence anisotropy measured at 440 nm for excitation at 340 nm for malignant (*open circles*) and normal (*filled circles*) human breast tissues as a function of tissue thickness [91]. The error bars represent standard deviation. The *solid and dashed curves* show theoretical fits for normal and malignant tissues, respectively (11.45). An expanded view of the dependencies of the anisotropy on tissue thickness for small thicknesses is shown in the insert

The fact that fluorescence anisotropy varies with tissue thickness is associated with the manifestation of various mechanisms of fluorescence depolarization which are caused by energy transfer and rotational diffusion in the fluorophores and by the scattering of excitation and emission light. Energy transfer and/or rotational diffusion of the fluorophores dominate in thin tissue sections, and these processes are faster in normal tissues than in malignant ones. In thicker sections, light scattering dominates with more contribution to depolarization during light transport within the malignant tissues.

As was already mentioned in the beginning of this section, the light scattering anisotropy factor g and, correspondingly, the reduced scattering coefficient, μ'_s, or the transport MFP, l_t, determine the characteristic depolarization depth in a scattering medium. Parameter A, characterizing the reduction of the fluorescence anisotropy per scattering event in the described model, depends on the value of the g-factor [91]. The theoretical analysis done by the authors of [91] has shown that, for an anisotropy parameter g ranging between 0.7 and 0.9, the value for A varies between 0.7 and 0.8.

The results presented suggest that fluorescence anisotropy measurements may be used for discriminating malignant sites from normal ones and may be especially useful for epithelial cancer diagnostics where superficial tissue layers are typically examined [94, 95].

11.11 Summary

We conclude that polarization-sensitive methods are promising tools for optical medical diagnostics and visualization, especially for in situ morphological analysis of living tissue. Additionally, polarization discrimination of scattered probe light, currently being integrated with traditional optical diagnostical methods, such as, diffuse reflectance spectroscopy and imaging with diffuse reflected or transmitted light, offers a possibility for improving the diagnostical potential of these methods. Another novel contribution to optical medical diagnostics should emerge from the morphological study of tissues with expressed structural anisotropy. Typically, almost all of the polarization-sensitive techniques that we considered in this chapter can be realized with inexpensive commercially available instrumentation; neither do they require sophisticated data processing algorithms. In other words, these methods are completely suitable for widespread implementation in clinical diagnostic practice. In addition, fluorescence polarization measurements that can provide information at the molecular level may also be useful for discriminating malignant sites from normal ones.

A

Appendix

Single-Scattering Mueller Matrix from Mie Theory

Mie theory describes the scattering of a vector plane wave by a homogeneous sphere. For a spherical particle, its single-scattering Jones matrix is

$$\mathbf{J} = \begin{pmatrix} S_2 & 0 \\ 0 & S_1 \end{pmatrix}, \tag{A.1}$$

where S_1 and S_2 are functions of the polar scattering angle and can be obtained from the Mie theory:

$$S_1(\theta) = \sum_{n=1}^{\infty} \frac{2n+1}{n(n+1)} \{a_n \pi_n(\cos\theta) + b_n \tau_n(\cos\theta)\},$$

$$S_2(\theta) = \sum_{n=1}^{\infty} \frac{2n+1}{n(n+1)} \{b_n \pi_n(\cos\theta) + a_n \tau_n(\cos\theta)\}. \tag{A.2}$$

The parameters π_n and τ_n represent

$$\pi_n(\cos\theta) = \frac{1}{\sin\theta} P_n^1(\cos\theta),$$

$$\tau_n(\cos\theta) = \frac{d}{d\theta} P_n^1(\cos\theta), \tag{A.3}$$

where $P_n^1(\cos\theta)$ is the associated Legendre polynomial. The following recursive relationships are used to calculate π_n and τ_n:

$$\pi_n = \frac{2n-1}{n-1} \pi_{n-1} \cos\theta - \frac{n}{n-1} \pi_{n-2},$$

$$\tau_n = n\pi_n \cos\theta - (n+1)\pi_{n-1}, \tag{A.4}$$

and the initial values are:

$$\begin{cases} \pi_1 = 1, \ \pi_2 = \cos\theta, \\ \tau_1 = \cos\theta, \ \tau_2 = 3\cos 2\theta. \end{cases} \tag{A.5}$$

The coefficients a_n and b_n are

$$\begin{cases} a_n = \dfrac{S'_n(y)S_n(x) - n_{\mathrm{rel}}S_n(y)S'_n(x)}{S'_n(y)\zeta_n(x) - n_{\mathrm{rel}}S_n(y)\zeta'_n(x)}, \\[3mm] b_n = \dfrac{n_{\mathrm{rel}}S'_n(y)S_n(x) - S_n(y)S'_n(x)}{n_{\mathrm{rel}}S'_n(y)\zeta_n(x) - S_n(y)\zeta'_n(x)}, \end{cases} \tag{A.6}$$

where

$$\begin{aligned} x &= 2\pi a n_{\mathrm{b}}/\lambda, \\ y &= 2\pi a n_{\mathrm{s}}/\lambda, \\ n_{\mathrm{rel}} &= n_{\mathrm{s}}/n_{\mathrm{b}}, \end{aligned} \tag{A.7}$$

where a is the radius of the scattering sphere, λ is the wavelength in vaccuo, n_{s} is the refractive index of the scattering spheres, and n_{b} is the refractive index of the background medium. ζ_n and S_n can be written in terms of Bessel functions:

$$\begin{cases} S_n(z) = \left(\frac{\pi z}{2}\right)^{0.5} J_{n+0.5}(z), \\ \zeta_n(z) = S_n(z) + iC_n(z), \\ C_n(z) = -\left(\frac{\pi z}{2}\right)^{0.5} N_{n+0.5}(z), \end{cases} \tag{A.8}$$

where $J_{n+0.5}(z)$ is the Bessel function of the 1st kind and $N_{n+0.5}(z)$ is the Bessel function of the 2nd kind. The derivatives of S_n and C_n can be obtained through

$$\begin{cases} S'_n(z) = \left(\frac{\pi}{8z}\right)^{0.5} J_{n+0.5}(z) + \left(\frac{\pi z}{2}\right)^{0.5} J'_{n+0.5}(z), \\ C'_n(z) = -\left(\frac{\pi}{8z}\right)^{0.5} N_{n+0.5}(z) - \left(\frac{\pi z}{2}\right)^{0.5} N'_{n+0.5}(z). \end{cases} \tag{A.9}$$

The single-scattering Mueller matrix can be derived from the Jones matrix (A.1):

$$\mathbf{M}(\theta) = \tfrac{1}{2}$$
$$\begin{bmatrix} |S_2|^2 + |S_1|^2 & |S_2|^2 - |S_1|^2 & 0 & 0 \\ |S_2|^2 - |S_1|^2 & |S_2|^2 + |S_1|^2 & 0 & 0 \\ 0 & 0 & S_2 S_1^* + S_1 S_2^* & -i(S_2 S_1^* - S_1 S_2^*) \\ 0 & 0 & i(S_2 S_1^* - S_1 S_2^*) & S_2 S_1^* + S_1 S_2^* \end{bmatrix}. \tag{A.10}$$

Coordinate Transformation in a Multiple Scattering Medium

In polarimetry, every Stokes vector and Mueller matrix are associated with a specific reference plane and coordinates. In the Mie theory, the Mueller

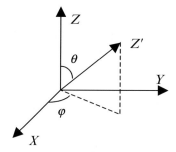

Fig. A.1. The coordinates transform of a single scattering event

matrix of a single scattering event is defined in the scattering plane that is formed by the incident light vector and the scattered light vector. For a general coordinate system associated with this scattering plane, the Z-axis is along the direction of photon propagation. The X-axis is within the reference plane and is perpendicular to the Z-axis. The Y-axis is perpendicular to both the Z-axis and the reference plane.

There is a local coordinate system associated with each incident photon packet, and its Stokes vector $\mathbf{S}_{\text{in}}$ is associated with this local coordinate system. As shown in Fig. A.1, the local coordinate system of the photon before scattering is (X, Y, Z). After the scattering event, the photon propagates along the Z'-axis; θ is the polar scattering angle and φ is the azimuth angle. The scattering plane is formed by the Z-axis and the Z'-axis, which is the new reference plane.

We use (9.1) to calculate the Stokes vector of the scattered light. As the Mueller matrix of the scattering event is defined in the reference plane, we first need to transform the Stokes vector of the incident light to the coordinate system associated with the reference plane. This transformation can be done by rotating the local coordinate system (X, Y, Z) by φ along the Z-axis, where the rotation matrix is

$$\mathbf{R}(\varphi) = \begin{bmatrix} 1 & 0 & 0 & 0 \\ 0 & \cos(2\varphi) & \sin(2\varphi) & 0 \\ 0 & -\sin(2\varphi) & \cos(2\varphi) & 0 \\ 0 & 0 & 0 & 1 \end{bmatrix}, \qquad (\text{A.11})$$

and the new Stokes vector is obtained by

$$\mathbf{S}'_{\text{in}} = \mathbf{R}(\varphi)\mathbf{S}_{\text{in}}. \qquad (\text{A.12})$$

The local coordinate system of the photon packet is tracked in the process. The transformation can be divided into two steps. From Fig. A.1, the first step is to rotate the (X, Y, Z) system by φ about the Z-axis, and the second step is to rotate the coordinate by θ about the rotated Y-axis to get (X', Y', Z'). After the transformation, the Z'-axis is aligned with the new light vector.

The transformation matrix is

$$
\begin{bmatrix} X' \\ Y' \\ Z' \end{bmatrix} = \begin{bmatrix} \cos\theta & 0 & -\sin\theta \\ 0 & 1 & 0 \\ \sin\theta & 0 & \cos\theta \end{bmatrix} \begin{bmatrix} \cos\phi & \sin\phi & 0 \\ -\sin\phi & \cos\phi & 0 \\ 0 & 0 & 1 \end{bmatrix} \begin{bmatrix} X \\ Y \\ Z \end{bmatrix}. \tag{A.13}
$$

After a photon packet passes through the turbid medium, its Stokes vector is recorded and accumulated. The local coordinate system was tracked in the simulation. In order to record the Stokes vector, the local coordinate system of each photon packet must be transformed into the laboratory coordinate system. In the laboratory coordinate system (e_1, e_2, e_3), the local photon coordinate can be written as

$$
\begin{bmatrix} x \\ y \\ z \end{bmatrix} = \begin{bmatrix} e_{1x} & e_{2x} & e_{3x} \\ e_{1y} & e_{2y} & e_{3y} \\ e_{1z} & e_{2z} & e_{3z} \end{bmatrix} \begin{bmatrix} e_1 \\ e_2 \\ e_3 \end{bmatrix}. \tag{A.14}
$$

To transform the photon Stokes vector from the local coordinate system into the laboratory coordinate system, the local coordinate system is rotated by its Z-axis so that the new X-axis lies within the (e_2, e_3) plane in the laboratory coordinate. The rotation angle is

$$
\varphi = \tan^{-1}(e_{1x}/e_{1y}). \tag{A.15}
$$

The rotation matrix and the new Stokes vector can be obtained from (A.11) and (A.12).

Glossary

This glossary was compiled using mostly [1-7] (Chap. 1).

absorption	the transformation of light (radiant) energy to some other form of energy, usually heat, as the light transverses matter
absorption spectrum	the spectrum formed by light that has passed through a medium in which light of certain wavelengths was absorbed
acquisition time	the period of time of acquiring experimental data
anisotropic scattering	a scattering process characterized by a clearly-apparent direction of photons that may be due to the presence of large scatterers
attenuation	a decrease in energy per unit area of a wave or beam of light; it occurs as the distance from the source increases and is caused by absorption or scattering
attenuation (extinction) coefficient	the reciprocal of the distance over which light of intensity I is attenuated to $I/e \approx 0.37I$; the units are typically cm^{-1}
autocorrelation	the correlation of an ordered series of observations with the same series in an altered order
autocorrelation function	the characteristic of the second-order statistics of a random process that shows how fast the random value changes from point to point, e.g., the autocorrelation function of intensity fluctuations caused by scattering of a laser beam by a rough surface characterizes the size and distribution of speckle sizes in the induced speckle pattern; the Fourier transform of the autocorrelation function represents the power spectrum of a random process
autofluorescence	natural tissue fluorescence
ballistic (coherent) photons	a group of unscattered and strictly straight-forward scattered photons
backscattering	the dispersion of a fraction of the incident radiation in a backward direction
bimodal distribution	a distribution having two modes

birefringence	the phenomenon exhibited by certain crystals in which an incident ray of light is split into two rays, called an ordinary ray and an extraordinary ray, which are plane- (linear) polarized in mutually orthogonal planes
chirality	in an object, describes the mirror-equal "right" or "left" modification; *optical activity* is one of the exhibitions of chirality, when the asymmetric structure of a molecule or crystal existing of two forms ("right" and "left") causes the substance (ensemble of these molecules or crystal) to rotate the plane of the incident linear polarized light; the pure "right" or "left" optically active substances have identical physical and chemical properties, but their biochemical and physiological properties can be quite different
chromophore	a chemical that absorbs light with a characteristic spectral pattern
coherence length	characterizes the degree of temporal coherence of a light source: $l_C = c\tau_C$, where c is the light speed and τ_C is the coherence time, which is approximately equal to the pulse duration of the pulse light source or inversely proportional to the frequency bandwidth of a continuous wave light source
coherent light	light in which the electromagnetic waves maintain a fixed phase relationship over a period of time and in which the phase relationship remains constant for various points that are perpendicular to the direction of propagation
constructive interference	the interference of two or more waves of equal frequency and phase, resulting in their mutual reinforcement and producing a single amplitude equal to the sum of the amplitudes of the individual waves
contrast of the intensity fluctuations	the relative difference between light and dark areas of a speckle pattern
correlation	the degree of correlation between two or more attributes or measurements on the same group of elements

correlation length	the length within which the degree of correlation between two measurements of a spatially dependent quantity is high (close to unity); for example, L_c is the correlation length of the scattering surface of the spatial inhomogeneities (random relief)
decorrelation of speckles	relates to statistics of the second order that characterize the size and distribution of speckle sizes and show how fast the intensity changes from point to point in the speckle pattern; decorrelation means that such changes of intensity tend to be faster
depolarization	depriving (destruction) of light polarization
destructive interference	the interference of two waves of equal frequency and opposite phase, resulting in their cancellation where the negative displacement of one always coincides with the positive displacement of the other
developed speckles	the speckles that are characterized by Gaussian statistics of the complex amplitude, the unity contrast of intensity fluctuations, and a negative exponential function of the intensity probability distribution (the most probable intensity value in the corresponding speckle pattern is equal to zero; i.e., destructive interference occurs with the highest probability)
diagnostic (or therapeutic) window	the spectral range from 600 to 1,600 nm within which the penetration depth of the light beams for most living tissues and blood is the highest; certain phototherapeutic and diagnostic modalities take advantage of this range for visible and NIR light
dichroism	a phenomenon related to pleochroism of a uniaxial crystal so that it exhibits two different colors when viewed from two different directions under transmitted light; *pleochroism* is the property possessed by certain crystals that exhibit different colors when viewed from different directions under transmitted light; this is one exhibition of the optical anisotropy caused by the anisotropy of absorption; the varieties of pleochroism are *circular dichroism*, different absorption for light with "right" and "left" circular polarization, and *linear dichroism*, different absorption for ordinary and extraordinary rays

diffuse photons	the photons that undertake multiple scatter with a broad variety of angles
diffusion wave spectroscopy (DWS)	the spectroscopy based on the study of dynamic light scattering in dense media with multiple scattering and related to the investigation of the dynamics of particles within very short time intervals
digital electronic autocorrelator	a device that reconstructs the time-domain autocorrelation function of intensity fluctuations
dispersion	the state of being dispersed, such as a photon trajectory (general); the variation of the index of refraction of a transparent substance, such as a glass, with the wavelength of light, the index of refraction increasing as the wavelength decreases (optics); the separation of white or compound light into its respective colors, as in the formation of a spectrum by a prism (optics); the scattering of values of a variable around the mean or median of a distribution (statistics); a system of dispersed particles suspended in a solid, liquid, or gas (chemistry)
Doppler effect	the apparent change in the frequency of a wave, such as a light wave or sound wave, resulting from a change in the distance between the source of the wave and the receiver
Doppler interferometry	the dynamic dual-beam interferometry when the reference beam pathlength is scanned with a constant speed; the Doppler signal induced is the measuring signal for depth profiling of an object placed in the measuring beam; the method is used in partially coherent interferometry or tomography of tissues
Doppler spectroscopy	the spectroscopy based on the study of the dynamic light scattering (*Doppler effect*) in media with single scattering and related to the investigation of the dynamics (velocity) of particles from the measurements of the Doppler shifts in the frequency of the waves scattered by the moving particles

dynamic light scattering — light scattering by a moving object that causes a Doppler shift in the frequency of the scattered wave relative to the frequency of the incident light

elastic (static) light scattering — light scattering by static (motionless) objects that occurs elastically, without changes of a photon energy or light frequency

emission spectrum — the emission obtained from a luminescent material at different wavelengths when it is excited by a narrow range of shorter wavelengths

excitation spectrum — the emission spectrum at one wavelength is monitored and the intensity at this wavelength is measured as a function of the exciting wavelength

far-field diffraction zone — the zone where Fraunhofer diffraction takes place; this is a type of diffraction in which the light source and the receiving screen are effectively at an infinite distance from the diffraction object, i.e., parallel beams of wave trains are used

form birefringence — birefringence that is caused by the structure of a medium; for example, a system of long dielectric cylinders made from an isotropic substance and arranged in a parallel fashion shows birefringence of form

forward scattering problem — the modeling of light propagation in a scattering medium by taking into account the experimental geometry, source, and detector characteristics and the known optical properties of a sample, and predicting the measurements and associated accuracies that result

fractal object — an object with a self-similar geometry, i.e., each arbitrary selected part of it is similar to the whole object

Fresnel diffraction — a type of diffraction in which the light source and the receiving screen are both at a finite distance from the diffraction object, i.e., divergent and convergent beams of wavetrains are used

Fresnel reflection — the reflection of a beam of radiation, such as light, which takes place at the interface between two media of different refractive indexes; not all the radiation is reflected, some may be refracted

Gaussian correlation function	the correlation function described by a bell-shaped (Gaussian) curve
Gaussian light beam	a light beam with a Gaussian shape for the transverse intensity profile; if the intensity at the center of the beam is I_0, then the formula for a Gaussian beam is $I = I_0 \exp(-2r^2/w^2)$, where r is the radial distance from the axis and w is the beam "waist;" the intensity profile of such a beam is said to be bell shaped; a laser beam is a Gaussian one; a single mode fiber also creates a Gaussian beam at its output
Gaussian statistics (normal statistics)	statistics when a bell-shaped (Gaussian) curve showing a distribution of probability associated with different values of a variate are valid
group refractive index	the refractive index associated with the group velocity of a train of waves traveling in a dispersive medium; the group velocity, and correspondingly the group refractive index, depends on the mean wavelength of a train of waves and on the rate of change of velocity with wavelength
hemoglobin spectrum	the main bands are the following: Soret band: 400–440 nm segment; Q bands: 540–580-nm segment
Henyey-Greenstein phase function (HG)	one of the practical semiempirical approximations of the scattering phase function
homogeneous medium	a medium that has common physical properties, including optical properties, throughout
image-carrying photons	a group of photons selected for producing an image of a certain macroinhomogeneity within a scattering medium
immersion medium (liquid)	a liquid that provides optical matching between an objective and a biological object; it enhances the numerical aperture of the objective and the microscope resolution; in addition, optical matching reduces surface reflection and scattering and consequently allows for receiving higher contrast images
immersion technique	the technique used for reduction of light scattering in a bulk inhomogeneous medium by matching of the refractive index of the scatterers and ground substance; immersion liquids with an appropriate refractive index and rate of diffusion are usually used

index of refraction	a number indicating the speed of light in a given medium as either the ratio of the speed of light in a vacuum to that in the given medium (absolute index of refraction) or the ratio of the speed of light in a specified medium to that in the given medium (relative index of refraction)
inhomogeneous medium	a medium with regular or irregular spatial distribution of physical properties, including optical properties
intensity probability density distribution function	a function that describes the distribution of probability over the values of the light intensity
interference of speckle fields (speckle-modulated fields)	the interference of the fields in which amplitudes and phases are randomly modulated due to their interaction (scattering) with inhomogeneous (scattering) media
inverse MC (IMC) method	the iterative method that is based on the statistical simulation of photon transport in the scattering media and that provides a tool for the most accurate solutions of inverse scattering problem; it takes into account the real geometry of the object, the measuring system, and light beams; the main disadvantage is the long computation time
inverse scattering problem	the attempt to take a set of measurements and error estimates, and only a limited set of parameters describing the sample and experiment, and to derive the remaining parameters; usually the geometry is known, intensities or their parameters are measured, and the optical properties or sizes of scatterers need to be derived; if these properties are considered to be spatially varying, then the resultant solutions can be presented as a 2 or 3D function of space, i.e., as an image
isotropic scattering	an equality of scattering properties along all axes
LASCA	acronym for laser speckle contrast analysis; the method uses the spatial statistics of time-integrated speckles; the full-field technique for visualizing capillary blood flow
latex	a suspension of micron-sized polystyrene spheres

light

ultraviolet (UV): UVC: 100 – 280 nm, UVB: 280 – 315 nm, UVA: 315 – 400 nm; visible: 400 – 780 nm (violet: 400 – 450 nm, blue: 450 – 480 nm, green: 510 – 560 nm, yellow: 560 – 590 nm, orange: 590 – 620 nm, red: 620 – 780 nm); infrared (IR) light: IRA (or NIR): 780 – 1400 nm, IRB (or middle IR (MIR)): 1400 – 3000 nm, IRC (or far IR (FIR)): 3 – 1000 μm

light scattering

change in direction of the propagation of light in a turbid medium caused by reflection and refraction by microscopic internal structures

low-step scattering

the scattering process in which on average each photon undertakes no more than a few scattering events (approximately less than five to ten)

LSM [light-scattering matrix (intensity or Mueller's matrix)]

the 4×4 matrix which connects the *Stokes vector* of the incident light with the *Stokes vector* of the scattered light; it describes the polarization state of the scattered light in the far zone that is dependent on the polarization state of the incident light and structural and optical properties of the object

LSM element

one of 16 elements of the light-scattering matrix; each element depends on the scattering angle and the wavelength, and geometrical and optical parameters of the scatterers and their arrangement

Mie or Lorenz–Mie scattering theory

exact solution of Maxwell's electromagnetic field equations for a homogeneous sphere

monodisperse model

a model presenting a disperse medium as a monodisperse one, such as an ensemble of scatterers with an equal size and refractive index for each scatterer

monodisperse system

a disperse system (medium) with a single value of characteristic parameter, such as an ensemble of scatterers with the equal size and refractive index for each scatterer; a healthy eye cornea is a good example of the monodisperse system, because it consists of dielectric rods with the same refractive index and radius dispersed in a homogeneous ground substance

Monte Carlo method	a numerical method of statistical modeling; in tissue studies it provides the most accurate simulation of photon transport in the samples with a complex geometry, accounting for the specificity of the measuring system and light beams configuration
multilayered tissue	a tissue that consists of many layers with different structural and optical properties, such as skin, blood vessel wall, and wall of bladder
multiple scattering	a scattering process in which on average each photon undertakes many scattering events (approximately more than five to ten)
non-Gaussian statistics	a statistically nonuniform process in which the statistical characteristics of the scattered light essentially depend on the observation angle and the degree of nonuniformity of an object
nonuniform medium	see **inhomogeneous medium**
objective speckles	the speckles formed in a free space and usually observed on a screen placed at a certain distance from an object
optical activity	the ability of a substance to rotate the plane of polarization of plane- (linear) polarized light (see **chirality**)
optical slicing	the process of extracting the optical image of a thin layer of tissue; the image is used for tomographic reconstruction of a whole body organ
optical path	the path of light through a medium, having a magnitude equal to the geometric distance through the system times the index of refraction of the medium
optical phantom	a medium that models the transport of visible and infrared light in tissue and is needed to evaluate techniques, to calibrate equipment, to optimize procedures, and for quality assurance
optical retarder	a device that provides an optical retardation: phase shift or optical path difference; such retarders as the half- or the quarter-wavelength plates provide, respectively, the half-wave or the quarter-wave phase difference

osmotic phenomenon   the tendency of a fluid to pass through a semipermeable membrane into a solution where its concentration is lower, thus equalizing the conditions on either side of the membrane

osmotic stress   the force that a dissolved substance exerts on a semipermeable membrane through which it cannot penetrate, when it is separated from a pure solvent by the membrane

paraxial region   the region where paraxial rays, lying close to the axis of an optical system, propagate

phase-contrast microscopy (phase microscopy)   a microscopy that translates the difference in the phase of light transmitted through or reflected by an object into difference of intensity in the image

phase fluctuations of the scattered field   the fluctuations that are induced by different optical paths for different parts or time periods of a wave front interacting with an inhomogeneous generally dynamic medium

phase object   an object that introduces the difference in phase of the light transmitted through or reflected by an object

$\lambda/4$-phase plate   see **optical retarder**; a device that provides an optical phase shift of 90° ($\pi/2$ radians) or an optical path difference equal to a quarter of the wavelength; a thin plate of birefringent substance, such as calcite or quartz, is cut parallel to the optical axis of the crystal and of a specific thickness that is calculated to give a phase difference of 90° ($\pi/2$ radians) between the emergent ordinary ray and the emergent extraordinary ray for light of a specified wavelength; quarter-wave plates are usually constructed for the wavelength of sodium light (589 nm); if the angle between the plane of polarization of light incident upon the plate and the optic axis of the plate is 45°, then circularly polarized light is produced and emerges from the plate; if the angle is other than 45°, elliptically polarized light is produced

phase shift (phase difference)	the difference in phase between two wave forms; phase difference is measured by the phase angle between the waves; when two waves have a phase shift (difference) of 90° (or $\pi/2$ radians), one wave is at maximum amplitude when the other wave is at zero amplitude; with a phase difference of 180° (π radians), both waves have zero amplitude at the same time, but one wave is at a crest when the other wave is at a trough
photon	a quantum of electromagnetic radiation, usually considered as an elementary particle that has its own antiparticle and that has zero rest mass and charge and a spin of 1
photon-correlation spectroscopy	a noninvasive method for studying the dynamics of particles on a comparatively large time scale; the implementation of the single-scattering regime and the use of coherent light sources are of fundamental importance in this case; the spatial scale of testing a colloid structure (an ensemble of biological particles) is determined by the inverse of the wave vector; *quasielastic light scattering* spectroscopy, *spectroscopy of intensity fluctuations*, and *Doppler spectroscopy* are synonymous terms related to *dynamic light scattering*
photon-counting system	a system that makes use of a specific method of photoelectron signal processing and provides sequential detection of single photons; photomultipliers (PMT) or avalanche photodetectors (APD) are usually used for photoncounting; the technique is applicable for detecting very weak signals
photon-density wave	a wave of progressively decaying intensity; microscopically, individual photons migrate randomly in a scattering medium, but collectively they form a photon-density wave at a modulation frequency that moves away from a radiation source
photosensitizer	a substance that increases the absorption of another substance at a particular wavelength band

photon transport	the process of photon travel in a homogeneous or inhomogeneous medium with possible macroinhomogeneities; a photon changes its direction due to reflection, refraction, diffraction or scattering and can be absorbed by an appropriate molecule on its way
pixel	the smallest element of an image that can be individually displayed
polarimetry	measurement of the polarization properties of light
polarization of light	a state, or the production of a state, in which rays of light exhibit different properties in different directions; *linear (plane)*: when the electric field vector oscillates in a single, fixed plane all along the beam, the light is said to be linearly (plane) polarized; *elliptical*: when the plane of the electric field rotates, the light is said to be elliptically polarized because the electric field vector traces out an ellipse at a fixed point in space as a function of time; *circular*: when the ellipse happens to be a circle, the light is said to be circularly polarized
polarization anisotropy	an inequality of polarization properties along different axes
polarizer	a device, often a crystal or prism, that produces polarized light from unpolarized light
probability density function (probability density distribution)	a function that describes the distribution of probability over the values of a variable
quasielastic light scattering	see **dynamic light scattering**
quasimonochromatic wave	a wave that has a very narrow but nonzero frequency (or wavelength) bandwidth; it can be presented as a group of monochromatic waves with a slightly different wavelength
quasiordered medium	a medium that has a structure very close to the ordered one, but nevertheless is not completely ordered which is caused by specific interactions between molecules and molecular structures; many of the natural media, including water and some living tissues, are examples of quasiordered media

radiation transfer theory (RTT)	the theory based on the integro-differential equation (the Boltzmann or linear transport equation), which is a balance equation describing the flow of particles (e.g., photons) in a given volume element that takes into account their velocity c, location $\bar{r}$, and changes due to collisions (i.e., scattering and absorption)
random medium	a specific state of a *nonuniform (inhomogeneous) medium* characterized by the irregular spatial distribution of its physical properties, including the optical properties
random phase screen (RPS)	a specific state of a *random medium* characterized by random spatial variations of the refractive index, which induce the corresponding variations in the phase shift of the optical wave transmitted through or reflected by the RPS
Rayleigh theory	the theory that addresses the problem of calculating scattering by small particles (with respect to the wavelength of the incident light) when individual particle scattering can be described as if it is a single dipole, the scattered irradiance is inversely proportional to λ^4 and increases as a^6, and the angular distribution of the scattered light is isotropic
Rayleigh–Debye–Gans theory (approximation)	the theory that addresses the problem of calculating the scattering by a special class of arbitrary shaped particles; it requires that the electric field inside the particle be close to that of the incident field and that the particle can be viewed as a collection of independent dipoles that are all exposed to the same incident field
reflectance (reflection coefficient)	the ratio of the intensity reflected from a surface to the incident intensity; it is a dimensional quantity
reflecting spectroscopy	the spectroscopy that is used for the spectral analysis of the light back-reflected (scattered) by an object

refractive index mismatch a difference in the index of refraction of two media being in contact; a scattering medium can be considered as a medium containing scattering particles whose index of refraction is mismatched relative to index of refraction of the ground substance

scatterer an inhomogeneity or a particle of a medium that refracts light or other electromagnetic radiation; light is diffused or deflected as a result of collisions between the wave and particles of the medium; sometimes it is a rough surface or a random-phase screen, also called scatterer

scattering medium a medium in which a wave or beam of particles is diffused or deflected by collisions with particles of this medium

scattering phase function the function that describes the scattering properties of the medium and is in fact the probability density function for scattering in the direction $\bar{s}'$ of a photon travelling in the direction $\bar{s}$; it characterizes an elementary scattering act; if scattering is symmetric relative to the direction of the incident wave, then the phase function depends only on the scattering angle θ (angle between directions $\bar{s}$ and $\bar{s}'$)

scattering spectrum the spectrum of scattered light; it can be differential, measured or calculated for a certain scattering angle, or integrated within an angle (field) of view of the measuring spectrometer

single-mode fiber a fiber in which only a single mode can be excited; for a fiber with a numerical aperture NA=0.1 and wavelength 633 nm the single mode can be excited if the core diameter is less than 4.8 μm

single-mode laser a laser that produces a light beam with a Gaussian shape of the transverse intensity profile without any spatial oscillations (see **Gaussian light beam**); in general, such lasers generate many optical frequencies (so-called longitudinal modes), which have the same transverse Gaussian shape

single scattering	the scattering process that occurs when a wave undertakes no more than one collision with particles of the medium in which it propagates
soft scattering particles	the refractive index of these particles, n_s, is close to the refractive index of the ground (interstitial) substance, n_0 ($n_s \geq n_0$)
spatial frequency	a spatial harmonic in the Fourier transform of a periodic or aperiodic (random) spatial distribution
spatial resolution	a measure of the ability of an optical imaging system to reveal the details of an image, i.e., to resolve adjacent elements
speckle	a single element of a speckle structure (pattern) that is produced as a result of the interference of a large number of elementary waves with random phases that arise when coherent light is reflected from a rough surface or when coherent light passes through a scattering medium
speckle correlometry	a technique that is based on the measurement of the intensity *autocorrelation function*, characterizing the size and the distribution of speckle sizes in a speckle pattern, caused, for example, by a scattering of coherent light beam from a rough surface; the statistical properties of the scattering object's structure can be deduced from such measurement
speckle photography	the measuring technique that uses a set of sequential photos of the speckle pattern taken at different moments or with different exposures; this is a full-field technique and can be used to study the dynamic properties of a scattering object (see **LASCA**); the updated instruments make use of computer controlled CCD cameras for averaging and storing the speckle patterns
speckle statistics of the first order	the statistics that define the properties of speckle fields at each point
speckle statistics of the second order	the statistics that show how fast the intensity changes from point to point in a speckle pattern, i.e., they characterize the size and distribution of speckle sizes in the pattern

specular	pertaining to or having the properties of a mirror
Stokes parameters	the four numbers I, Q, U, and V presenting an arbitrary polarization of light: I refers to the irradiance or intensity of the light; the parameters Q, U, and V represent the extent of horizontal liner, 45° linear, and circular polarization, respectively
Stokes vector	the vector that is formed by the four *Stokes parameters*
structure function	the function that describes the second-order statistics of a random process and is proportional to the difference between values of the autocorrelation function for zero and arbitrary values of the argument; the structure function is more sensitive to small-scale oscillations
subjective speckles	the speckles produced in the image space of an optical system (including an eye)
time-of-flight	the mean time of photon travel between two points which account for refractive index and scattering properties of the medium
tissue optical parameters (properties) control	any kind of physical or chemical action, such as mechanical stress or changes in osmolarity, which induces reversible or irreversible changes in the optical properties of a tissue [see **immersion medium (liquid)** and **immersion technique**]
tomographic reconstruction	obtaining 3D images by which the size, shape, and position of a hidden object can be determined
transmittance	ratio of the intensity transmitted through a sample to the incident intensity; it is a dimensionless quantity
two-photon fluorescence microscopy	the microscopy that employs both the ballistic and scattered photons at the wavelength of the second harmonic of incident radiation coming to a wide-aperture photodetector exactly from the focal area of the excitation beam

References

Chapter 1

1. *Medical Optical Tomography: Functional Imaging and Monitoring* **IS11**, Eds. G. Mueller, B. Chance, R. Alfano et al., SPIE, Bellingham (1993)
2. *Selected Papers on Tissue Optics Applications in Medical Diagnostics and Therapy* **MS 102**, Ed. V.V. Tuchin, SPIE, Bellingham (1994)
3. *Selected Papers on Optical Tomography, Fundamentals and Applications in Medicine* **MS 147**, Eds. O. Minet, G. Mueller, and J. Beuthan, SPIE, Bellingham (1998)
4. V.V. Tuchin, *Tissue Optics: Light Scattering Methods and Instruments for Medical Diagnosis,* SPIE Tutorial Texts in Optical Engineering, **TT38**, SPIE, Bellingham (2000)
5. *Handbook of Optical Biomedical Diagnostics* **PM107**, Ed. V.V. Tuchin, SPIE, Bellingham, USA (2002)
6. L.-H.V. Wang, G.L. Cote', and S.L. Jacques Eds., "Special section on Tissue Polarimetry," *J. Biomed. Opt.* **7**(3), 278–397 (2002)
7. *Biomedical Photonics Handbook*, Ed. Tuan Vo-Dinh, CRC, Boca Raton (2003)
8. C. Chandrasekhar, *Radiative Transfer*, Dover, Toronto, Ontario (1960)
9. R.G. Newton, *Scattering Theory of Waves and Particles*, McGraw-Hill, New York (1966)
10. M. Kerker, *The Scattering of Light and Other Electromagnetic Radiation*, Academic, New York (1969)
11. V.V. Sobolev, *Light Scattering in Planetary Atmospheres*, Pergamon, Oxford (1974)
12. A. Ishimaru, *Wave Propagation and Scattering in Random Media*, Academic, New York (1978)
13. H.C. van de Hulst, *Multiple Light Scattering. Tables, Formulas, and Applications*, Academic, New York (1980)
14. H.C. van de Hulst, *Light Scattering by Small Particles*, Wiley, New York (1957) [reprint, Dover, New York (1981)]
15. L.D. Barron, *Molecular Light Scattering and Optical Activity*, Cambridge University Press, London (1982)
16. C.F. Bohren and D.R. Huffman, *Absorption and Scattering of Light by Small Particles*, Wiley, New York (1983)
17. E.P. Zege, A.P. Ivanov, and I.L. Katsev, *Image Transfer through a Scattering Medium*, Springer, Berlin Heidelberg New York (1991)

18. A.Z. Dolginov, Yu.N. Gnedin, and N.A. Silant'ev, *Propagation and Polarization of Radiation in Cosmic Media*, Gordon and Breach, Basel (1995)

19. *Light Scattering by Nonspherical Particles*, Eds. M.I. Mishchenko, J.W. Hovenier, and L.D. Travis, Academic, San Diego (2000)

20. M.I. Mishchenko, L.D. Travis, and A.A. Lacis, *Scattering, Absorption, and Emission of Light by Small Particles*, Cambridge University Press, Cambridge (2002)

21. W.A. Shurcliff, *Polarized Light. Production and Use*, Harvard University Press, Cambridge, MA (1962)

22. W.A. Shurcliff and S.S. Ballard, *Polarized Light*, Van Nostrand, Princeton (1964)

23. E.L. O'Neill, *Introduction to Statistical Optics*, Addison-Wesley, Reading, MA (1963)

24. M. Born and E. Wolf, *Principles of Optics*, 7th ed., Cambridge University Press, Cambridge (1999)

25. D.S. Kliger, J.W. Lewis, and C.E. Randall, *Polarized Light in Optics and Spectroscopy*, Academic, Boston (1990)

26. E. Collet, *Polarized Light. Fundamentals and Applications*, Dekker, New York (1993)

27. R.M.A. Azzam and N.M. Bashara, *Ellipsometry and Polarized Light*, Elsevier, Amsterdam (1994)

28. C. Brosseau, *Fundamentals of Polarized Light: A Statistical Optics Approach*, Wiley, New York (1998)

29. S. Jiao, G. Yao, and L.-H.V. Wang, "Depth-resolved two-dimensional Stokes vectors of backscattered light and Mueller matrices of biological tissue measured with optical coherence tomography," *Appl. Opt.* **39**(34), 6318–6324 (2000)

Chapter 2

1. *Medical Optical Tomography: Functional Imaging and Monitoring* **IS11**, Eds. G. Mueller, B. Chance, R. Alfano et al., SPIE, Bellingham (1993)

2. *Selected Papers on Tissue Optics Applications in Medical Diagnostics and Therapy* **MS 102**, Ed. V.V. Tuchin, SPIE, Bellingham (1994)

3. *Selected Papers on Optical Tomography, Fundamentals and Applications in Medicine* **MS 147**, Eds. O. Minet, G. Mueller, and J. Beuthan, SPIE, Bellingham (1998)

4. V.V. Tuchin, *Tissue Optics: Light Scattering Methods and Instruments for Medical Diagnosis*, SPIE Tutorial Texts in Optical Engineering, **TT38**, SPIE, Bellingham (2000)

5. *Handbook of Optical Biomedical Diagnostics* **PM107**, Ed. V.V. Tuchin, SPIE, Bellingham (2002)

6. L.-H.V. Wang, G.L. Cote', and S.L. Jacques Eds., "Special section on Tissue Polarimetry", *J. Biomed. Opt.* **7**(3), 278–397 (2002)

7. *Biomedical Photonics Handbook*, Ed. Tuan Vo-Dinh, CRC, Boca Raton (2003)

8. C. Chandrasekhar, *Radiative Transfer*, Dover, Toronto, Ontario (1960)

9. R.G. Newton, *Scattering Theory of Waves and Particles*, McGraw-Hill, New York (1966)

10. M. Kerker, *The Scattering of Light and Other Electromagnetic Radiation*, Academic, New York (1969)

11. V.V. Sobolev, *Light Scattering in Planetary Atmospheres*, Pergamon, Oxford (1974)
12. A. Ishimaru, *Wave Propagation and Scattering in Random Media*, IEEE, New York (1997)
13. H.C. van de Hulst, *Multiple Light Scattering. Tables, Formulas, and Applications*, Academic, New York (1980)
14. H.C. van de Hulst, *Light Scattering by Small Particles*, Wiley, New York (1957) [reprint, Dover, New York (1981)].
15. L.D. Barron, *Molecular Light Scattering and Optical Activity*, Cambridge University Press, London (1982)
16. C.F. Bohren and D.R. Huffman, *Absorption and Scattering of Light by Small Particles*, Wiley, New York (1983)
17. E.P. Zege, A.P. Ivanov, and I.L. Katsev, *Image Transfer through a Scattering Medium*, Springer, Berlin Heidelberg New York (1991)
18. A.Z. Dolginov, Yu.N. Gnedin, and N.A. Silant'ev, *Propagation and Polarization of Radiation in Cosmic Media*, Gordon and Breach, Basel (1995)
19. *Light Scattering by Nonspherical Particles*, Eds. M.I. Mishchenko, J.W. Hovenier, and L.D. Travis, Academic, San Diego (2000)
20. M.I. Mishchenko, L.D. Travis, A.A. Lacis, *Scattering, Absorption, and Emission of Light by Small Particles*, Cambridge University Press, Cambridge (2002)
21. W.A. Shurcliff, *Polarized Light. Production and Use*, Harvard University Press, Cambridge, MA (1962)
22. W.A. Shurcliff and S.S. Ballard, *Polarized Light*, Van Nostrand, Princeton (1964)
23. E.L. O'Neill, *Introduction to Statistical Optics*, Addison-Wesley, Reading, MA (1963)
24. M. Born and E. Wolf, *Principles of Optics*, 7th ed., Cambridge University Press, Cambridge (1999)
25. D.S. Kliger, J.W. Lewis, and C.E. Randall, *Polarized Light in Optics and Spectroscopy*, Academic, Boston (1990)
26. E. Collet, *Polarized Light. Fundamentals and Applications*, Dekker, New York (1993)
27. R.M.A. Azzam and N.M. Bashara, *Ellipsometry and Polarized Light*, Elsevier, Amsterdam (1994)
28. C. Brosseau, *Fundamentals of Polarized Light: A Statistical Optics Approach*, Wiley, New York (1998)
29. J.M. Schmitt and G. Kumar, "Turbulent nature of refractive-index variations in biological tissue," *Opt. Lett.* **21**, 1310–1312 (1996)
30. D.A. Zimnyakov, V.V. Tuchin, and A.A. Mishin, "Spatial speckle correlometry in applications to tissue structure monitoring," *Appl. Opt.* **36**, 5594–5607 (1997)
31. J.M. Schmitt and G. Kumar, "Optical scattering properties of soft tissue: a discrete particle model," *Appl. Opt.* **37**(13), 2788–2797 (1998)
32. J.W. Goodman, *Statistical Optics*, Wiley, New York (1985)
33. S.M. Rytov, Y.A. Kravtsov, and V.I. Tatarskii, *Wave Propagation Through Random Media*, Vol. 4 of *Principles of Statistical Radiophysics*, Springer Berlin Heidelberg New York (1989)
34. S.Ya. Sid'ko, V.N. Lopatin, and L.E. Paramonov, *Polarization Characteristics of Solutions of Biological Particles*, Nauka, Novosibirsk (1990)

35. R.D. Dyson, *Cell Biology: A Molecular Approach*, Allyn and Bacon, Boston (1974)

36. P. Latimer, "Light scattering and absorption as methods of studying cell population parameters," *Ann. Rev. Biophys. Bioeng.* **11**(1), 129–150 (1982)

37. R. Drezek, M. Guillaud, T. Collier, I. Boiko, A. Malpica, C. Macaulay, M. Follen, and R. Richards-Kortum, "Light scattering from cervical cells throughout neoplastic progression: influence of nuclear morphology, DNA content, and chromatin texture," *J. Biomed. Opt.* **8**, 7–16 (2003)

38. J.R. Mourant, J.P. Freyer, A.H. Hielscher, A.A. Eick, D. Shen, and T.M. Johnson, "Mechanisms of light scattering from biological cells relevant to noninvasive optical-tissue diagnostics," *Appl. Opt.* **37**, 3586–3593 (1998)

39. J.R. Mourant, M. Canpolat, C. Brocker, O. Esponda-Ramos, T. Johnson, A. Matanock, K. Stetter, and J.P. Freyer, "Light scattering from cells: the contribution of the nucleous and the effects of proliferative status," *J. Biomed. Opt.* **5**, 131–137 (2000)

40. A. Dunn, C. Smithpeter, A.J. Welch, and R. Richards-Kortum, "Finite-difference time-domain simulation of light scattering from single cells," *J. Biomed. Opt.* **2**(3), 262–266 (1997)

41. R. Drezek, A. Dunn, and R. Richards-Kortum, "Light scattering from cells: finite-difference time-domain simulations and goniometric measurements," *Appl. Opt.* **38**(16), 3651–3661 (1999)

42. K. Sokolov, R. Drezek, K. Gossagee, and R. Richards-Kortum, "Reflectance spectroscopy with polarized light: Is it sensitive to cellular and nuclear morphology," *Opt. Express* **5**, 302–317 (1999)

43. A.N. Yaroslavsky, I.V. Yaroslavsky, T. Goldbach, and H.-J. Schwarzmaier, "Influence of the scattering phase function approximation on the optical properties of blood determined from the integrating sphere measurements," *J. Biomed. Opt.* **4**(1), 47–53 (1999)

44. G. Kumar and J.M. Schmitt, "Micro-optical properties of tissue," *Proc. SPIE* **2679**, 106–116 (1996)

45. J.R. Mourant, T.M. Johnson, S. Carpenter, A. Guerra, T. Aida, and J.P. Freyer, "Polarized angular dependent spectroscopy of epithelial cells and epithelial cell nuclei to determine the size scale of scattering structures," *J. Biomed. Opt.* **7**(3), 378–387 (2002)

46. M.J. Hogan, J.A. Alvardo, and J. Weddel, *Histology of the Human Eye*, Saunders, Philadelphia (1971)

47. Q. Zhou and R.W. Knighton, "Light scattering and form birefringence of parallel cylindrical arrays that represent cellular organelles of the retinal nerve fiber layer," *Appl. Opt.* **36**(10), 2273–2285 (1997)

48. G. Videen and D. Ngo, "Light scattering multipole solution for a cell," *J. Biomed. Opt.* **3**, 212–220 (1998)

49. K.S. Shifrin, *Physical Optics of Ocean Water*, American Institute of Physics, New York (1988)

50. V.V. Bakutkin, I.L. Maksimova, P.I. Saprykin, V.V. Tuchin, and L.P. Shubochkin, "Light scattering by human eye sclera," *J. Appl. Spectrosc. (USSR)* **46**, 104–107 (1987)

51. V.V. Tuchin, I.L. Maksimova, D.A. Zimnyakov, I.L. Kon, A.H. Mavlutov, and A.A. Mishin, "Light propagation in tissues with controlled optical properties," *J. Biomed. Opt.* **2**, 401–417 (1997)

52. A. Brunsting and P.F. Mullaney, "Differential light scattering from spherical mammalian cells," *Biophys. J.* **10**, 439–453 (1974)
53. J. Beuthan, O. Minet, J. Helfmann, M. Herring, and G. Mueller, "The spatial variation of the refractive index in biological cells," *Phys. Med. Biol.*, **41**(3), 369–382 (1996)
54. P.S. Tuminello, E.T. Arakawa, B.N. Khare, J.M. Wrobel, M.R. Querry, and M.E. Milham, "Optical properties of Bacillus subtilis spores from 0.2 to 2.5 µm," *Appl. Opt.* **36**, 2818–2824 (1997)
55. F.A. Duck, *Physical Properties of Tissue: A Comprehensive Reference Book*, Academic, San Diego (1990)
56. K.W. Keohane and W.K. Metcalf, "The cytoplasmic refractive index of lymphocytes, its significance and its changes during active immunisatioin," *Q. J. Exp. Physiol. Cogn. Med. Sci.* **44**, 343–346 (1959)
57. F.P. Bolin, L.E. Preuss, R.C. Taylor, and R.J. Ference, "Refractive index of some mammalian tissues using a fiber optic cladding method," *Appl. Opt.* **28**, 2297–2303 (1989)
58. *Laser–Induced Interstitial Thermotherapy*, Eds. G. Müller and A. Roggan, SPIE, Bellingham (1995)
59. F.H. Silver, *Biological Materials: Structure, Mechanical Properties, and Modeling of Soft Tissues*, New York University Press, New York (1987)
60. R.G. Kessel, *Basic Medical Histology: The Biology of Cells, Tissues, and Organs*, Oxford University Press, New York (1998)
61. F. Bettelheim, "On the optical anisotropy of lens fibre cells," *Exp. Eye Res.* **21**, 231–234 (1975)
62. J.Y.T. Wang and F.A. Bettelheim, "Comparative birefringence of cornea," *Comp. Biochem. Physiol., Part A: Mol. Integr. Physiol.* **51**, 89–94 (1975)
63. R.P. Hemenger, "Birefringence of a medium of tenuous parallel cylinders," *Appl. Opt.* **28**(18), 4030–4034 (1989)
64. D.J. Maitland and J.T. Walsh, "Quantitative measurements of linear birefringence during heating of native collagen," *Laser Surg. Med.* **20**, 310–318 (1997)
65. H.B. Klein Brink, "Birefringence of the human crystalline lens in vivo," *J. Opt. Soc. Am. A* **8**, 1788–1793 (1991)
66. R.P. Hemenger, "Refractive index changes in the ocular lens result from increased light scatter," *J. Biomed. Opt.* **1**, 268–272 (1996)
67. E.J. Naylor, "The structure of the cornea as revealed by polarized light," *Q. J. Microsc. Sci.* **94**, 83–88 (1953)
68. E.P. Chang, D.A. Keedy, and C.W. Chien, "Ultrastructures of rabbit corneal stroma: Mapping of optical and morphological anisotropies," *Biochim. Biophys. Acta* **343**, 615–626 (1974)
69. V.F. Izotova, I.L. Maksimova, I.S. Nefedov, and S.V. Romanov, "Investigation of Mueller matrices of anisotropic nonhomogeneous layers in application to optical model of cornea," *Appl. Opt.* **36**(1), 164–169 (1997)
70. J.S. Baba, B.D. Cameron, S. Theru, and G.L. Coté, "Effect of temperature, pH, and corneal birefringence on polarimetric glucose monitoring in the eye," *J. Biomed. Opt.* **7**(3), 321–328 (2002)
71. G.J. van Blokland, "Ellipsometry of the human retina in vivo: preservation of polarization," *J. Opt. Soc. Am. A* **2**, 72–75 (1985)
72. H.B. Klein Brink and G.J. van Blokland, "Birefringence of the human foveal area assessed in vivo with Mueller-matrix ellipsometry," *J. Opt. Soc. Am. A* **5**, 49–57 (1988)

73. R.C. Haskell, F.D. Carlson, and P.S. Blank, "Form birefringence of muscle," *Biophys. J.* **56**, 401–413 (1989)
74. S. Bosman, "Heat-induced structural alterations in myocardium in relation to changing optical properties," *Appl. Opt.* **32**(4), 461–463 (1993)
75. G.V. Simonenko, T.P. Denisova, N.A. Lakodina, and V.V. Tuchin, "Measurement of an optical anisotropy of biotissues," *Proc. SPIE* **3915**, 152–157 (2000)
76. G.V. Simonenko, V.V. Tuchin, N.A. Lakodina, "Measurement of the optical anisotropy of biological tissues with the use of a nematic liquid crystal cell," *J. Opt. Technol.* **67**(6), 559–562 (2000)
77. O.V. Angel'skii, A.G. Ushenko, A.D. Arkhelyuk, S.B. Ermolenko, and D.N. Burkovets, "Scattering of laser radiation by multifractal biological structures," *Opt. Spectrosc.* **88**(3), 444–447 (2000)
78. M.R. Hee, D. Huang, E.A. Swanson, and J.G. Fujimoto, "Polarization-sensitive low-coherence reflectometer for birefringence characterization and ranging," *J. Opt. Soc. Am. B* **9**, 903–908 (1992)
79. J.F. de Boer, T.E. Milner, M.J.C. van Gemert, and J.S. Nelson, "Two-dimensional birefringence imaging in biological tissue by polarization-sensitive optical coherence tomography," *Opt. Lett.* **22**(12), 934–936 (1997)
80. M.J. Everett, K. Schoenerberger, B.W. Colston, Jr., and L.B. Da Silva, "Birefringence characterization of biological tissue by use of optical coherence tomography," *Opt. Lett.* **23**(3), 228–230 (1998)
81. J.F. de Boer, T.E. Milner, and J.S. Nelson, "Determination of the depth resolved Stokes parameters of light backscattered from turbid media using polarization sensitive optical coherence tomography," *Opt. Lett.* **24**, 300–302 (1999)
82. J.F. de Boer and T.E. Milner, "Review of polarization sensitive optical coherence tomography and Stokes vector determination," *J. Biomed. Opt.* **7**(3), 359–371 (2002)
83. C.K. Hitzenberger, E. Gotzinger, M. Sticker, M. Pircher, and A.F. Fercher, "Measurement and imaging of birefringence and optic axis orientation by phase resolved polarization sensitive optical coherence tomography," *Opt. Express* **9**, 780–790 (2001)
84. S. Jiao and L.-H.V. Wang, "Jones-matrix imaging of biological tissues with quadruple-channel optical coherence tomography," *J. Biomed. Opt.* **7**(3), 350–358 (2002)
85. M.G. Ducros, J.F. de Boer, H. Huang, L. Chao, Z. Chen, J.S. Nelson, T.E. Milner, and H.G. Rylander, "Polarization sensitive optical coherence tomography of the rabbit eye," *IEEE J. Sel. Top. Quantum Electron.* **5**, 1159–1167 (1999)
86. M.G. Ducros, J.D. Marsack, H.G. Rylander III, S.L. Thomsen, and T.E. Milner, "Primate retina imaging with polarization-sensitive optical coherence tomography," *J. Opt. Soc. Am. A* **18**, 2945–2956 (2001)
87. C.E. Saxer, J.F. de Boer, B.H. Park, Y. Zhao, C. Chen, and J.S. Nelson, "High speed fiber based polarization sensitive optical coherence tomography of in vivo human skin," *Opt. Lett.* **26**, 1069–1071 (2001)
88. B.H. Park C.E. Saxer, S.M. Srinivas, J.S. Nelson, and J.F. de Boer, "*In vivo* burn depth determination by high-speed fiber-based polarization sensitive optical coherence tomography," *J. Biomed. Opt.* **6**, 474–479 (2001)
89. X.J. Wang, T.E. Milner, J.F. de Boer, Y. Zhang, D.H. Pashley, and J.S. Nelson, "Characterization of dentin and enamel by use of optical coherence tomography," *Appl. Opt.* **38**, 2092–2096 (1999)

90. A. Baumgartner, S. Dichtl, C.K. Hitzenberger, H. Sattmann, B. Robl, A. Moritz, A.F. Fercher, and W. Sperr, "Polarization-sensitive optical coherence tomography of dental structures," *Caries Res.* **34**(1), 59–69 (2000)

91. R.C.N. Studinski and I.A. Vitkin, "Methodology for examining polarized light interactions with tissues and tissuelike media in the exact backscattering direction," *J. Biomed. Opt.* **5**(3), 330–337 (2000)

92. K.C. Hadley and I.A. Vitkin, "Optical rotation and linear and circular depolarization rates in diffusively scattered light from chiral, racemic, and achiral turbid media," *J. Biomed. Opt.* **7**(3), 291–299 (2002)

93. J. Applequist, "Optical activity: Biot's bequest," *Am. Sci.* **75**, 59–67 (1987)

94. *Theory and Practice of Histological Techniques*, Eds. J.D. Bancroft and A. Stevens, Churchill Livingstone, Edinburgh (1990)

95. D.M. Maurice, *The Cornea and Sclera. The Eye*, Ed. H. Davson, Academic, Orlando, pp. 1–158 (1984)

96. R.W. Hart and R.A. Farrell, "Light scattering in the cornea," *J. Opt. Soc. Am.* **59**(6), 766–774 (1969)

97. R.L. McCally, and R.A. Farrell,"Light scattering from cornea and corneal transparency," in *Noninvasive Diagnostic Techniques in Ophthalmology*, Ed. B.R. Master, Springer, Berlin Heidelberg New York, pp. 189–210 (1990)

98. I.L. Maksimova, V.V. Tuchin, and L.P. Shubochkin, "Polarization features of eye's cornea," *Opt. Spectrosc.* **60**(4), 801–807 (1986)

99. I.L. Maksimova and L.P. Shubochkin, "Light-scattering matrices for a close-packed binary system of hard spheres," *Opt. Spectrosc.* **70**(6), 745–748 (1991)

100. V. Shankaran, M.J. Everett, D.J. Maitland, and J.T. Walsh, Jr., "Polarized light propagation through tissue phantoms containing densely packed scatterers," *Opt. Lett.* **25**(4), 239–241 (2000)

101. V. Shankaran, J.T. Walsh, Jr., D.J. Maitland, "Comparative study of polarized light propagation in biological tissues," *J. Biomed. Opt.* **7**(3), 300–306 (2002)

102. I.L. Maksimova, V.V. Tuchin, and L.P. Shubochkin "Light scattering matrix of crystalline lens," *Opt. Spectrosc.* **65**(3), 615–619 (1988)

103. I.L. Maksimova, D.A. Zimnyakov, and V.V. Tuchin, "Control of optical properties of biotissues: I. Spectral properties of the eye sclera," *Opt. Spectrosc.* **89**(1), 78–86 (2000)

104. I.L. Maksimova, "Scattering of radiation by regular and random systems comprised of parallel long cylindrical rods," *Opt. Spectrosc.* **93**(4), 610–619 (2002)

105. Y. Kamai and T. Ushiki, "The three-dimensional organization of collagen fibrils in the human cornea and sclera," *Invest. Ophthalmol. Visual Sci.* **32**, 2244–2258 (1991)

106. M.S. Borcherding, L.J. Blasik, R.A. Sittig, J.W. Bizzel, M. Breen, and H.G. Weinstein, "Proteoglycans and collagen fiber organization in human corneoscleral tissue," *Exp. Eye Res.* **21**, 59–70 (1975)

107. P. Rol, P. Niederer, U. Dürr, P.-D. Henchoz, and F. Fankhauser, "Experimental investigation on the light scattering properties of the human sclera," *Laser Light Ophthalmol.* **3**, 201–212 (1990)

108. P.O. Rol, "Optics for transscleral laser applications," Ph.D dissertation, Swiss Federal Institute of Technology, Zurich, Switzerland, 1992, ETH No. 9655.

109. M. Spitznas, "The fine structure of human scleral collagen," *Am. J. Ophthalmol.* **71**(1), 68–75 (1971)

110. Y. Huang and K.M. Meek, "Swelling studies on the cornea and sclera: the effect of pH and ionic strength," *Biophys. J.* **77**, 1655–1665 (1999)

111. S. Vaezy and J.I. Clark, "A quantitative analysis of transparency in the human sclera and cornea using Fourier methods," *J. Microsc.* **163**, 85–94 (1991)

112. S. Vaezy and J.I. Clark, "Quantitative analysis of the microstructure of the human cornea and sclera using 2-D Fourier methods," *J. Microsc.* **175**(2), 93–99 (1994)

113. D.E. Freund, R.L. McCally, R.A. Farrell, S.M. Cristol, N.L. L'Hernault, and H.F. Edelhauser, "Ultrastructure in anterior and posterior stroma of perfused human and rabbit corneas: relation to transparency," *Invest. Ophthalmol. Vis. Sci.* **36**, 1508–1523 (1995)

114. D.W. Leonard and K.M. Meek, "Refractive indices of the collagen fibrils and extrafibrillar material of the corneal stroma," *Biophys. J.* **72**, 1382–1387 (1997)

115. Z.S. Sacks, R.M. Kurtz, T. Juhasz, and G.A. Mourau, "High precision subsurface photodisruption in human sclera," *J. Biomed. Opt.* **7**(3), 442–450 (2002)

116. V.V. Tuchin, "Lasers light scattering in biomedical diagnostics and therapy," *J. Laser Appl.* **5**(2/3), 43–60 (1993)

117. V.V. Tuchin, "Light scattering study of tissues," *Physics – Uspekhi* **40**(5), 495–515 (1997)

118. V.V. Tuchin, "Coherence-domain methods in tissue and cell optics," *Laser Phys.* **8**(4), 807–849 (1998)

119. F.A. Bettelheim, "Physical basis of lens transparency," in: *The Ocular Lens: Structure, Function and Pathology*, Ed. H. Maisel, Dekker, New York (1985)

120. S. Zigman, G. Sutliff, and M. Rounds, "Relationships between human cataracts and environmental radiant energy. Cataract formation, light scattering and fluorescence," *Lens Eye Toxicity Res.* **8**, 259–280 (1991)

121. J. Xu, J. Pokorny, and V.C. Smith, "Optical density of the human lens," *J. Opt. Soc. Am. A.* **14**(5), 953–960 (1997)

122. B.K. Pierscionek and R.A. Weale, "Polarising light biomicroscopy and the relation between visual acuity and cataract," *Eye* **9**, 304–308 (1995)

123. B.K. Pierscionek, "Aging changes in the optical elements of the eye," *J. Biomed. Opt.* **1**(3), 147–156 (1996)

124. J.A. van Best, E.V.M.J. Kuppens, "Summary of studies on the blue – green autofluorescence and light transmission of the ocular lens," *J. Biomed. Opt.* **1**(3), 243–250 (1996)

125. F.A. Bettelheim, A.C. Churchill, W.G. Robinson, Jr., and J.S. Zigler, Jr., "Dimethyl sulfoxide cataract: a model for optical anisotropy fluctuations," *J. Biomed. Opt.* **1**(3), 273–279 (1996)

126. N.-T. Yu, B.S. Krantz, J.A. Eppstein, K.D. Ignotz, M.A. Samuels, J.R. Long, and J.F. Price, "Development of noninvasive diabetes screening device using the ratio of fluorescence to Rayleigh scattered light," *J. Biomed. Opt.* **1**(3), 280–288 (1996)

127. M.J. Costello, T.N. Oliver, and L.M. Cobo, "Cellular architecture in aged-related human nuclear cataracts," *Invest. Ophthalmol. Vis. Sci.* **3**(11), 2244–2258 (1992)

128. X. Wang, T.E. Milner, M.C. Chang, J.S. Nelson, "Group refractive index measurement of dry and hydrated type I collagen films using optical low-coherence reflectometry," *J. Biomed. Opt.* **1**(2) 212–216 (1996)

129. Y. Ozaki, "Medical application of Raman spectroscopy," *Appl. Spectrosc. Rev.* **24**(3), 259–312 (1988)

130. G.B. Benedek, "Theory of transparency of the eye," *Appl. Opt.* **10**(3), 459–473 (1971)

131. A. Tardieu and M. Delaye, "Eye lens proteins and transparency from light transmission theory to solution X-ray structural analysis," *Ann. Rev. Biophys. Chem.* **17**, 47–70 (1988)

132. A.V. Krivandin, "On the supramolecular structure of eye lens crystallins. The study by small-angle X-ray scattering," *Biophysica* **46**(6), 1274–1278 (1997)

133. S. Vaezy and J.I. Clark, "Characterization of the cellular microstructures of ocular lens using 2-D power law analysis," *Ann. Biomed. Eng.* **23**, 482–490 (1995)

134. J. Feder, *Fractals*, Plenum, New York (1988)

135. H.-O. Peitgen and D. Saupe, *The Science of Fractal Images*, Springer Berlin Heidelberg New York (1988)

136. *Fractal Frontiers*, Eds. M. Novak and T.G. Deway, World Scientific, Singapore (1999)

137. *Optics of Nanostructured Materials*, Eds. V.A. Markel and Th.F. George, Wiley, New York (2000)

138. B.R. Masters, "Fractal analysis of human retinal blood vessel patterns: developmental and diagnostic aspects," in: *Noninvasive Diagnostic Techniques in Ophthalmology*, Ed. B.R. Master, Springer Berlin Heidelberg New York, pp. 515–527 (1990)

139. H. Shimizu, T. Fujii, and T. Yokoyama, "Fractal dimension of the spatial structure of the human liver vascular network-computer analysis from serial tissue section," *Forma* **4**, 135–139 (1989)

140. Y. Sakurada, J. Uozumi, and T. Asakura, "Scaling properties of the Fresnel's diffraction field produced by one dimensional regular fractals," *Pure Appl. Opt.* **3**(3), 374–380 (1994)

141. S.S. Ulyanov, D.A. Zimnyakov, and V.V. Tuchin, "Fundamentals and applications of dynamic speckles induced by focused laser beam scattering," *Opt. Eng.* **33**(10), 3189–3201 (1994)

142. D.A. Zimnyakov, V.V. Tuchin, and S.R. Uttz, "A study of statistical properties of partially developed speckle fields as applied to the diagnostic of structural changes in human skin," *Opt. Spectrosc.* **76**(5), 838–844 (1994)

143. O.V. Angelsky and P.P. Maksimyak, "Optical correlations diagnostics of random field and objects," *Opt. Eng.* **34**, 937–981 (1995)

144. K. Ishii, T. Iwai, J. Uozumi, and T. Asakura, "Optical free-path-length distribution in a fractal aggregate and its effect on enhanced backscattering," *Appl. Opt.* **37**, 5014–5018 (1998)

145. K. Ishii, T. Iwai, and T. Asakura, "Correlation properties of light backscattered multiply from fractal aggregates of particles under Brownian motion," *J. Biomed. Opt.* **4**(2) 230–235 (1999)

146. A.V. Priezzhev, O.M. Ryaboshapka, N.N. Firsov, and I.V. Sirko, "Aggregation and disaggregation of erythrocytes in whole blood: study by backscattering technique," *J. Biomed. Opt.* **4**(1), 76–84 (1999)

147. M.Y. Lin, H.M. Lindsay, D.A. Weitz, R.C. Ball, R. Klein, and P. Meakin, "Universality of fractal aggregates as probed by light scattering," *Proc. R. Soc. London Ser. A* **423**(1864), 71–87 (1989)

148. N.G. Khlebtsov and A.G. Melnikov, "Structure factor and exponent of scattering by polydisperse fractal colloidal aggregates," *J. Colloid Interface Sci.* **163**(1), 145–151 (1994)

149. R. Botet, P. Rannou, and M. Cabane, "Sensitivity of some optical properties of fractals to the cut-off functions," *J. Phys. A: Math. Gen.* **28**, 297–316 (1995)

150. N.G. Khlebtsov, "Spectroturbidimetry of fractal clusters: test of density correlation cutoff," *Appl. Opt.* **35**(21), 4261–4270 (1996)
151. J. Cai, N. Lu, and C.M. Sorensen, "Analysis of fractal cluster morphology parameters: structural coefficient and density correlation function cutoff," *J. Colloid Interface Sci.* **171**(2), 470–473 (1995)
152. P. Meakin, "Models for colloidal aggregation," *Ann. Rev. Phys. Chem.* **39**, 237–267 (1988)

Chapter 3

1. *Medical Optical Tomography: Functional Imaging and Monitoring* **IS11**, Eds. G. Mueller, B. Chance, R. Alfano et al., SPIE, Bellingham (1993)
2. *Selected Papers on Tissue Optics Applications in Medical Diagnostics and Therapy* **MS 102**, Ed. V.V. Tuchin, SPIE, Bellingham (1994)
3. *Selected Papers on Optical Tomography, Fundamentals and Applications in Medicine* **MS 147**, Eds. O. Minet, G. Mueller, and J. Beuthan, SPIE, Bellingham (1998)
4. V.V. Tuchin, *Tissue Optics: Light Scattering Methods and Instruments for Medical Diagnosis*, SPIE Tutorial Texts in Optical Engineering, **TT38**, SPIE, Bellingham (2000)
5. *Handbook of Optical Biomedical Diagnostics* **PM107**, Ed. V.V. Tuchin, SPIE, Bellingham (2002)
6. L.-H.V. Wang, G.L. Cote, and S.L. Jacques Eds., "Special section on Tissue Polarimetry", *J. Biomed. Opt.* **7**(3), 278–397 (2002)
7. *Biomedical Photonics Handbook*, Ed. Tuan Vo-Dinh, CRC, Boca Raton (2003)
8. C. Chandrasekhar, *Radiative Transfer*, Dover, Toronto, Ontario (1960)
9. R.G. Newton, *Scattering Theory of Waves and Particles*, McGraw-Hill, New York (1966)
10. M. Kerker, *The Scattering of Light and Other Electromagnetic Radiation*, Academic, New York (1969)
11. V.V. Sobolev, *Light Scattering in Planetary Atmospheres*, Pergamon, Oxford (1974)
12. A. Ishimaru, *Wave Propagation and Scattering in Random Media*, IEEE, New York (1997)
13. H.C. van de Hulst, *Multiple Light Scattering. Tables, Formulas, and Applications*, Academic, New York (1980)
14. H.C. van de Hulst, *Light Scattering by Small Particles*, Wiley, New York (1957) [reprint, Dover, New York (1981)]
15. L.D. Barron, *Molecular Light Scattering and Optical Activity*, Cambridge University Press, London (1982)
16. C.F. Bohren and D.R. Huffman, *Absorption and Scattering of Light by Small Particles*, Wiley, New York (1983)
17. E.P. Zege, A.P. Ivanov, and I.L. Katsev, *Image Transfer through a Scattering Medium*, Springer Berlin Heidelberg New York (1991)
18. A.Z. Dolginov, Yu.N. Gnedin, and N.A. Silant'ev, *Propagation and Polarization of Radiation in Cosmic Media*, Gordon and Breach, Basel (1995)
19. *Light Scattering by Nonspherical Particles*, Eds. M.I. Mishchenko, J.W. Hovenier, and L.D. Travis, Academic, San Diego (2000)

20. M.I. Mishchenko, L.D. Travis, and A.A. Lacis, *Scattering, Absorption, and Emission of Light by Small Particles*, Cambridge University Press, Cambridge (2002)

21. W.A. Shurcliff, *Polarized Light. Production and Use*, Harvard University Press, Cambridge, MA (1962)

22. W.A. Shurcliff and S.S. Ballard, *Polarized Light*, Van Nostrand, Princeton (1964)

23. E.L. O'Neill, *Introduction to Statistical Optics*, Addison-Wesley, Reading, MA (1963)

24. M. Born and E. Wolf, *Principles of Optics*, 7th ed., Cambridge University Press, Cambridge (1999)

25. D.S. Kliger, J.W. Lewis, and C.E. Randall, *Polarized Light in Optics and Spectroscopy*, Academic, Boston (1990)

26. E. Collet, *Polarized Light. Fundamentals and Applications*, Dekker, New York (1993)

27. R.M.A. Azzam and N.M. Bashara, *Ellipsometry and Polarized Light*, Elsevier, Amsterdam (1994)

28. C. Brosseau, *Fundamentals of Polarized Light: A Statistical Optics Approach*, Wiley, New York (1998)

29. S. Jiao, G. Yao, and L.-H.V. Wang, "Depth-resolved two-dimensional Stokes vectors of backscattered light and Mueller matrices of biological tissue measured with optical coherence tomography," *Appl. Opt.* **39**(34), 6318–6324 (2000)

30. R.G. Johnston, S.B. Singham, and G.C. Salzman, "Polarized light scattering," *Comments Mol. Cell. Biophys.* **5**(3), 171–192 (1988)

31. G.C. Salzmann, S.B. Singham, R.G. Johnston, and C.F. Bohren, "Light scattering and cytometry," in: *Flow Cytometry and Sorting*, 2nd ed., Eds. M.R. Melamed, T. Lindmo, and M.L. Mendelsohn, Wiley, New York, pp. 81–107 (1990)

32. S. Jiao and L.-H.V. Wang, "Jones-matrix imaging of biological tissues with quadruple-channel optical coherence tomograthy," *J. Biomed. Opt.* **7**(3), 350–358 (2002)

33. G.W. Kattawar and E.S. Fry, "Inequalities between the elements of the Mueller scattering matrix: Comments," *Appl. Opt.* **21**, 18 (1982)

34. V.F. Izotova, I.L. Maksimova, and S.V. Romanov, "Utilization of relations between elements of the Mueller matrices for estimating properties of objects and the reliability of experiments," *Opt. Spectrosc.* **80**(5), 753–759 (1996)

35. I.L. Maksimova, S.N. Tatarintsev, and L.P. Shubochkin, "Multiple scattering effects in laser diagnostics of bioobjects," *Opt. Spectrosc.* **72**, 1171–1177 (1992)

36. A. Dunn, C. Smithpeter, A.J. Welch, and R. Richards-Kortum, "Finite-difference time-domain simulation of light scattering from single cells," *J. Biomed. Opt.* **2**(3), 262–266 (1997)

37. R. Drezek, A. Dunn, and R. Richards-Kortum, "Light scattering from cells: finite-difference time-domain simulations and goniometric measurements," *Appl. Opt.* **38**(16), 3651–3661 (1999)

38. K. Sokolov, R. Drezek, K. Gossagee, and R. Richards-Kortum, "Reflectance spectroscopy with polarized light: Is it sensitive to cellular and nuclear morphology," *Opt. Express* **5**, 302–317 (1999)

39. R.W. Hart and R.A. Farrell, "Light scattering in the cornea," *J. Opt. Soc. Am.* **59**(6), 766–774 (1969)

40. J.M. Ziman, *Models of Disorder: The Theoretical Physics of Homogeneously Disordered Systems*, Cambridge University Press, London (1979)

41. M.S. Wertheim, "Exact solution of the Percus-Yevick integral equation for hard spheres," *Phys. Rev. Lett.* **10**(8), 321–323 (1963)

42. J.L. Lebovitz, "Exact solution of generalized Percus-Yevick equation for a mixture of hard spheres," *Phys. Rev.* **133**(4A), 895–899 (1964)

43. R.J. Baxter, "Ornstein-Zernike relation and Percus-Yevick approximation for fluid mixtures," *J. Chem. Phys.* **52**(9), 4559–4562 (1970)

44. A.P. Ivanov, V.A. Loiko, and V.P. Dik, *Light Propagation in Densely Packed Disperse Media*, Nauka i Tekhnika, Minsk (1988)

45. N.G. Khlebtsov, I.L. Maksimova, V.V. Tuchin, and L.-H.V Wang, "Introduction to light scattering by biological objects" in: *Handbook of Optical Biomedical Diagnostics* **PM107**, Ed. V.V. Tuchin, SPIE, Bellingham, pp. 31–167 (2002)

46. S.M. Rytov, Y.A. Kravtsov, and V.I. Tatarskii, *Wave Propagation Through Random Media*, Vol. 4 of *Principles of Statistical Radiophysics*, Springer, Berlin Heidelberg New York (1989)

47. R.P. Hemenger, "Birefringence of a medium of tenuous parallel cylinders," *Appl. Opt.* **28**(18), 4030–4034 (1989)

48. R.P. Hemenger, "Refractive index changes in the ocular lens result from increased light scatter," *J. Biomed. Opt.* **1**, 268–272 (1996)

49. V.F. Izotova, I.L. Maksimova, I.S. Nefedov, and S.V. Romanov, "Investigation of Mueller matrices of anisotropic nonhomogeneous layers in application to optical model of cornea," *Appl. Opt.* **36**(1), 164–169 (1997)

50. R.L. McCally and R.A. Farrell, "Light scattering from cornea and corneal transparency," in: *Noninvasive Diagnostic Techniques in Ophthalmology*, Ed. B.R. Master, Springer, Berlin Heidelberg New York, pp. 189–210 (1990)

51. I.L. Maksimova, V.V. Tuchin, and L.P. Shubochkin, "Polarization features of eye's cornea," *Opt. Spectrosc.* **60**(4), 801–807 (1986)

52. I.L. Maksimova and L.P. Shubochkin, "Light-scattering matrices for a close-packed binary system of hard spheres," *Opt. Spectrosc.* **70**(6), 745–748 (1991)

53. I.L. Maksimova, V.V. Tuchin, and L.P. Shubochkin "Light scattering matrix of crystalline lens," *Opt. Spectrosc.* **65**(3), 615–619 (1988)

54. I.L. Maksimova, "Scattering of radiation by regular and random systems comprised of parallel long cylindrical rods," *Opt. Spectrosc.* **93**(4), 610–619 (2002)

55. D.E. Freund, R.L. McCally, R.A. Farrell, S.M. Cristol, N.L. L'Hernault, and H.F. Edelhauser, "Ultrastructure in anterior and posterior stroma of perfused human and rabbit corneas: relation to transparency," *Invest. Ophthalmol. Visual Sci.* **36**, 1508–1523 (1995)

56. V.V. Tuchin, "Lasers light scattering in biomedical diagnostics and therapy," *J. Laser Appl.* **5**(2/3), 43–60 (1993)

57. V.V. Tuchin, "Light scattering study of tissues," *Physics – Uspekhi* **40**(5), 495–515 (1997)

58. V.V. Tuchin, "Coherence-domain methods in tissue and cell optics," *Laser Phys.* **8**(4), 807–849 (1998)

59. F.A. Bettelheim, "Physical basis of lens transparency," in: *The Ocular Lens: Structure, Function and Pathology*, Ed. H. Maisel, Marcel Dekker, New York (1985)

60. J. Xu, J. Pokorny, and V.C. Smith, "Optical density of the human lens," *J. Opt. Soc. Am. A.* **14**(5), 953–960 (1997)

61. G.B. Benedek, "Theory of transparency of the eye," *Appl. Opt.* **10**(3), 459–473 (1971)
62. A. Tardieuand and M. Delaye, "Eye lens proteins and transparency from light transmission theory to solution X-ray structural analysis," *Ann. Rev. Biophys. Chem.* **17**, 47–70 (1988)

Chapter 4

1. *Medical Optical Tomography: Functional Imaging and Monitoring* **IS11**, Ed. G. Mueller, B. Chance, R. Alfano et al., SPIE, Bellingham (1993)
2. *Selected Papers on Tissue Optics Applications in Medical Diagnostics and Therapy* **MS 102**, Ed. V.V. Tuchin, SPIE, Bellingham (1994)
3. *Selected Papers on Optical Tomography, Fundamentals and Applications in Medicine* **MS 147**, Eds. O. Minet, G. Mueller, and J. Beuthan, SPIE, Bellingham (1998)
4. V.V. Tuchin, *Tissue Optics: Light Scattering Methods and Instruments for Medical Diagnosis,* SPIE Tutorial Texts in Optical Engineering, **TT38**, SPIE, Bellingham (2000)
5. *Handbook of Optical Biomedical Diagnostics* **PM107**, Ed. V.V. Tuchin, SPIE, Bellingham (2002)
6. L.-H.V. Wang, G.L. Cote', and S.L. Jacques Eds., "Special section on Tissue Polarimetry," *J. Biomed. Opt.* **7**(3), 278–397 (2002)
7. *Biomedical Photonics Handbook*, Ed. Tuan Vo-Dinh, CRC, Boca Raton (2003)
8. *Light Scattering by Nonspherical Particles*, Eds. M. I. Mishchenko, J. W. Hovenier, and L.D. Travis, Academic, San Diego (2000)
9. M.I. Mishchenko, L.D. Travis, A.A. Lacis, *Scattering, Absorption, and Emission of Light by Small Particles,* Cambridge University Press, Cambridge (2002)
10. H.H. Tynes, G.W. Kattawar, E.P. Zege, I.L. Katsev, A.S. Prikhach, and L.I. Chaikovskaya, "Monte Carlo and multicomponent approximation methods for vector radiative transfer by use of effective Mueller matrix calculations," *Appl. Opt.* **40**(3), 400–412 (2001)
11. C. Chandrasekhar, *Radiative Transfer,* Dover, Toronto, Ontario (1960)
12. V. V. Sobolev, *Light Scattering in Planetary Atmospheres,* Pergamon, Oxford (1974)
13. A. Ishimaru, *Wave Propagation and Scattering in Random Media,* IEEE, New York (1997)
14. H.C. van de Hulst, *Multiple Light Scattering. Tables, Formulas, and Applications,* Academic, New York (1980)
15. A.Z. Dolginov, Yu.N. Gnedin, and N.A. Silant'ev, *Propagation and Polarization of Radiation in Cosmic Media,* Gordon and Breach, Basel (1995)
16. J. Lenoble, *Radiative Transfer in Scattering and Absorbing Atmospheres: Standard Computational Procedures,* Deepak, Hampton, VA (1985)
17. E.J. Yanovitskij, *Light Scattering in Inhomogeneous Atmospheres,* Springer, Berlin Heidelberg New York (1997)
18. G.E. Thomas and K. Stamnes, *Radiative Transfer in the Atmosphere and Ocean,* Cambridge University Press, New York (1999)
19. A. Ishimaru, "Diffusion of light in turbid material," *Appl. Opt.* **28**, 2210–2215 (1989)

20. M. Motamedi, S. Rastegar, G. LeCarpentier, and A.J. Welch, "Light and temperature distribution in laser irradiated tissue the influence of anisotropic scattering and refractive index," *Appl. Opt.* **28**, 2230–2237 (1989)

21. S.R. Arridge, M. Schweiger, M. Hiraoka, and D.T. Delpy, "A finite element approach for modelling photon transport in tissue," *Med. Phys.* **20**, 299–309 (1993)

22. V.V. Tuchin, S.R. Utz, and I.V. Yaroslavsky, "Tissue optics, light distribution, and spectroscopy," *Opt. Eng.* **33**, 3178–3188 (1994)

23. G. Yoon, A.J. Welch, M. Motamedi, and M.C.J. Van Gemert, "Development and application of three–dimensional light distribution model for laser irradiated tissue," *IEEE J. Quantum Electron.* **23**(10), 1721–1733 (1987)

24. I.M. Stockford, S.P. Morgan, P.C.Y. Chang, and J.G. Walker, "Analysis of the spatial distribution of polarized light backscattering," *J. Biomed. Opt.* **7**(3), 313–320 (2002)

25. *Monte Carlo Method in Statistical Physics*, Ed. K. Binder, Springer, Berlin Heidelberg New York (1979)

26. B.G. Wilson and G.A. Adam, "Monte Carlo model for the absorption and flux distributions of light in tissue," *Med. Phys.* **10**, 824–830 (1983)

27. S.L. Jacques, "Monte Carlo modeling of light transport in tissue," in: *Tissue Optics*, Eds. A. J. Welch and M. C. J. van Gemert, Academic, New York (1992)

28. D. Bicout, C. Brosseau, A.S. Martinez, and J.M. Schmitt, "Depolarization of multiply scattering waves by spherical diffusers: influence of the size parameter," *Phys. Rev. E.* **49**, 1767–1770 (1994)

29. C.J. Hourdakis and A. Perris, "A Monte Carlo estimation of tissue optical properties for the use in laser dosimetry," *Phys. Med. Biol.* **40**, 351–364 (1995)

30. A. Yodh, B. Tromberg, E. Sevick-Muraca, and D. Pine, "Diffusing photons in turbid media," *J.Opt. Soc. Am. A* **14**, 136–342 (1997)

31. A.H. Hielsher, J.R. Mourant, and I.J. Bigio, "Influence of particle size and concentration on the diffuse backscattering of polarized light from tissue phantoms and biological cell suspensions," *Appl. Opt.* **36**, 125–135 (1997)

32. M.J. Racovic, G.W. Kattavar, M. Mehrubeoglu, B.D. Cameron, L.-H.V. Wang, S. Rasteger, and G.L. Cote, "Light backscattering polarization patterns from turbid media: theory and experiment," *Appl. Opt.* **38**, 3399–3408 (1999)

33. S. Bartel and A.H. Hielscher, "Monte Carlo simulations of the diffuse backscattering Mueller matrix for highly scattering media," *Appl. Opt.* **39**, 1580–1588 (2000)

34. M. Moscoso, J.B. Keller, and G. Papanicolaou, "Depolarization and blurring of optical images by biological tissue," *J. Opt. Soc. Am.* **18**(4), 948–960 (2001)

35. I.L. Maksimova, S.V. Romanov, and V.F. Izotova, "The effect of multiple scattering in disperse media on polarization characteristics of scattered light," *Opt. Spectrosc.* **92**(6), 915–923 (2002)

36. X. Wang and L.-H.V. Wang, "Propagation of polarized light in birefringent turbid media: A Monte Carlo study," *J. Biomed. Opt.* **7**(3), 279–290 (2002)

37. S. Ya. Sid'ko, V. N. Lopatin, and L. E. Paramonov, *Polarization Characteristics of Solutions of Biological Particles,* Nauka, Novosibirsk (1990)

38. V. Shankaran, J.T. Walsh, Jr., D.J. Maitland, "Comparative study of polarized light propagation in biological tissues," *J. Biomed. Opt.* **7**(3), 300–306 (2002)

39. I.L. Maksimova, S.N. Tatarintsev, and L.P. Shubochkin, "Multiple scattering effects in laser diagnostics of bioobjects," *Opt. Spectrosc.* **72**, 1171–1177 (1992)

40. V.F. Izotova, I.L. Maksimova, and S.V. Romanov, "Utilization of relations between elements of the Mueller matrices for estimating properties of objects and the reliability of experiments," *Opt. Spectrosc.* **88**(5), 753–759 (1996)

41. M.J. Rakovic and G.W. Kattawar, "Theoretical analysis of polarization patterns from incoherent backscattering of light," *Appl. Opt.* **37**(15), 3333–3338 (1998)

42. S.M. Rytov, Y.A. Kravtsov, and V.I. Tatarskii, *Wave Propagation Through Random Media*, Vol. 4 of *Principles of Statistical Radiophysics*, Springer, Berlin Heidelberg New York (1989)

43. V.G. Vereshchagin and A.N. Ponyavina, "Statistical characteristic and transparency of thin closely packed disperse layer," *J. Appl. Spectrosc. (USSR)* **22**(3), 518–524 (1975)

44. V. Twersky "Interface effects in multiple scattering by large, low refracting, absorbing particles," *J. Opt. Soc. Am.* **60**(7), 908–914 (1970)

45. M. Lax, "Multiple scattering of waves II. The effective field in dense system," *Phys. Rev.* **85**(4), 621–629 (1952)

46. L. Tsang, J.A. Kong, and R.T. Shin, *Theory of Microwave Remote Sensing*, Wiley, New York (1985)

47. K.M. Hong, "Multiple scattering of electromagnetic waves by a crowded monolayer of spheres: Application to migration imaging films," *J. Opt. Soc. Am.* **70**(7), 821–826 (1980)

48. A. Ishimaru and Y. Kuga "Attenuation constant of a coherent field in a dense distribution of particles," *J. Opt. Soc. Am.* **72**(10), 1317–1320 (1982)

49. A.N. Ponyavina, "Selection of optical radiation in scattering by partially ordered disperse media," *J. Appl. Spectrosc.* **65**(5), 721–733 (1998)

50. T.R. Smith, "Multiple scattering in the cornea," *J. Mod. Opt.* **35**(1), 93–101 (1988)

51. V. Twersky, "Absorption and multiple scattering by biological suspensions," *J. Opt. Soc. Am.* **60**, 1084–1093 (1970)

52. J.M. Steinke and A. P. Shepherd, "Diffusion model of the optical absorbance of whole blood," *J. Opt. Soc. Am.A* **5**, 813–822 (1988)

53. J.M. Schmitt and G. Kumar, "Optical scattering properties of soft tissue: a discrete particle model," *Appl. Opt.* **37**(13), 2788–2797 (1998)

54. I.F. Cilesiz and A.J. Welch, "Light dosimetry: effects of dehydration and thermal damage on the optical properties of the human aorta," *Appl. Opt.* **32**, 477–487 (1993)

55. W.-C. Lin, M. Motamedi, and A.J. Welch, "Dynamics of tissue optics during laser heating of turbid media," *Appl. Opt.* **35**(19), 3413–3420 (1996)

56. G. Vargas, E.K. Chan, J.K. Barton, H.G. Rylander III, and A.J. Welch, "Use of an agent to reduce scattering in skin," *Laser. Surg. Med.* **24**, 138–141 (1999)

57. R.M.P. Doornbos, R. Lang, M.C. Aalders, F.W. Cross, and H.J.C.M. Sterenborg, "The determination of in vivo human tissue optical properties and absolute chromophore concentrations using spatially resolved steady-state diffuse reflectance spectroscopy," *Phys. Med. Biol.* **44**, 967–981 (1999)

Chapter 5

1. *Medical Optical Tomography: Functional Imaging and Monitoring* **IS11**, Eds. G. Mueller, B. Chance, R. Alfano et al., SPIE, Bellingham (1993)

260   References

2. *Selected Papers on Tissue Optics Applications in Medical Diagnostics and Therapy* **MS 102**, Ed. V.V. Tuchin, SPIE, Bellingham (1994)
3. *Selected Papers on Optical Tomography, Fundamentals and Applications in Medicine*MS **147**, Eds. O. Minet, G. Mueller, and J. Beuthan, SPIE, Bellingham, (1998)
4. V.V. Tuchin, *Tissue Optics: Light Scattering Methods and Instruments for Medical Diagnosis*, SPIE Tutorial Texts in Optical Engineering, **TT38,** SPIE, Bellingham (2000)
5. *Handbook of Optical Biomedical Diagnostics* **PM107**, Ed. V.V. Tuchin, SPIE, Bellingham (2002)
6. L.-H.V. Wang, G.L. Cote', and S.L. Jacques Eds. "Special section on Tissue Polarimetry," *J. Biomed. Opt.* **7**(3), 278–397 (2002)
7. *Biomedical Photonics Handbook*, Eds. Tuan Vo-Dinh, CRC, Boca Raton (2003)
8. R.G. Johnston, S.B. Singham, and G.C. Salzman, "Polarized light scattering," *Comments Mol. Cell. Biophys.* **5**(3), 171–192 (1988)
9. G.C. Salzmann, S.B. Singham, R.G. Johnston, and C.F. Bohren, "Light scattering and cytometry," in: *Flow Cytometry and Sorting*, 2nd ed., Ed. M.R. Melamed, T. Lindmo, and M.L. Mendelsohn, Wiley, New York, pp. 81–107 (1990)
10. V.F. Izotova, I.L. Maksimova, and S.V. Romanov, "Utilization of relations between elements of the Mueller matrices for estimating properties of objects and the reliability of experiments," *Opt. Spectrosc.* **88** (5), 753–759 (1996)
11. C.F. Bohren and D.R. Huffman, *Absorption and Scattering of Light by Small Particles,* Wiley, New York (1983)
12. R.M.A. Azzam and N.M. Bashara, *Ellipsometry and Polarized Light,* Elsevier, Amsterdam (1994)
13. R.M.A. Azzam, "Photopolarimetric measurement of the Muellermatrix by Fourier analysis of a single detected signal," *Opt. Lett.* **2**, 148–150 (1987)
14. V.V. Tuchin, "Lasers light scattering in biomedical diagnostics andtherapy," *J. Laser Appl.* **5**(2/3), 43–60 (1993)
15. V.F. Izotova, I.L. Maksimova, and S.V. Romanov,"Simulation of polarization characteristics of the crystallinelens during protein aggregation with regard to multiplescattering," *Opt. Spectrosc.* **86**(6) 902–908 (1999)
16. A.V. Priezzhev, V.V. Tuchin, and L.P. Shubochkin,"Laser microdiagnostics of eye optical tissues and form elementsof blood," *Bull. USSR Acad. Sci., Phys. Ser.* **53**(8) 1490–1495 (1989)
17. I.L. Maksimova, A.P. Mironychev, S.V. Romanov, S.N. Tatarintsev, V.V. Tuchin, and L.P. Shubochkin, "Methods and equipment for laser diagnostics inophthalmology," *Bull. USSR Acad. Sci., Phys. Ser.* **54**(10), 1918–1923 (1990)
18. A.N. Korolevich, A.Ya. Khairulina, and L.P. Shubochkin, "Scattering matrix of amonolayer of optically "soft" particles at their densepackage," *Opt. Spectrosc. (USSR)* **68**(2), 403–409 (1990)
19. V.F. Izotova, I.L.Maksimova, and S.V. Romanov, "Analysis of accuracy of laserpolarization nephelometer," *Opt. Spectrosc.* **80**,1001–1007 (1996)
20. P.S. Hauge, "Recent developments ininstruments in ellipsometry," *Surf. Sci.* **96**, 108–140 (1980)
21. J.R. Mourant, T.M. Johnson, S. Carpenter, A. Guerra, T. Aida, and J.P. Freyer, "Polarizedangular dependent spectroscopy of epithelial cells and epithelialcell nuclei to determine the size scale of scatteringstructures," *J. Biomed. Opt.* **7**(3), 378–387 (2002)

22. G.J. van Blokland, "Ellipsometry of the humanretina *in vivo*: preservation of polarization," *J. Opt. Soc. Am. A* **2**, 72–75 (1985)
23. H.B. Klein Brink and G.J. van Blokland, "Birefringence of thehuman foveal area assessed in vivo with Mueller-matrixellipsometry," *J. Opt. Soc. Am. A* **5**, 49–57 (1988)
24. R.C.N. Studinski and I.A. Vitkin, "Methodology forexamining polarized light interactions with tissues and tissuelikemedia in the exact backscattering direction," *J. Biomed. Opt.* **5**(3), 330–337 (2000)
25. K.C. Hadley and I.A. Vitkin, "Optical rotation and linear and circular depolarizationrates in diffusively scattered light from chiral, racemic, andachiral turbid media," *J. Biomed. Opt.* **7**(3),291–299 (2002)
26. V. Shankaran, M.J. Everett, D.J. Maitland,and J.T. Walsh, Jr., "Polarized light propagation through tissuephantoms containing densely packed scatterers," *Opt.Lett.* **25**(4), 239–241 (2000)
27. V. Shankaran, J.T. Walsh, Jr., D.J. Maitland, "Comparative study of polarized lightpropagation in biological tissues," *J. Biomed. Opt.***7**(3), 300–306 (2002)
28. J.F. Bille, A.W. Dreher, and G. Zinser, Scanning laser tomography of the living human eye, in: *Noninvasive Diagnostic Techniques in Ophthalmology,* Ed. B.R. Master, Springer, Berlin, Heidelberg, New York, pp. 528–547 (1990)
29. A.W. Dreher and K. Reiter, "Retinal laserellipsometry: A new method for measuring the retinal nerve fiberlayer thickness distributions," *Clin. Vis. Sci.* **7**(6), 481–485 (1992)
30. F. Delplancke, "Automatedhigh-speed Mueller matrix scatterometer," *Appl. Opt.* **36**(22), 5388–5395 (1997)
31. M.H. Smith, "Optimizing adual-rotating-retarder Mueller matrix polarimeter," in*Polarization Analysis and Measurement IV*, Eds. D. Goldstein, D. Chenault, *Proc. SPIE* **4481** (2001)
32. M.H. Smith, "Interpreting Mueller matrix images of tissues," in *Laser-Tissue Interaction XII*, Eds. D. Duncan, S. Jacques, P. Johnson, *Proc. SPIE* **4257**, 82–89 (2001)
33. J.S. Baba, J.-R. Chung, A.H. DeLaughter, B.D. Cameron, and G.L. Cote', "Development and calibration of an automated Mueller matrixpolarization imaging system," *J. Biomed. Opt.***7**(3), 341–349 (2002)
34. X.-R. Huang and R.W. Knighton, " Linear birefringence of the retinal nerve fiber layermeasured in vitro with a multispectral imaging micropolarimeter," *J. Biomed. Opt.* **7**(2), 199–204 (2002)
35. I.M. Stockford, S.P. Morgan, P.C.Y. Chang, and J.G. Walker, "Analysis of the spatial distribution of polarized lightbackscattering," *J. Biomed. Opt.* **7**(3), 313–320 (2002)
36. S.L. Jacques, J.C. Ramella-Roman, and K. Lee,"Imaging skin pathology with polarized light," *J.Biomed. Opt.* **7**(3), 329–340 (2002)
37. X. Gan and M. Gu, "Image reconstruction through turbid media under atransmission-mode microscope," *J. Biomed. Opt.***7**(3), 372–377 (2002)
38. J.S. Tyo, M.P. Rowe, E.N. Pugh, and N. Engheta, "Target detection in optically scatteringmedia by polarization-difference imaging," *Appl. Opt.* **35**, 1855–1870 (1996)
39. D.H. Goldstein and R.A.Chipman, "Error analysis of a Mueller matrix polarimeter," *J. Opt. Soc. Am. A* **7**, 693–700 (1990)

40. D.S. Sabatke, M.R. Descour, E.L. Dereniak, W.C. Sweatt, S.A. Kemme, and G.S. Phipps, "Optimization of retardance for acomplete Stokes polarimeter," *Opt. Lett.* **25**, 802–804 (2000)

41. J.S. Tyo, "Noise equalization in Stokesparameter images obtained by use of variable-retardancepolarimeters," *Opt. Lett.* **25**, 1198–1200 (2000)

42. Y. Ichihashi, M.H. Khin, K. Ishikawa, and T. Hatada, Birefringence effect of the in vivo cornea," *Opt. Eng.***34**(3), 693–700 (1995)

43. R.L. McCally, R.A. Farrell, "Light scattering from cornea and cornealtransparency," in: *Noninvasive Diagnostic Techniques inOphthalmology,* Ed. B.R. Master, Springer, Berlin Heidelberg New York, pp. 189-210 (1990)

44. S.Ya. Sid'ko, V.N. Lopatin, and L.E. Paramonov, *Polarization Characteristics of Solutions of Biological Particles,* Nauka, Novosibirsk (1990)

45. V.V. Tuchin, "Light scattering study of tissues," *Physics – Uspekhi* **40**(5), 495–515 (1997)

46. V.V. Tuchin,"Coherence-domain methods in tissue and cell optics," *Laser Phys.* **8**(4), 807–849 (1998)

47. N.G. Khlebtsov, I.L. Maksimova, V.V. Tuchin, and L.-H.V. Wang, "Introduction to light scattering by biological objects" in: *Handbook of Optical Biomedical Diagnostics* **PM107**, Ed. V.V. Tuchin, SPIE, Bellingham, pp. 31–167 (2002)

48. A.G. Hoekstra and P.M.A. Sloot, "Biophysicaland biomedical applications of nonspherical scattering," in: *Light Scattering by Nonspherical Particles*, Eds. M.I. Mishchenko, J.W. Hovenier, and L.D. Travis, Academic, San Diego, pp. 585-602 (2002)

49. W.S. Bickel, J.F. Davidson, D.R. Huffman, and R. Kilkson, "Application of polarization effectsin light scattering: a new biophysical tool," *Proc. Natl Acad. Sci. USA* **73**, 486–490 (1976)

50. K.J. Voss and E.S. Fry, "Measurement of the Mueller matrix for ocean water," *Appl. Opt.* **23**, 4427–4439 (1984)

51. K.D. Lofftus, M.S. Quinby-Hunt, A.J. Hunt, F. Livolant, and M. Maestre, "Light scattering by *Prorocentrum micans*: A newmethod and results," *Appl. Opt.* **31**, 2924–2931 (1992)

52. W.S. Bickel and M.E. Stafford, "Biological particles as irregularly shaped particles," in: *LightScattering by Irregularly Shaped Particles*, Ed. D. Schuerman, Plenum, New York, pp. 299–305 (1980)

53. A.G. Hoekstra and P.M.A. Sloot, "Dipolar unit size in coupled-dipole calculationsof the scattering matrix elements," *Appl. Opt.* **18**, 1211–1213 (1993)

54. B.V. Bronk, S.D. Druger, J. Czege, and W.P. van de Merwe, "Measuring diameters ofrod-shaped bacteria in vivo with polarized light scattering," *Biophys. J.* **69**, 1170–1177 (1995)

55. W.P. Van de Merwe, Z.-Z. Li, B.V. Bronk, and J. Czege, "Polarizedlight scattering for rapid observation of bacterial sizechanges," *Biophys. J.* **73**, 500–506 (1997)

56. W.P. Van de Merwe, D.R. Huffman, and B.V. Bronk, "Reproducibility and sensitivity of polarized light scatteringfor identifying bacterial suspensions," *Appl. Opt.***28**(23), 5052–5057 (1989)

57. A.N. Korolevich, A.Ya. Khairullina, and L.P. Shubochkin, "Effects of aggregation oflarge biological particles on the scattering matrix elements," *Opt. Spectrosc.* **77**(2) 278–282 (1994)

58. B.G. de Grooth, L.W.M.M. Terstappen, G.J. Puppels, and J. Greve, "Light-scattering polarization measurements as a new parameter inflow cytometry," *Cytometry* **8**, 539–544 (1987)

59. R.M.P. Doornbos, A.G. Hoekstra, K.E.I. Deurloo, B.G. deGrooth, P.M.A. Sloot, and J. Greve, "Lissajous-like patterns inscatter plots of calibration beads," *Cytometry* **16**, 236–242 (1994)

60. A.G. Ushenko, S.B. Ermolenko, D.N. Burkovets, and Yu.A. Ushenko, "Polarizationmicrostructure of laser radiation scattered by optically activebiotissues," *Opt. Spectrosc.* **87**(3), 434–438 (1999)

61. A.G. Ushenko, "Stokes-correlometry of biotissues, " *Laser Phys.* **10**, 1–7, (2000)

62. V.V. Tuchin, I.L. Maksimova, A.N. Yaroslavskaya, T.N. Semenova, S.N. Tatarintsev, V.I. Kochubey, and V.F. Izotova, "Human eyelens spectroscopy and modeling of its transmittance, " *Proc. SPIE* **2126**, 393–406 (1994)

63. M. Kerker, *The Scattering of Light and Other Electromagnetic Radiation*, Academic, New York (1969)

64. A. Ishimaru, "Diffusion of light in turbid material," *Appl. Opt.* **28**, 2210–2215 (1989)

65. J.A. van Best, E.V.M.J. Kuppens, "Summary of studies on the blue – greenautofluorescence and light transmission of the ocular lens," *J. Biomed. Opt.* **1**(3), 243–250 (1996)

66. Q. Zhou and R.W. Knighton, "Light scattering and form birefringenceof parallel cylindrical arrays that represent cellular organellesof the retinal nerve fiber layer," *Appl. Opt.* **36**(10), 2273–2285 (1997)

67. C. Bustamante, M.F. Maestre, D. Keller, and I. Tinoco, Jr., "Differential scattering (CIDS) of circularly polarized light by dense particles," *J. Chem. Phys.* **80**, 4817–4823 (1984)

68. S. Zietz, A. Belmont, and C. Nicolini, "Differential scattering ofcircularly polarized light as an unique probe of polynucleosomesuperstructures," *Cell Biophys.* **5**, 163–187(1983)

69. C.T. Gross, H. Salamon, A.J. Hunt, R.I. Macey, F. Orme, and A.T. Quintanilha, "Hemoglobin polymerization in sicklecells studied by circular polarized light scattering," *Biochim. Biophys. Acta* **1079**(2), 152–160 (1991)

70. V.F. Izotova, I.L. Maksimova, I.S. Nefedov, and S.V. Romanov, "Investigation of Mueller matrices of anisotropicnonhomogeneous layers in application to optical model of cornea," *Appl. Opt.* **36**(1), 164–169 (1997)

71. A. Ishimaru, *Wave Propagation and Scattering in Random Media* IEEE, New York (1997)

72. R.J. McNichols and G.L. Coté, "Optical glucose sensing in biological fluids: anoverview," *J. Biomed. Opt.* **5**(1), 5–16 (2000)

73. J.S. Baba, B.D. Cameron, S. Theru, and G.L. Coté, "Effect of temperature, *p*H, and corneal birefringence onpolarimetric glucose monitoring in the eye," *J. Biomed.Opt.* **7**(3), 321–328 (2002)

74. I.L. Maksimova, S.N. Tatarintsev, and L.P. Shubochkin, "Multiple scattering effects inlaser diagnostics of bioobjects," *Opt. Spectrosc.* **72**, 1171–1177 (1992)

75. D. Bicout, C. Brosseau, A.S. Martinez, and J.M. Schmitt, "Depolarization of multiplyscattering waves by spherical diffusers: influence of the sizeparameter," *Phys. Rev. E.* **49**, 1767–1770 (1994)

76. I.L. Maksimova, S.V. Romanov, and V.F. Izotova, "The effectof multiple scattering in disperse media on polarizationcharacteristics of scattered light," *Opt. Spectrosc.* **92**(6), 915–923 (2002)

77. M.J. Racovic, G.W. Kattavar, M. Mehrubeoglu, B.D. Cameron, L.-H.V. Wang, S. Rasteger, and G.L. Cote, "Light backscattering polarization patterns from-turbid media: theory and experiment," *Appl. Opt.***38**, 3399– 3408 (1999)

78. L.O. Svaasand and Ch.J. Gomer, "Optics of tissue," in: *Dosimetry of Laser-Radiation in Medicine and Biology*, **IS5**, SPIE, Bellingham, WA, pp. 114–132 (1989)

79. A.N. Yaroslavsky, I.V. Yaroslavsky, T. Goldbach, and H.-J. Schwarzmaier, "Influence ofthe scattering phase function approximation on the opticalproperties of blood determined from the integrating spheremeasurements," *J. Biomed. Opt.* **4**(1), 47–53 (1999)

80. A.H. Hielsher, J.R. Mourant, and I.J. Bigio, "Influence of particle size and concentration on the diffusebackscattering of polarized light from tissue phantoms andbiological cell suspensions," *Appl. Opt.* **36**, 125–135 (1997)

81. H. Horinaka, K. Hashimoto, K. Wada, and Y. Cho, "Extraction of quasi-straightforward propagating photonsfrom diffused light transmitting through a scattering medium bypolarization modulation," *Opt. Lett.* **20**, 1501–1503 (1995)

82. S.P. Morgan, M.P. Khong, and M.G. Somekh,"Effects of polarization state and scatterer concentrationoptical imaging through scattering media," *Appl. Opt.* **36**, 1560–1565 (1997)

83. A.B. Pravdin, S.P. Chernova,and V.V. Tuchin, "Polarized collimated tomography for biomedicaldiagnostics," *Proc. SPIE* **2981**, 230–234 (1997)

84. M.R. Ostermeyer, D.V. Stephens, L.-H.V. Wang, and S.L. Jacques, "Nearfield polarization effects on light propagation in randommedia," *OSA TOPS* **3**, 20–25 (1996)

85. R.R. Anderson,"Polarized light examination and photography of the skin," *Arch. Dermatol.* **127**, 1000–1005 (1991)

86. N. Kollias, "Polarized light photography of human skin," in*Bioengineering of the Skin: Skin Surface Imaging and Analysis*, Eds. K.-P. Wilhelm, P. Elsner, E. Berardesca, and H.I. Maibach, CRC, Boca Raton, pp. 95–106 (1997)

87. G.V. Simonenko, T.P. Denisova, N.A. Lakodina, and V.V. Tuchin "Measurement of an optical anisotropy of biotissues" *Proc. SPIE* **3915**, 152–157 (2000)

88. G.V. Simonenko, V.V. Tuchin, and N.A. Lakodina, "Measurement of theoptical anisotropy of biological tissues with the use of a nematicliquid crystal cell" *J. Opt. Technol.* **67**(6), 559–562 (2000)

89. O.V. Angelski, A.G. Ushenko, A.D. Arkhelyuk, S.B. Ermolenko, and D.N. Burkovets, "Scattering oflaser radiation by multifractal biological structures," *Opt. Spectrosc.* **88**(3), 444–447 (2000)

90. M.R. Hee, D. Huang, E.A. Swanson, and J.G. Fujimoto, "Plarization-sensitivelow-coherence reflectometer for birefringence characterization an-dranging," *J. Opt. Soc. Am. B* **9**, 903–908 (1992)

91. J.F. de Boer, T.E. Milner, M.J.C. van Gemert, and J.S. Nelson, ' 'Two-dimensional birefringence imaging in biologicaltissue by polarization-sensitive optical coherence tomography," *Opt. Lett.* **22**(12), 934–936 (1997)

92. M.J. Everett, K. Schoenerberger, B.W. Colston, Jr., and L.B. Da Silva,"Birefringence characterization of biological tissue by use ofoptical coherence tomography," *Opt. Lett.* **23**(3), 228–230 (1998)

93. J.F. de Boer, T.E. Milner, and J.S. Nelson, "Determination of the depth resolved Stokes parameters oflight backscattered from turbid media using

polarization sensitiveoptical coherence tomography," *Opt. Lett.* **24**, 300–302 (1999)

94. J.F. de Boer and T.E. Milner, "Review ofpolarization sensitive optical coherence tomography and Stokesvector determination," *J. Biomed. Opt.* **7**(3), 359–371 (2002)

95. C.K. Hitzenberger, E. Gotzinger, M. Sticker, M. Pircher, and A.F. Fercher, "Measurement and imagingof birefringence and optic axis orientation by phase resolvedpolarization sensitive optical coherence tomography," *Opt. Express* **9**, 780–790 (2001)

96. S. Jiao and L.-H.V. Wang, "Jones-matrix imaging of biological tissues withquadruple-channel optical coherence tomograthy," *J. Biomed. Opt.* **7**(3), 350–358 (2002)

97. M.G. Ducros, J.F. de Boer, H. Huang, L. Chao, Z. Chen, J.S. Nelson, T.E. Milner, and H.G. Rylander, "Polarization sensitive opticalcoherence tomography of the rabbit eye," *IEEE J. Sel.Top. Quantum Electron.* **5**, 1159–1167 (1999)

98. M.G. Ducros, J.D. Marsack, H.G. Rylander III, S.L. Thomsen, and T.E. Milner, "Primate retina imaging with polarization-sensitiveoptical coherence tomography," *J. Opt. Soc. Am. A* **18**, 2945–2956 (2001)

99. C.E. Saxer, J.F. de Boer, B.H. Park, Y. Zhao, C. Chen, and J.S. Nelson, "High speed fiberbased polarization sensitive optical coherence tomography of invivo human skin," *Opt. Lett.* **26**, 1069–1071 (2001)

100. B.H. Park, C.E. Saxer, S.M. Srinivas, J.S. Nelson, and J.F. de Boer, "*In vivo* burn depthdetermination by high-speed fiber-based polarization sensitiveoptical coherence tomography," *J. Biomed. Opt.* **6**, 474–479 (2001)

101. X.J. Wang, T.E. Milner, J.F. deBoer, Y. Zhang, D.H. Pashley, and J.S. Nelson, "Characterizationof dentin and enamel by use of optical coherence tomography," *Appl. Opt.* **38**, 2092–2096 (1999)

102. A.Baumgartner, S. Dichtl, C.K. Hitzenberger, H. Sattmann, B. Robl, A. Moritz, A.F. Fercher, and W. Sperr, "Polarization-sensitiveoptical coherence tomography of dental structures," *Caries Res.* **34**(1), 59–69 (2000)

103. M. Moscoso, J.B. Keller, and G. Papanicolaou, "Depolarization andblurring of optical images by biological tissue," *J. Opt. Soc. Am.* **18** (4), 948–960 (2001)

104. X. Wang and L.-H.V. Wang, "Propagation of polarized light in birefringent turbidmedia: A Monte Carlo study," *J. Biomed. Opt.* **7**(3), 279–290 (2002)

105. S.G. Demos and R.R. Alfano,"Optical polarization imaging," *Appl. Opt.* **36**, 150–155 (1997)

106. G. Yao and L.-H.V. Wang, "Two-dimensionaldepth-resolved Mueller matrix characterization of biologicaltissue by optical coherence tomography," *Opt. Lett.* **2**4, 537–539 (1999)

107. E.E. Gorodnichev and D.B. Rogozkin, "Small angle multiple scattering in randominhomogeneous media, *JETP* **107**, 209–235 (1995)

108. V.L. Kuzmin and V.P. Romanov, "Coherent effects at lightscattering in disordered systems," *Usp. Fiz. Nauk* **166**(3), 247–278 (1996)

109. K.M. Yoo, F. Liu, and R.R. Alfano, "Biological materials probed by the temporal andangular profiles of the backscattered ultrafast laser pulses," *J. Opt. Soc. Am. B* **7**, 1685–1693 (1990)

110. I. Freund, M. Kaveh, R. Berkovits, and M. Rosenbluh, "Universalpolarization correlations and microstatistics of optical waves inrandom media," *Phys. Rev. B* **42**(4), 2613–2616 (1990)

Chapter 6

1. V.L. Kuz'min and V.P. Romanov, "Coherent phenomena in light scattering from disordered systems," *Soviet Phys. Usp.* **166**, 247–277 (1996)
2. D. Bicout and C. Brosseau, "Multiply scattered waves through a spatially random medium: entropy production and depolarization," *J. Phys. I*, **2**, 2047–2063 (1992)
3. D. Bicout, C. Brosseau, A.S. Martinez, and J.M. Schmitt, "Depolarization of multiply scattering waves by spherical diffusers: Influence of size parameter," *Phys. Rev. E* **49**, 1767–1770 (1994)
4. C. Brosseau, *Fundamentals of Polarized Light: a Statistical Optics Approach*, Wiley, New York (1998)
5. A. Ishimaru, *Wave Propagation and Scattering in Random Media*, Academic, New York (1978)
6. A.A. Golubentsev, "On the suppression of the interference effects under multiple scattering of light," *Zh. Eksp. Teor. Fiz.* **86**, 47–59 (1984)
7. M.P. Van Albada and A. Lagendijk, "Observation of weak localization of light in a random medium," *Phys. Rev. Lett.* **55**, 2692–2695 (1985)
8. P. Wolf and G.Maret, "Weak localization and coherent backscattering of photons in disordered media," *Phys. Rev. Lett.* **55**, 2696–2699 (1985)
9. E. Akkermans, P.E. Wolf, and R. Maynard, "Coherent backscattering of light by disordered media: Analysis of the peak line shape," *Phys. Rev. Lett.* **56**, 1471–1474 (1986)
10. M.J. Stephen and G. Cwillich, "Rayleigh scattering and weak localization: effects of polarization," *Phys. Rev. B* **34**, 7564–7572 (1986)
11. E. Akkermans, P.E. Wolf, R. Maynard, and G. Maret, "Theoretical study of the coherent backscattering of light by disordered media," *J. Phys. France*, **49**, 77–98 (1988)
12. H. Kaveh, M. Rosenbluh, I. Edrei, and I. Edrei, "Weak localization and light scattering from disordered solids," *Phys. Rev. Lett.* **57**, 2049–2052 (1986)
13. F.C. MacKintosh and S. John, "Coherent backscattering of light in the presence of time-reversal-noninvariant and non-parity conversing media," *Phys. Rev. B* **37**, 884–897 (1988)
14. F.C. MacKintosh, J.X. Zhu, D.J. Pine, and D.A. Weitz, "Polarization memory of multiply scattered light," *Phys. Rev. B* **40**, 9342 – 9345 (1989)
15. F.C. MacKintosh and S. John, "Diffusing-wave spectroscopy and multiple scattering of light in correlated random media," *Phys. Rev. B* **40**, 2383–2406 (1989)
16. V.I. Kuz'min and V.P. Romanov, "Multiply scattered light correlations in an expanded temporal range," *Phys. Rev. E* **56**, 6008–6019 (1997)
17. S. Sanyal, A.K. Sood, S. Ramkumar, S. Ramaswamy, and N. Kumar, "Novel polarization dependence in diffusing-wave spectroscopy of crystallizing colloidal suspensions," *Phys. Rev. Lett.* **72**, 2963–2966 (1994)
18. *Photon Correlation and Light Beating Spectroscopy*, Eds. H. Z. Cummins and E. R. Pike, Plenum, New York (1974)

19. D.A. Zimnyakov, V.V. Tuchin, and A.G. Yodh, "Characteristic scales of optical field depolarization and decorrelation for multiple scattering media and tissues," *J. Biomed. Opt.* **4**, 157–163 (1999)

20. J.W. Goodman, *Statistical Optics*, Wiley, New York (1985)

21. R. Barakat, "The statistical properties of partially polarized light," *Opt. Acta* **32**, 295–312 (1985)

22. D. Eliyahu, "Vector statistics of correlated Gaussian fields," *Phys. Rev. B* **47**, 2881–2892 (1993)

23. I. Freund, M. Kaveh, R. Berkovits, and M. Rosenbluh, "Universal polarization correlations and microstatistics of optical waves in random media," *Phys. Rev. B* **42**, 2613–2616 (1991)

24. S. M. Cohen, D. Eliyahu, I. Freund, and M. Kaveh, "Vector statistics of multiply-scattered waves in random systems," *Phys. Rev. A* **43**, 5748–5751 (1991)

25. D. Eliyahu, M. Rosenbluh, and I. Freund, "Angular intensity and polarization dependence of diffuse transmission through random media," *JOSA A* **10**, 477–491 (1993)

26. I.I. Tarhan and G.H. Watson, "Polarization microstatistics of laser speckle," *Phys. Rev. A* **45**, 6013–6018 (1992)

27. I. Freund and M. Kaveh, "Comment on Polarization memory of multiply scattered light", *Phys. Rev. B* **45**, 8162–8164 (1992)

28. F.C. MacKintosh, J.X. Zhu, D.J. Pine, and D.A. Weitz, "Reply to "Comment on 'Polarization memory of multiply scattered light'," *Phys. Rev. B* **45**, 8165 (1992)

Chapter 7

1. A.A. Golubentsev, "On the suppression of the interference effects under multiple scattering of light," *Zh. Eksp. Teor. Fiz.* **86**, 47–59 (1984)

2. M.J. Stephen, "Temporal fluctuations in wave propagation in random media," *Phys. Rev. B.* **37**, 1–5 (1988)

3. S. John and M. Stephen, "Wave propagation and localization in a long-range correlated random potential," *Phys. Rev. B.* **28**, 6358–6380 (1983)

4. S. John, "Electromagnetic absorption in a disordered medium near a photon mobility edge," *Phys. Rev. Lett.* **53**, 2169–2172 (1984)

5. E. Akkermans, P.E. Wolf, R. Maynard, and G. Maret, "Theoretical study of the coherent backscattering of light by disordered media," *J. Phys. France* **49**, 77–98 (1988)

6. F.C. MacKintosh and S. John, "Diffusing-wave spectroscopy and multiple scattering of light in correlated random media," *Phys. Rev. B* **40**, 2383–2406 (1989)

7. M.J. Stephen and G. Cwillich, "Rayleigh scattering and weak localization: effects of polarization," *Phys. Rev. B* **34**, 7564–7572 (1986)

8. F.C. MacKintosh, J.X. Zhu, D.J. Pine, and D.A. Weitz, "Polarization memory of multiply scattered light," *Phys. Rev. B* **40**, 9342 – 9345 (1989)

9. V.L. Kuz'min and V.P. Romanov, "Coherent phenomena in light scattering from disordered systems," *Soviet Phys. Usp.* **166**, 247–277 (1996)

10. D.A. Zimnyakov, "On some manifestations of similarity multiple scattering of coherent light," *Waves Random Media* **10**, 417–434 (2000)

11. D.A. Zimnyakov, "Effects of similarity in the case of multiple scattering of coherent light: phenomenology and experiments," *Opt. Spectrosc.* **89**, 494–504 (2000)

12. D.A. Zimnyakov, "Coherence phenomena and statistical properties of multiply scattered light," in: *Handbook of Optical Biomedical Diagnostics*, Ed. V. Tuchin, SPIE, Belligham, pp. 265–310 (2002)

13. G. Maret and P.E. Wolf, "Multiple light scattering from disordered media. The effect of Brownian motion of scatterers," *Z. Phys. B* **65**, 409–413 (1987)

14. D.J. Pine, D.A. Weitz, P.M. Chaikin, and E. Herbolzheimer, "Diffusing wave spectroscopy," *Phys. Rev. Lett.* **60**, 1134–1137 (1988)

15. I. Freund, M. Kaveh, and M. Rosenbluh, "Dynamic light scattering: ballistic photons and the breakdown of the photon-diffusion approximation," *Phys. Rev. Lett.* **60**, 1130–1133 (1988)

16. P.-A. Lemieux, M.U. Vera, and D.J. Durian, "Diffusing-light spectroscopies beyond the diffusion limit: The role of ballistic transport and anisotropic scattering," *Phys. Rev. E* **57**, 4498–4515 (1998)

17. D. Bicout, C. Brosseau, A.S. Martinez, and J.M. Schmitt, "Depolarization of multiply scattering waves by spherical diffusers: Influence of size parameter," *Phys. Rev. E* **49**, 1767–1770 (1994)

18. A. Dogariu, C. Kutsche, P. Likamwa, G. Boreman, and B. Moudgil, "Time-domain depolarization of waves retroreflected from dense colloidal media," *Opt. Lett.* **22**, 585–587 (1997)

19. D.A. Zimnyakov and V.V. Tuchin, "About interrelations of distinctive scales of depolarization and decorrelation of optical fields in multiple scattering," *JETP Lett.* **67**, 455–460 (1998)

20. D.A. Zimnyakov, V.V. Tuchin, and A.G. Yodh, "Characteristic scales of optical field depolarization and decorrelation for multiple scattering media and tissues," *J. Biomed. Opt.* **4**, 157–163 (1999)

21. D.A. Zimnyakov, Yu.P. Sinichkin, P.V. Zakharov, and D.N. Agafonov, "Residual polarization of noncoherently backscattered linearly polarized light: the influence of the anisotropy parameter of the scattering medium," *Waves Random Media* **11**, 395–412 (2001)

22. P. Wolf and G. Maret, "Weak localization and coherent backscattering of photons in disordered media," *Phys. Rev. Lett.* **55**, 2696–2699 (1985)

23. D.A. Zimnyakov and Yu.P. Sinichkin, "Ultimate degree of residual polarization of incoherently backscattered light of multiple scattering of linearly polarized light," *Opt. Spectrosc.* **91**, 103–108 (2001)

24. E.E. Gorodnichev and D.B. Rogozkin, "Small-angle multiple scattering of light in a random medium," *JETP* **80**, 112–126 (1995)

25. E.E. Gorodnichev, A.I. Kuzovlev, and D.B. Rogozkin, "Depolarization of light in small-angle multiple scattering in random media," *Laser Phys.* **9**, 1210–1227 (1999)

26. S.L. Jacques, M.R. Ostermeyer, L.-H.V. Wang, and D. Stephens, "Polarized light transmission through skin using videoreflectometry: Toward optical tomography of superficial tissue layers," *Proc. SPIE* **2671**, 199–210 (1996)

27. S.L. Jacques and K. Lee, "Polarized video imaging of skin," *Proc. SPIE* **3245**, 356–362 (1998)

28. S.L. Jacques, R.J. Roman, and K. Lee, "Imaging superficial tissues with polarized light," *Lasers Surg. Med.* **26**, 119–129 (2000)

29. S.L. Jacques, J.C. Ramella-Roman, and K. Lee, "Imaging skin pathology with polarized light," *J. Biomed. Opt.* **7**, 329–340 (2002)

30. V. Sankaran, M.J. Everett, D.J. Maitland, and J.T. Walsh, "Comparison of polarized light propagation in biologic tissue and phantoms," *Opt. Lett.* **24**, 1044–1046 (1999)

31. V. Sankaran, J.T. Walsh, and D.J. Maitland, "Polarized light propagation through tissue phantoms containg densely packed scatterers," *Opt. Lett.* **25**, 239–241 (2000)

32. V. Sankaran, J.T. Walsh, and D.J. Maitland, "Comparative study of polarized light propagation in biologic tissues," *J. Biomed. Opt.* **7**, 300–306 (2002)

33. A.H. Hielscher, J.R. Mourant, and I.J. Bigio, "Influence of particle size and concentration on the diffuse backscatteringof polarized light from tissue phantoms and biological cell suspensions," *Appl. Opt.* **36**, 125–135 (1997)

34. R.C. Studinski and I.A. Vitkin, "Methodology for examining polarized light interactionwith tissues and tissue-like media in the exact backscattering direction," *J. Biomed. Opt.* **5**, 330–337 (2000)

35. G. Jarry, E. Steiner, V. Damaschini, M. Epifanie, M. Jurczak, and R. Kaizer, "Coherence and polarization of light propagating through scattering media and biological tissues," *Appl. Opt.* **37**, 7357–7367 (1998)

36. D.A. Zimnyakov and Yu.P. Sinichkin, "Ultimate degree of residual polarization of incoherently backscattered light of multiple scattering of linearly polarized light," *Opt. Spectrosc.* **91**, 103–108 (2001)

37. P.M. Saulnier, M.P. Zinkin, and G.H. Watson, "Scatterer correlation effects on photon transport in dense random media," *Phys. Rev. B* **42**, 2621–2623 (1992)

38. D.A. Zimnyakov, Yu. P. Sinichkin, I.V. Kiseleva, and D.N. Agafonov, "Effect of absorption of multiply scattering media on the degree of residual polarization of backscattered light," *Opt. Spectrosc.* **92**, 765–771 (2002)

39. A. Ishimaru, *Wave Propagation and Scattering in Random Media*, Academic, New York (1978)

40. T. Durduran, A.G. Yodh, B. Chance, and D.A. Boas, "Does the photon-diffusion coefficient depend on absorption?," *JOSA A* **14**, 3358–3365 (1997)

41. K. Furutsu and Y. Yamada, "Diffusion approximation for a dissipative random medium and the applications," *Phys. Rev. E.* **50**, 3634–3640 (1994)

42. K. Furutsu, "Pulse wave scattering by an absorber and integrated attenuation in the dffusion approximation," *JOSA A* **14**, 267–274 (1997)

43. M. Bassani, F. Martelli, G. Zaccanti, and D. Contini, "Independence of the diffusion coefficient from absorption: experimental and numerical evidence," *Opt. Lett.* **22**, 853–855 (1997)

44. G.G. Tinekov and V.G. Tinekov, *Microstructure of Milk and Dairy Produce*, Pishchevaya Promyshlennost', Moscow (1972)

45. S.T. Flock, S.L. Jacques, and B.C. Wilson, "Optical properties of Intralipid: a phantom medium for light propagation studies," *Lasers Surg. Med.* **12**, 510–519 (1992)

46. Yu.P. Sinichkin, N. Kollias, G. Zonios, S.R. Utz, and V.V. Tuchin. "Back reflectance and fluorescence spectroscopy of the human skin in vivo," in: *Handbook of Optical Biomedical Diagnostics*, Ed. V.V. Tuchin, SPIE, Bellingham, pp. 725–785 (2002)

47. Yu.P. Sinichkin, S.R. Utz, L.E. Dolotov, H.A Pilipenko, and V.V. Tuchin, "Technique and device for evaluation of erythema degree and melanin pigmentation of the human skin," *Radiotechnika* **4**, 77–81 (1997)

48. V.V. Tuchin, S. R. Utz, and I.V. Yaroslavsky, "Tissue optics, light distribution and spectroscopy," *Opt. Eng.* **33**, 3178–3188 (1994)

49. S. Prahl. http://omlc.ogi.edu.

50. A.P. Sviridov, D.A. Zimnyakov, Yu.P. Sinichkin, L.N. Butvina, A.I. Omel'chenko, G.Sh. Makhmutova, and V.N. Bagratashvili, "IR Fourier spectroscopy of in-vivo human skin and polarization of backscattered light in the case of skin ablation by IAG:Nd laser radiation," *J. Appl. Spectrosc.* **69**, 484–488 (2002)

51. A. Dogariu, M. Dogariu, K. Richardson, and G.D. Boreman, "Polarization asymmetry in waves backscattering from highly absorbant random media," *Appl. Opt.* **36**, 8159–8164 (1997)

52. D. Eliyahu, M. Rosenbluh, and I. Freund, "Angular intensity and polarization dependence of diffuse transmission through random media," *JOSA A* **10**, 477–491 (1993)

Chapter 8

1. J.W. Goodman, Statistics properties of laser speckle patterns, in: *Laser Speckle and Related Phenomenon*, Ed. J. C. Dainty, Springer, Berlin Heidelberg New york, pp. 9–75 (1975)

2. A.F. Fercher and P.F. Steeger, "First-order statistics of Stokes parameters in speckle fields," *Optica Acta* **28**, 443–448 (1981)

3. P.F. Steeger and A.F. Fercher, "Experimental investigation of the first-order statistics of Stokes parameters in speckle fields," *Optica Acta* **29**, 1395–1400 (1982)

4. R. Barakat, "The statistical properties of partially polarized light," *Optica Acta* **32**, 295–312 (1985)

5. C. Brosseau, "Statistics of the normalized Stokes parameters for a Gaussian stochastic plane wave field," *Appl. Opt.* **34**, 4788–4793 (1995)

6. I. Freund, M. Kaveh, R. Berkovits, and M. Rosenbluh, "Universal polarization correlations and microstatistics of optical waves in random media," *Phys. Rev. B* **42**, 2613–2616 (1990)

7. I.I. Tarhan and G.H. Watson, "Polarization microstatistics of laser speckle," *Phys. Rev. A* **45**, 6013–6018 (1992)

8. P. Elies, B. LeJeune, F. LeRoyBrehonnet, J. Cariou, and J. Lotrian, "Experimental investigation of the speckle polarization for a polished aluminium sample," *J. Phys. D: Appl. Phys.* **30**, 29–39 (1997)

9. J. Li, G. Yao, and L.-H.V. Wang, "Degree of polarization in laser speckles from turbid media: Implications in tissue optics," *J. Biomed. Opt.* **7**, 307–312 (2002)

10. J.P. Mathieu, *Optics*, Pergamon, New York (1975)

11. J.C. Dainty, Introduction, in: *Laser Speckle and Related Phenomenon*, Ed. J. C. Dainty, Springer, Berlin Heidelberg New York, pp. 1–7 (1975)

12. A.E. Ennos, Speckle interferometry, in: *Laser Speckle and Related Phenomenon*, Ed. J. C. Dainty, Springer, Berlin, Heidelberg New York, pp. 203–253 (1975)

13. S. Jiao, G. Yao, and L.-H.V. Wang, "Depth-resolved two-dimensional Stokes vectors of backscattered light and Mueller matrices of biological tissue measured by optical coherence tomography," *Appl. Opt.* **39**, 6318–6324 (2000)

Chapter 9

1. B.C. Wilson and G. Adam, "A Monte Carlo model for the absorption and flux distribution of light in tissue," *Med. Phys.* **10**, 824–830 (1983)
2. M. Keijzer, S.L. Jacques, S.A. Prahl, and A.J. Welch, "Light distributions in artery tissue: Monte Carlo simulation for finite-diameter laser beams," *Lasers Surg. Med.* **9**, 148–154 (1989)
3. L.-H.V. Wang and S.L. Jacques, "Hybrid model of Monte Carlo simulation and diffusion theory for light reflectance by turbid media," *J. Opt. Soc. Am. A* **10**, 1746–1752 (1993)
4. L.-H.V. Wang, S.L. Jacques, and L.-Q. Zheng, "MCML – Monte Carlo modeling of photon transport in multi-layered tissue," *Comput. Methods Programs Biomed.* **47**, 131–146 (1995)
5. L.-H.V. Wang, G.L. Coté, and S.L. Jacques, "Special Section Guest Editorial – Special section on tissue polarimetry," *J. Biomed. Opt.* **7**(3), 278 (2002)
6. J.F. de Boer, S.M. Srinivas, A. Malekafzali, Z. Chen and J.S. Nelson, "Imaging thermally damaged tissue by polarization sensitive optical coherence tomography," *Opt. Exp.* **3**, 212–215 (1998)
7. G.L. Coté, M.D. Fox, and R.B. Northrop, "Noninvasive optical polarimetric glucose sensing using a true phase measurement technique," *IEEE Trans. Biomed Eng.* **39**, 752–756 (1992)
8. D. Bicout, C. Brosseau, A.S. Martinez, and J.M. Schmitt, "Depolarization of multiply scattered waves by spherical diffusers: influence of the size parameter," *Phy. Rev. E* **49**, 1767–1770 (1994)
9. G. Yao and L.-H.V. Wang, "Propagation of polarized light in turbid media: an animated simulation study," *Opt. Express* **7**, 198–203 (2000)
10. G. Yao, *Ultrasound-Modulated Optical Tomography*, Ph.D. dissertation, Texas A&M University (2000)
11. H.C. van de Hulst, *Light Scattering by Small Particles*, Dover, New York (1981)
12. W.S. Bickel and W.M. Bailey, "Stokes vectors, Mueller matrices, and polarized scattered light," *Am. J. Phys.* **53**, 468–478 (1995)
13. I.L. Maksimova, S.V. Romanov, and V.F. Izotova, "The effect of multiple scattering in disperse media on polarization characteristics of scattered light," *Opt. Spectrosc.* **92**(6), 915–923 (2002)
14. M.J. Rakovic, G.W. Kattawar, M. Mehrubeoglu, B.D. Cameron, L.-H.V. Wang, S. Rastegar, and G.L. Coté, "Light backscattering polarization patterns from turbid media: theory and experiments," *Appl. Opt.* **38**, 3399–3408 (1999)
15. S. Bartel and A.H. Hielscher, "Monte Carlo simulation of the diffuse backscattering Mueller matrix for highly scattering media," *Appl. Opt.* **39**, 1580–1588 (2000)
16. S. Jiao, G. Yao, and L.-H.V. Wang, "Depth-resolved two-dimensional Stokes vectors of backscattered light and Mueller matrices of biological tissue measured by optical coherence tomography," *Appl. Opt.* **39**(34), 6318–6324 (2000)
17. X. Liang, L. Wang, P.P. Ho, and R.R. Alfano, "Time-resolved polarization shadowgrams in turbid media," *Appl. Opt.* **36**, 2984–2989 (1997)

Chapter 10

1. W. Drexler, U. Morgner, F.X. Kärtner, C. Pitris, S.A. Boppart, X.D. Li, E.P. Ippen, and J.G. Fujimoto, "In vivo ultrahigh-resolution optical coherence tomography," *Opt. Lett.* **24**(17), 1221–1223 (1999)

2. I. Hartl, X.D. Li, C. Chudoba, R.K. Ghanta, T.H. Ko, and J.G. Fujimoto, "Ultrahigh-resolution optical coherence tomography using continuum generation in an air-silica microstructure optical fiber," *Opt. Lett.* **29**(9), 608–610 (2001)

3. G.J. Tearney, B.E. Bouma, and J.G. Fujimoto, "High-speed phase- and group-delay scanning with a grating-based phase control delay line," *Opt. Lett.* **22**(23), 1811–1813 (1997)

4. A.M. Rollins, M.D. Kulkarni, S. Yazdanfar, R. Ung-arunyawee, and J.A. Izatt, "In vivo video rate optical coherence tomography," *Opt. Express* **3**(6), 219–229 (1998)

5. J.M. Schmitt, "Restoration of optical coherence images of living tissues using the CLEAN algorithm," *J. Biomed. Opt.* **3**(1), 66–75 (1998)

6. M. Baskansky and J. Reintjes, "Statistics and reduction of speckle in optical coherence tomography," *Opt. Lett.* **25**(8), 545–547 (2000)

7. J.F. de Boer, T.E. Milner, M.J.C. van Gemert, and J.S. Nelson, "Two-dimensional birefringence imaging in biological tissue by polarization-sensitive optical coherence tomography," *Opt. Lett.* **22**(12), 934–936 (1997)

8. M.J. Everett, K. Schoenerberger, B.W. Colston, Jr., and L.B. Da Silva, "Birefringence characterization of biological tissue by use of optical coherence tomography," *Opt. Lett.* **23**(3), 228–230 (1998)

9. J.F. de Boer, T.E. Milner, and J.S. Nelson, "Determination of the depth-resolved Stokes parameters of light backscattered from turbid media by use of polarization-sensitive optical coherence tomography," *Opt. Lett.* **24**(5), 300–302 (1999)

10. G. Yao and L.-H.V. Wang, "Two-dimensional depth-resolved Mueller matrix characterization of biological tissue by optical coherence tomography," *Opt. Lett.* **24**(8), 537–539 (1999)

11. S. Jiao, G. Yao, and L.-H.V. Wang, "Depth-resolved two-dimensional Stokes vectors of backscattered light and Mueller matrices of biological tissue measured with optical coherence tomography," *Appl. Opt.* **39**(34), 6318–6324 (2000)

12. C.K. Hitzenberger, E. Gotzinger, M. Sticker, M. Pircher, and A.F. Fercher, "Measurement and imaging of birefringence and optic axis orientation by phase resolved polarization sensitive optical coherence tomography," *Opt. Express* **9**, 780–790 (2001)

13. J.E. Roth, J.A. Kozak, S. Yazdanfar, A.M. Rollins, and J.A. Izatt, "Simplified method for polarization-sensitive optical coherence tomography," *Opt. Lett.* **26**, 1069–1071 (2001)

14. H.W. Ren, Z.H. Ding, Y.H. Zhao, J.J. Miao, J.S. Nelson, and Z.P. Chen, "Phase-resolved functional optical coherence tomography: simultaneous imaging of in situ tissue structure, blood flow velocity, standard deviation, birefringence, and Stokes vectors in human skin," *Opt. Lett.* **27**, 1702–1704 (2002)

15. R.V. Kuranov, V.V. Sapozhnikova, I.V. Turchin, E.V. Zagainova, V.M. Gelikonov, V.A. Kamen-sky, L.B. Snopova, and N.N. Prodanetz, "Complementary use of cross-polarization and standard OCT for differential diagnosis of pathological tissues," *Opt. Express* **10**, 707–713 (2002)

16. Y. Yang, L. Wu, Y.Q. Feng, and R.K. Wang, "Observations of birefringence in tissues from optic-fibre-based optical coherence tomography," *Meas. Sci. Technol.* **14**, 41–46 (2003)

17. S. Jiao and L.-H.V. Wang, "Two-dimensional depth-resolved Mueller matrix of biological tissue measured with double-beam polarization-sensitive optical coherence tomography," *Opt. Lett.* **27**(2), 101–103 (2002)

18. S. Jiao and L.-H.V. Wang, "Jones-matrix imaging of biological tissues with quadruple-channel optical coherence tomography," *J. Biomed. Opt.* **7**(3), 350–358 (2002)

19. J.J. Gil and E. Bernabeu, "Obtainment of the polarizing and retardation parameters of a non-depolarizing optical system from the polar decomposition of its Mueller matrix," *Optik* **76**(2), 67–71 (1987)

20. E. Collett, *Polarized Light Fundamentals and Applications*, Dekker, New York (1993), Chap. 10.

21. F. Le Roy-Brehonnet and B. Le Jeune, "Utilization of Mueller matrix formalism to obtain optical targets depolarization and polarization properties," *Prog. Quantum. Electron.* **21**(2), 109–151 (1997)

22. N. Vansteenkiste, P. Vignolo, and A. Aspect, "Optical reversibility theorems for polarization: application to remote control of polarization," *J. Opt. Soc. Am. A* **10**(10), 2240–2245 (1993)

23. G. Yao and L.-H.V. Wang, "Monte Carlo simulation of an optical coherence tomography signal in homogenous turbid media," *Phys. Med. Biol.* **44**, 2307–2320 (1999)

24. Y. Zhao, Z. Chen, C. Saxer, S. Xiang, J.F. de Boer, and J.S. Nelson, "Phase-resolved optical coherence tomography and optical Doppler tomography for imaging blood flow in human skin with fast scanning speed and high velocity sensitivity," *Opt. Lett.* **25**(2), 114–116 (2000)

25. Alexander D. Poularikas, *The Transforms and Applications Handbook*, CRC Boca Raton, FL (1996), Chap. 7.

26. R.A. Chipman, Chap. 22. Polarimetry, in: *The Handbook of Optics* **II**, (OSA/McGraw-Hill, New york (1995)

27. C.E. Saxer, J.F. de Boer, B.H. Park, Y.H. Zhao, Z.P. Chen, and J.S. Nelson, "High-speed fiber-based polarization-sensitive optical coherence tomography of in vivo human skin," *Opt. Lett.* **25**, 1355–1357 (2000)

28. S. Jiao, W. Yu, G. Stoica, and L.-H.V. Wang, "Optical-fiber-based Mueller optical coherence tomography," *Opt. Lett.* **28**, 1206–1208 (2003)

29. Y. Yasuno, S. Makita, Y. Sutoh, M. Itoh, and T. Yatagai, "Birefringence imaging of human skin by polarization-sensitive spectral interferometric optical coherence tomography," *Opt. Lett.* **27**, 1803–1805 (2002)

Chapter 11

1. R.R. Anderson, "Polarized light examination and photography of the skin," *Arch. Dermatol.* **127**, 1000–1005 (1991)

2. P.F. Bilden, S.B. Phillips, N. Kollias, J.A. Muccini, and L.A. Drake, "Polarized light photography of acne vulgaris," *J. Invest. Dermatol.* **98**, 606 (1992)

3. S.G. Demos, W.B. Wang, and R.R. Alfano, "Imaging objects hidden in scattering media with fluorescence polarization preservation of contrast agents," *Appl. Opt.* **37**, 792–797 (1998)

4. N. Kollias, "Polarized light photorgaphy of human skin," in *Bioengineering of the Skin: Skin Surface Imaging and Analysis*, Eds. K.-P. Wilhelm, P. Elsner, E. Berardesca, H. I. Maibach, CRC, New York, pp. 95–106 (1997)

5. S.G. Demos, W.B. Wang, J. Ali, and R.R. Alfano, "New optical difference approaches for subsurface imaging of tissues," *OSA TOPS* **21**, in *Advances in Optical Imaging and Photon Migration* Eds. J.G. Fujimoto and M.S. Patterson, pp. 405–410 (1998)

6. S.L. Jacques, J.R. Roman, and K. Lee, "Imaging superficial tissues with polarized light," *Lasers Surg. Med.* **26**, 119–129 (2000)

7. A. Muccini, N. Kollias, S.B. Phillips, R.R. Anderson, A.J. Sober, M.J. Stiller, and L.A. Drake, "Polarized light photography in the evaluation of photoaging," *J. Am. Acad. Dermatol.* **33**, 765–769 (1995)

8. O. Emile, F. Bretenaker, and A. LeFloch, "Rotating polarization imaging in turbid media," *Opt. Lett.* **21**, 1706–1709 (1996)

9. S.L. Jacques, J.C. Ramella-Roman, and K. Lee, "Imaging skin pathology with polarized light," *J. Biomed. Opt.* **7**(3), 329–340 (2002)

10. J.S. Tyo, "Enhancement of the point-spread function for imaging in scattering media by use of polarization-difference imaging," *J. Opt. Soc. Am. A* **17**, 1–10 (2000)

11. H.-J. Schnorrenberg, R. Haβner, M. Hengstebeck, K. Schlinkmeier, and W. Zinth, "Polarization modulation can improve resolutionin diaphanography," *Proc. SPIE* **2326**, 459–464 (1995)

12. D.A. Zimnyakov and Yu. P. Sinichkin, "A study of polarization decay as applied to improved imaging in scattering media," *J. Opt. A: Pure Appl. Opt.* **2**, 200–208 (2000)

13. Yu. P. Sinichkin, D.A. Zimnyakov, D.N. Agafon ov, and L.V. Kuznetsova, "Visualization of scattering media upon backscattering of a linearly polarized nonmonochromatic light," *Opt. Spectrosc.* **93**, 110–116 (2002)

14. D.A. Boas and A.G. Yodh, "Spatially varying dynamical properties of turbid media probed with diffusing temporal light correlation," *J. Opt. Soc. Am. A* **14**, 192–215 (1997)

15. S.L. Jacques, K. Lee, and J. Roman, "Scattering of polarized light by biological tissues," *Proc. SPIE* **4001**, 14–28 (2000)

16. G.A. Wagnieres, W.M. Star, and B.C. Wilson, "In vivo fluorescence spectroscopy and imaging for oncological applications," *Photochem. Photobiol.* **68**(5), 603–632 (1998)

17. J.R. Mourant, T. Fuselier, J. Boyer, T.M. Johnson, and I.J. Bigio, "Predictions and measurements of scattering and absorption over broad wavelength ranges in tissue phantoms," *Appl. Opt.* **36**, 949–957 (1997)

18. L.T. Perelman, V. Backman, M. Wallace, G. Zonios, R. Manoharan, A. Nustar, S. Shields, M. Seiler, C. Lima, T. Hamano, I. Itzkan, J. Van Dam, J.M. Crawford, and M.S. Feld, "Observation of periodic fine structure in reflectance from biological tissue: A new technique for measuring nuclear size distribution," *Phys. Rev. Lett.* **80**, 627–630 (1998)

19. K. Sokolov, R. Drezek, K. Gossage, and R. Richards-Kortum, "Reflectance spectroscopy with polarized light: Is it sensitive to cellular and nuclear morphology," *Opt. Express* **5**, 302–317 (1999)

20. V. Backman, R. Gurjar, K. Badizadegan, I. Itzkan, R. Dasari, L.T. Perelman, and M.S. Feld, "Polarized light scattering spectroscopy for quantitative measurements of epithelial cellular structures in situ," *IEEE J. Sel. Top. Quantum Electron.* **5**, (1999)

21. A. Myakov, L. Nieman, L. Wicky, U. Utzinger, R. Richards-Kortum, and K. Sokolov, "Fiber optic probe for polarized reflectance spectroscopy in vivo: Design and performance," *J. Biomed. Opt.* **7**(3), 388–397 (2002)

22. B. Rabinovitch, W.F. March, and R.L. Adams, "Noninvasive glucose monitoring of the aqueous humor of the eye: Part I. Measurement of very small optical rotations," *Diabetes Care* **5**(3), 254–258 (1982)

23. W.F. March, B. Rabinovitch, and R.L. Adams, "Noninvasive glucose monitoring of the aqueous humor of the eye: Part II. Animal studies and the scleral lens," *Diabetes Care* **5**(3), 259–265 (1982)

24. G.L. Coté, M.D. Fox, and R.B. Northrup, "Noninvasive optical glucose sensing using a true phase measurement technique," *IEEE Trans. Biomed. Eng.* **39**(7), 752–756 (1992)

25. J.A. Tamada, N.J.V. Bohannon, and R.O. Potts, "Measurement of glucose in diabetic subjects using noninvasive transdermal extraction," *Nat. Med.* (N.Y.) **1**(11), 1198–1201 (1995)

26. T. Mitsui and K. Sakurai, "Precise measurement of the refractive and optical rotatory power of a suspension by a delayed optical heterodyne technique," *Appl. Opt.* **35**, 2253–2258 (1996)

27. A. Ishimaru, *Wave Propagation and Scattering in Random Media*, IEEE New York (1997)

28. D.S. Klonoff, "Noninvasive blood glucose monitoring," *Diabetes Care* **20**(3), 433–237 (1997)

29. G.L. Coté, "Noninvasive optical glucose sensing – an overview," *J. Clin. Eng.* **22**(4), 253–259 (1997)

30. G.L. Coté and B.D. Cameron, "Noninvasive polarimetric measurement of glucose in cell culture media," *J. Biomed. Opt.* **2**(3), 275–281 (1997)

31. C. Chou, Y.C. Huang, C.M. Feng, and M. Chang, "Amplitude sensitive optical heterodyne and phase lock-in technique on small optical rotation angle detection of chiral liquid," *Jpn. J. Appl. Phys.* **36**, 356–359 (1997)

32. C. Chou, C.Y. Han, W.C. Kuo, Y.C. Huang, C.M. Feng, and J.C. Shyu, "Noninvasive glucose monitoring *in vivo* with an optical heterodyne polarimeter," *Appl. Opt.* **37**(16), 3553–3557 (1998)

33. J.N. Roe and D.R. Smoller, "Bloodless glucose measurements," *Crit. Rev. Ther. Drug Carrier Syst.*, **15**(3), 199–241 (1998)

34. R.J. McNichols and G.L. Coté, "Development of a noninvasive polarimetric glucose sensor," *IEEE-LEOS Newslett.* **12**(2), 30–31 (1998)

35. J.R. Lakowicz, I. Gryczynski, Z. Gryczynski, L. Tolosa, L. Randers-Eichhorn, and G. Rao, "Polarization-based sensing of glucose using an oriented reference film," *J. Biomed. Opt.* **4**(4), 443–449 (1999)

36. O.S. Khalil, "Spectroscopic and clinical aspects of noninvasive glucose measurements," *Clin. Chem.* **45**(2), 165–177 (1999)

37. R.J. McNichols and G.L. Coté, "Optical glucose sensing in biological fluids: an overview," *J. Biomed. Opt.* **5**(1), 5–16 (2000)

38. J.S. Baba, B.D. Cameron, S. Theru, and G.L. Coté, "Effect of temperature, pH, and corneal birefringence on polarimetric glucose monitoring in the eye," *J. Biomed. Opt.* **7**(3), 321–328 (2002)

39. S. Böckle, L. Rovati, and R.R. Ansari, "Glucose sensing using Brewster-reflection: Polarimetric ray-tracing based upon an anatomical eye model," *Proc. SPIE* 4965–21 (2003)

40. R.O. Esenaliev, K.V. Larin, I.V. Larina, and M. Motamedi, "Noninvasive monitoring of glucose concentration with optical coherence tomography," *Opt. Lett.* **26**(13), 992–994 (2001)

41. V.V. Tuchin, R.K. Wang, E.I. Galanzha, N.A. Lakodina, and A.V. Solovieva, "Monitoring of glycated hemoglobin in a whole blood by refractive index measurement with OCT, Conference Program CLEO/QELS (2003) Baltimore, June 1–6 2003, p. 120

42. R.C.N. Studinski and I. A. Vitkin, "Methodology for examining polarized light interactions with tissues and tissuelike media in the exact backscattering direction," *J. Biomed. Opt.* **5**(3), 330–337 (2000)

43. K. C. Hadley and I. A. Vitkin, "Optical rotation and linear and circular depolarization rates in diffusively scattered light from chiral, racemic, and achiral turbid media," *J. Biomed. Opt.* **7**(3), 291–299 (2002)

44. R.L. Stamper, "Aqueous humor: secretion and dynamics," in *Physiology of the Human Eye and Visual System*, Ed. R.E. Records, Harper & Row, Hagerstown, MD, pp. 156–182, 1979

45. A.N. Bashkatov, E.A. Genina, Y.P. Sinichkin, V.I. Kochubey, N.A. Lakodina, and V.V. Tuchin, "Glucose and manitol diffusion in human *dura mater*," *Biophys. J.* **79** (2003)

46. V.V. Tuchin, *Tissue Optics: Light Scattering Methods and Instruments for Medical Diagnosis*, SPIE Tutorial Texts in Optical Engineering, **TT38**, SPIE Bellingham (2000)

47. W.S. Bickel, J.F. Davidson, D.R. Huffman, and R. Kilkson, "Application of polarization effects in light scattering: a new biophysical tool," *Proc. Natl Acad. Sci. USA* **73**, 486–490 (1976)

48. W.S. Bickel and M.E. Stafford, "Polarized light scattering from biological systems: A technique for cell differentiation," *J. Biol. Phys.* **9**, 53–66 (1981)

49. R.G. Johnston, S.B. Singham, and G.C. Salzman, "Polarized light scattering," *Comments Mol. Cell. Biophys.* **5**(3), 171–192 (1988)

50. G.C. Salzmann, S.B. Singham, R.G. Johnston, and C.F. Bohren, "Light scattering and cytometry," in *Flow Cytometry and Sorting*, 2nd ed., Eds. M.R. Melamed, T. Lindmo, and M.L. Mendelsohn, Wiley-Liss Inc., New York, pp. 81–107 (1990)

51. S. Ya. Sid'ko, V.N. Lopatin, and L.E. Paramonov, *Polarization Characteristics of Solutions of Biological Particles* Nauka, Novosibirsk (1990)

52. W.P. Van de Merwe, D.R. Huffman, and B.V. Bronk, "Reproducibility and sensitivity of polarized light scattering for identifying bacterial suspensions," *Appl. Opt.* **28**(23), 5052–5057 (1989)

53. B.V. Bronk, S.D. Druger, J. Czege, and W.P. van de Merwe, "Measuring diameters of rod-shaped bacteria in vivo with polarized light scattering," *Biophys. J.* **69**, 1170–1177 (1995)

54. W.P. Van de Merwe, Z.-Z. Li, B.V. Bronk, and J. Czege, "Polarized light scattering for rapid observation of bacterial size changes," *Biophys. J.* **73**, 500–506 (1997)

55. B.G. de Grooth, L.W.M.M. Terstappen, G.J. Puppels, and J. Greve, "Light-scattering polarization measurements as a new parameter in flow cytometry," *Cytometry* **8**, 539–544 (1987)

56. R.M.P. Doornbos, A.G. Hoekstra, K.E.I. Deurloo, B.G. de Grooth, P.M.A. Sloot, and J. Greve, "Lissajous-like patterns in scatter plots of calibration beads," *Cytometry* **16**, 236–242 (1994)

57. V.V. Tuchin, "Lasers light scattering in biomedical diagnostics and therapy," *J. Laser Appl.* **5**(2/3), 43–60 (1993)

58. V.V. Tuchin, "Coherence-domain methods in tissue and cell optics," *Laser Phys.* **8**(4), 807–849 (1998)

59. A.G. Hoekstra and P.M.A. Sloot, "Biophysical and biomedical applications of nonspherical scattering," in *Light Scattering by Nonspherical Particles*, Eds. M. I. Mishchenko, J. W. Hovenier, and L. D. Travis, Academic Press, San Diego, pp. 585–602 (2000)

60. N.G. Khlebtsov, I.L. Maksimova, V.V. Tuchin, and L.-H.V. Wang, "Introduction to light scattering by biological objects" in *Handbook of Optical Biomedical Diagnostics* **PM107**, Ed. V.V. Tuchin, SPIE, Bellingham, pp. 31–167 (2002)

61. A.G. Hoekstra and P.M.A. Sloot, "Dipolar unit size in coupled-dipole calculations of the scattering matrix elements," *Appl. Opt.* **18**, 1211–1213 (1993)

62. C.F. Bohren and D.R. Huffman, *Absorption and Scattering of Light by Small Particles*, Wiley, New York (1983)

63. R.G. Johnston, S.B. Singham, and G.C. Salzman, "Phase differential scattering from microspheres," *Appl. Opt.* **25**, 3566–3572 (1986)

64. R. Doornboss, E.J. Hennink, C.A.J. Putman, et al., "White blood cell differentiation using a solid state flow cytometry," *Cytometry* **14**, 589–594 (1993)

65. R.H. Newton, J.Y. Brown, and K.M. Meek, "Polarised light microscopy technique for quantitative mapping collagen fibril orientation in cornea," *Proc. SPIE*, **2926**, 278–284 (1996)

66. S. Inoué, "Video imaging processing greatly enhance contrast, quality, and speed in polarization-based microscopy," *J. Cell Biol.* **89**, 346–356 (1981)

67. R. Oldenbourg and G. Mei, "New polarized light microscope with precision universal compensator," *J. Microsc.* **180**(2), 140–147 (1995)

68. Q. Zhou and R.W. Knighton, "Light scattering and form birefringence of parallel cylindrical arrays that represent cellular organelles of the retinal nerve fiber layer," *Appl. Opt.* **36**(10), 2273–2285 (1997)

69. X.-R. Huang and R.W. Knighton, " Linear birefringence of the retinal nerve fiber layer measured *in vitro* with a multispectral imaging micropolarimeter," *J. Biomed. Opt.* **7**(2), 199–204 (2002)

70. G. Mazarevica, T. Freivalds, and A. Jurka, Properties of erythrocyte light refraction in diabetic patients," *J. Biomed. Opt.* **7**(2), 244–247 (2002)

71. P.S. Hauge, "Recent developments in instruments in ellipsometry," *Surf. Sci.* **96**, 108–140 (1980)

72. T.T. Tower and R.T.Tranquillo, "Alignment maps of tissues: I. Microscopic elliptical polarimetry," *Biophys. J.* **81**, 2954–2963 (2001)

73. T.T. Tower and R.T.Tranquillo, "Alignment maps of tissues: II. Fast harmonic analysis for imaging," *Biophys. J.* **81**, 2964–2971 (2001)

74. D.A. Yakovlev, S.P. Kurchatkin, A.B. Pravdin, E.V. Gurianov, M.Yu. Kasatkin, and D.A. Zimnyakov, "Polarization monitoring of structure and optical properties of the heterogenous birefringent media: application in the study of liquid crystals and biological tissues," *Proc. SPIE*, **5067**, 64–72 (2003)

75. I. Freund, M. Deutsch, and A. Sprecher, "Connective tissue polarity: Optical second-harmonic microscopy, crossed-beam summation, and small-angle scattering in rat-tail tendon," *Biophys. J.* **50**, 693–512 (1986)

76. P. Stoller, B.-M. Kim, A.M. Rubenchik, "Polarization-dependent optical second-harmonic imaging of a rat-tail tendon," *J. Biomed. Opt.* **7**(2), 205–214 (2002)

77. X. Gan, S.P. Schilders, and M. Gu, "Image enhancement through turbid media under a microscope by use of polarization gating methods," *J. Opt. Soc. Am. A* **16**, 2177–2184 (1999)
78. X. Gan and M. Gu, "Image reconstruction through turbid media under a transmission-mode microscope," *J. Biomed. Opt.* **7**(3), 372–377 (2002)
79. V.V. Tuchin, "Coherent optical techniques for the analysis of tissue structure and dynamics," *J. Biomed. Opt.* **4**, 106–124 (1999)
80. G.V. Simonenko, T.P. Denisova, N.A. Lakodina, and V.V. Tuchin "Measurement of an optical anisotropy of biotissues," *Proc. SPIE* **3915**, 152–157 (2000)
81. G.V. Simonenko, V.V. Tuchin, N.A. Lakodina, "Measurement of the optical anisotropy of biological tissues with the use of a nematic liquid crystal cell," *J. Opt. Technol.* **67**(6), 559–562 (2000)
82. N. Huse, A. Schönle, and S.W. Hell, "Z-polarized confocal microscopy," *J. Biomed. Opt.* **6**(3), 273–276 (2001)
83. A.A. Caputo and J.P. Standlee, *Biomechanics in Clinical Dentistry*, Quintessence, Chicago, pp. 19–28 (1987)
84. A. Asundi, "Phase shifting in photoelasticity," *Exp. Tech.* **17**, 19–23 (1993)
85. A. Asundi and A. Kishen, "Digital photoelastic investigations on the tooth-bone interface," *J. Biomed. Opt.* **6**, 224–230 (2001)
86. A. Kishen and A. Asundi, "Photomechanical investigations on post endodontically rehabilitated teeth," *J. Biomed. Opt.* **7**(2), 262–270 (2002)
87. J.W. Dally and W.F. Riley, *Experimental Stress Analysis*, 3rd ed., McGraw–Hill, New York, pp. 425–429 (1991)
88. J. Lackowicz, *Principles of Fluorescence Spectroscopy*, 2nd ed., Kluwer Academic/Plenum, New York (1999)
89. D.B. Tata, M. Foresti, J. Cordero, P. Tomashefsky, M.A. Alfano, and R.R. Alfano, "Fluorescence polarization spectroscopy and time-resolved fluorescence kinetics of native cancerous and normal rat kidney tissues," *Biophys. J.* **50**, 463–469 (1986)
90. A. Pradhan, S.S. Jena, B.V. Laxmi, and A. Agarwal, "Fluorescence depolarization of normal and diseased skin tissues," *Proc. SPIE* **3250**, 78–82 (1998)
91. S.K. Mohanty, N. Ghosh, S.K. Majumder, and P.K. Gupta, "Depolarization of autofluorescence from malignant and normal human breast tissues," *Appl. Opt.* **40**(7), 1147–1154 (2001)
92. F.W.J. Teale, "Fluorescence depolarization by light scattering in turbid solutions," *Photochem. Photobiol.* **10**, 363–374 (1969)
93. G.A. Wagnieres, W.M. Star, and B.C. Wilson, "In-vivo fluorescence spectroscopy and imaging for oncological applications," *Photochem. Photobiol.* **68**, 603–632 (1998)
94. P.K. Gupta, S.K. Majumder, and A. Uppal, "Breast cancer diagnosis using N_2 laser excited autofluorescence spectroscopy," *Lasers Surg. Med.* **21**, 417–422 (1997)
95. N. Ghosh, S.K. Majumder, and P.K. Gupta, "Polarized fluorescence spectroscopy of human tissue," *Opt. Lett.* **27**, 2007–2009 (2002)

Index

Printing: Krips bv, Meppel
Binding: Stürtz, Würzburg